Urban Infrastructure and Resilient, Sustainable Buildings

Guest Editors

Shenghua Zhou
Tiantian Gu
Mun On Wong

MDPI

Basel • Beijing • Wuhan • Barcelona • Belgrade • Novi Sad • Cluj • Manchester

Guest Editors
Shenghua Zhou
School of Civil Engineering
Southeast University
Nanjing
China

Tiantian Gu
School of Mechanics and Civil Engineering
China University of Mining and Technology
Xuzhou
China

Mun On Wong
Department of Civil and Environmental Engineering
University of Macau
Macau
China

Editorial Office
MDPI AG
Grosspeteranlage 5
4052 Basel, Switzerland

This is a reprint of the Special Issue, published open access by the journal *Buildings* (ISSN 2075-5309), freely accessible at: https://www.mdpi.com/journal/buildings/special_issues/3GM32P38L0.

For citation purposes, cite each article independently as indicated on the article page online and as indicated below:

Lastname, A.A.; Lastname, B.B. Article Title. *Journal Name* **Year**, *Volume Number*, Page Range.

ISBN 978-3-7258-5571-1 (Hbk)
ISBN 978-3-7258-5572-8 (PDF)
https://doi.org/10.3390/books978-3-7258-5572-8

Contents

About the Editors

Shenghua Zhou

Shenghua Zhou is an Assistant Professor at Southeast University. He earned his BEng and MPhil from Southeast University, completed a PhD in Civil Engineering at the University of Hong Kong, and was a China Scholarship Council-funded visiting scholar at the University of Cambridge. He has been selected for Jiangsu Province's "Dual-Innovation Doctor" talent program and is a council member of the Engineering Construction Management Committee of the Chinese Society for Technology Economics. His research focuses on AI-enabled infrastructure management, including cascading failures and the resilience of lifeline systems. As of 2025, he has led projects funded by the National Natural Science Foundation of China, the Jiangsu Social Science Fund, and central university grants. He has published more than 50 peer-reviewed papers and serves as a reviewer for dozens of SCI/SSCI journals.

Tiantian Gu

Tiantian Gu is an Associate Professor of engineering management at the school of mechanics and civil engineering of China University of Mining and Technology, Xuzhou, China. She has been recognized as a "Dual-Innovation Doctor" in Jiangsu Province, a "Twelfth Batch of Outstanding Young Teachers" at China University of Mining and Technology, and a participant in the "Sailing Program" for young teachers at the same university. Her primary research interests focus on resilient city development, public engagement, urban renewal, and smart community construction. To date, she has conducted over 30 projects and has published over 50 peer-reviewed journal papers and conference papers.

Mun On Wong

Mun On Wong is an Assistant Professor of construction project management at the Department of Civil and Environmental Engineering of the University of Macau, Macau SAR, China. He received a B.Eng. in Civil Engineering from Shanghai Jiao Tong University in 2017 and a Ph.D. in Construction Engineering and Management from The University of Hong Kong in 2021. His research focuses on developing collaborative technologies to integrate human beings, physical environments, and digital twins for safer and more sustainable construction. His work has been published in leading journals, including Automation in Construction, Advanced Engineering Informatics, and the ASCE Journal of Construction Engineering and Management. He has served as the PI and Co-I for projects funded by the Science and Technology Development Fund (FDCT) of the Macao SAR and Universitas 21 alliance. His research has achieved several awards, including the metaverse hackathon prize in the 22nd international CONVR conference and the construction innovation award from the Hong Kong Construction Industry Council (CIC).

Preface

The subject of this Reprint is the evolving landscape of urban infrastructures and sustainable buildings under the pressures of carbon reduction, disaster resilience, and social uncertainty. Its scope includes intelligent monitoring, digital modeling, low-carbon adoption, and participatory governance strategies. The aim is to offer an integrated perspective that moves beyond isolated technical solutions toward cohesive strategies linking design, operation, and policy.

The motivation for compiling this work is the urgent need to confront cascading hazards, resource constraints, and uncertainties facing cities worldwide. By presenting 12 contributions across the pillars of intelligence, decarbonization, and resilience, the Reprint demonstrates how technological innovation, community practice, and governance frameworks can be mobilized to create more adaptive and sustainable urban systems.

This Reprint is intended for academics, engineers, planners, and decision-makers seeking to transform the built environment into a resilient foundation for future generations. By combining theoretical insights with applied case studies, it bridges research and practice.

Shenghua Zhou, Tiantian Gu, and Mun On Wong
Guest Editors

buildings

MDPI

Editorial

Editorial on Urban Infrastructure and Resilient, Sustainable Buildings

Shenghua Zhou [1,*], Tiantian Gu [2] and Mun On Wong [3]

1 School of Civil Engineering, Southeast University, Nanjing 210018, China
2 School of Mechanics and Civil Engineering, China University of Mining and Technology, Xuzhou 221116, China
3 Department of Civil and Environmental Engineering, University of Macau, Macau 999078, China; mowong@um.edu.mo
* Correspondence: shenghua@seu.edu.cn

1. Introduction

Urban infrastructures and buildings today face the intertwined pressures of extreme disasters, cascading effects, carbon constraints, and social uncertainty [1,2]. Addressing these challenges demands a technological, environmental, and societal synthesis that evolves beyond siloed fixes into integrated capabilities spanning design, operation, and governance [3,4]. This Special Issue brings together 12 papers organized around three foundational pillars of sustainable buildings and infrastructures: Intelligence [5,6], Decarbonization [7,8], and Resilience [9,10], each providing sharpened insights into how cities can harness digital sensing and twins, accelerate low-carbon adoption and community practices, and strengthen data governance, social responsibility, and spatial design to evolve under future challenges.

2. Intelligence: Smart-Sensing Digital Twins and Scalable Monitoring

This theme captures the shift from periodic inspection to continuous, explainable, and scalable monitoring. Huang et al. (contribution 1) present a deep learning framework based on Inception v3 that achieves high-accuracy corrosion classification for steel, offering a lightweight route to automated condition assessment. Yang et al. (contribution 2) integrate Building Information Modeling (BIM) at Level of Development (LOD) 400 with a digital twin and live Internet of Things (IoT) feeds to deliver real-time anomaly detection and dashboarding for a noise barrier tunnel. Jiang et al. (contribution 3) combine unmanned aerial vehicle (UAV)-borne Light Detection and Ranging (LiDAR) camera fusion with learning-based segmentation to recover bridge geometry at scale. Wang et al. (contribution 4) provide a review of wireless sensor networks for viaduct structural health monitoring (SHM). Together, these works sketch a new paradigm of multi-sensor intelligence, digital twins, and machine learning that closes the loop from sensing to action.

3. Decarbonization: Adoption Incentives and Community Practice

This theme examines the drivers and levers that turn carbon intent into measurable outcomes. Jin et al. (contribution 5) develop and validate an integrated Technology Acceptance Model and Technology Organization Environment (TAM-TOE) framework for hospital construction. Li et al. (contribution 6) provide a bibliometric analysis of human settlement improvement from 2012 to 2022. Gu et al. (contribution 7) classify and quantify resident engagement in sponge-style old community renewal, linking micro-level participation to performance and maintenance of low-impact development (LID) facilities.

Collectively, these contributions form a complementary evidence chain across technology adoption, knowledge landscapes, and community governance.

4. Resilience: Data Governance Participation and Spatial Design

Resilience here spans engineering, data, social, and spatial dimensions. Xie et al. (contribution 8) trace how social risks in urban renewal evolve among stakeholders. Wu et al. (contribution 9) show that megaproject social responsibility behaviors improve performance, with resource integration capacity acting as a mediator. Hua et al. (contribution 10) link data governance capabilities to project organization resilience. Li and Lin (contribution 11) use multisource data and machine learning to map the impact of streetscape features on social sensing. Meknaci et al. (contribution 12) apply space syntax to reveal how connectivity and accessibility enhance sustainability in a Saharan city. The through line is clear: data governance, responsible action, participatory co-management, and spatial optimization must be intertwined to deliver comprehensive resilience.

5. Conclusions

Urban infrastructures and resilient, sustainable buildings are entering a decisive decade where digital intelligence, low-carbon pathways, and systemic resilience will no longer be optional, but essential. Looking ahead, four directions are particularly critical for advancing urban infrastructures and buildings. First, infrastructures across transport, energy, water, and buildings must be integrated to deliver systemic efficiency and resilience. Second, large-scale AI models and cross-modal learning should drive predictive maintenance and intelligent governance at the city scale. Third, low-carbon mechanisms must couple technology adoption with policy and economic incentives to secure tangible emission reductions. Finally, participatory and risk-sensitive governance needs to embed social responsibility and stakeholder co-management into urban renewal and megaprojects.

Funding: This work was supported by National Natural Science Foundation of China (No. 72201057); Social Science Foundation of Jiangsu Province (No. 23GLC020); National Natural Science Foundation of China (No. 72104233); China Postdoctoral Science Foundation (No. 2023M743767); and Science and Technology Development Fund, Macao S.A.R (FDCT project No. 0022/2024/ITP1).

Acknowledgments: The Guest Editors gratefully acknowledge the authors for their high-quality submissions and the anonymous reviewers whose rigorous feedback enhanced every manuscript. We also thank the Buildings editorial team for their professional support.

Conflicts of Interest: The authors declare that they have no conflicts of interest.

List of Contributions:

1. Huang, X.; Duan, Z.; Hao, S.; Hou, J.; Chen, W.; Cai, L. A Deep Learning Framework for Corrosion Assessment of Steel Structures Using Inception v3. *Buildings* **2025**, *15*, 512. https://doi.org/10.3390/buildings15040512.
2. Yang, S.W.; Lee, Y.; Kim, S.A. Design and Validation of a Real-Time Maintenance Monitoring System Using Building Information Modeling and Digital Twin Integration. *Buildings* **2025**, *15*, 1312. https://doi.org/10.3390/buildings15081312.
3. Jiang, S.; Yang, Y.; Gu, S.; Li, J.; Hou, Y. Bridge Geometric Shape Measurement Using LiDAR–Camera Fusion Mapping and Learning-Based Segmentation Method. *Buildings* **2025**, *15*, 1458. https://doi.org/10.3390/buildings15091458.
4. Wang, T.; Cui, T.; Qi, X.; Huan, X. Urban Viaduct Structural Health Monitoring: A Review of Wireless Sensor Approaches. *Buildings* **2025**, *15*, 1619. https://doi.org/10.3390/buildings15101619.

5. Jin, L.; Liang, M.; Zhu, H.; Ni, X.; Zhang, J. Determinants of Low-Carbon Technology Adoption in Hospitals: Integrating Technology–Organization–Environment and Technology Acceptance Model Perspectives. *Buildings* **2025**, *15*, 2703. https://doi.org/10.3390/buildings15152703.
6. Li, C.; Miao, J.; Zhai, X.; Liu, Q. A Bibliometric Analysis of Research on Human Settlements Environment Improvement. *Buildings* **2025**, *15*, 1805. https://doi.org/10.3390/buildings15111805.
7. Gu, T.; Xie, M.; Hao, E.; Wang, Y.; Zhou, S. Classifying and Quantifying Resident Engagement in the Sponge-Style Old Community Renewal of China. *Buildings* **2024**, *14*, 3773. https://doi.org/10.3390/buildings14123773.
8. Xie, H.; Zhang, L.; Cui, P.; Yuan, J.; Li, Q. Exploring the Evolution Mechanisms of Social Risks Associated with Urban Renewal from the Perspective of Stakeholders. *Buildings* **2024**, *14*, 1470. https://doi.org/10.3390/buildings14051470.
9. Wu, Y.; Zhou, Z.; Xie, L.; Xia, B.; Huang, M. A Bayesian Network Model of Megaproject Social Responsibility Behavior and Project Performance. *Buildings* **2024**, *14*, 1143. https://doi.org/10.3390/buildings14041143.
10. Hua, Y.; Kang, M.; Yao, H.; Fu, Y. How to Foster Project Organization Resilience: The Mechanism of Data Governance Capabilities. *Buildings* **2025**, *15*, 1219. https://doi.org/10.3390/buildings15081219.
11. Li, K.; Lin, Y. Mapping the Impact of Spontaneous Streetscape Features on Social Sensing in the Old City of Quanzhou, China. *Buildings* **2025**, *15*, 1522. https://doi.org/10.3390/buildings15091522.
12. Meknaci, M.E.F.; Wang, X.; Brown, R.W. A Space Syntax Analysis of the Urban Spatial Structure of Bechar, Algeria and How It Affects Sustainability. *Buildings* **2024**, *14*, 2103. https://doi.org/10.3390/buildings14072103.

References

1. Li, J.; Shirowzhan, S.; Pignatta, G.; Sepasgozar, S.M. Data-driven net-zero carbon monitoring: Applications of geographic information systems, building information modelling, remote sensing, and artificial intelligence for sustainable and resilient cities. *Sustainability* **2024**, *16*, 6285. [CrossRef]
2. Omrany, H.; Mehdipour, A.; Oteng, D. Digital twin technology and social sustainability: Implications for the construction industry. *Sustainability* **2024**, *16*, 8663. [CrossRef]
3. Zhou, S.; Liu, K.; Li, D.; Fu, C.; Ning, Y.; Ji, W.; Liu, X.; Xiao, B.; Wei, R. Augmenting general-purpose large-language models with domain-specific multimodal knowledge graph for question-answering in construction project management. *Adv. Eng. Inform.* **2025**, *65*, 103142. [CrossRef]
4. Wu, Y.; Zhou, Z.; Xie, L.; Xia, B.; Huang, M. A Bayesian Network Model of Megaproject Social Responsibility Behavior and Project Performance. *Buildings* **2024**, *14*, 1143. [CrossRef]
5. Omar, O. Digital Twins for Climate-Responsive Urban Development: Integrating Zero-Energy Buildings into Smart City Strategies. *Sustainability* **2025**, *17*, 6670. [CrossRef]
6. Tan, Z.; Li, Z. Digital twins for sustainable design and management of smart city buildings and municipal infrastructure. *Sustain. Energy Technol. Assess.* **2024**, *64*, 103682. [CrossRef]
7. Zhou, S.; Wang, H.; Li, D.; Ng, S.T.; Wei, R.; Zhao, Y.; Zhou, Y. Revealing public attitudes toward mobile cabin hospitals during COVID-19 pandemic: Sentiment and topic analyses using social media data in China. *Sustain. Cities Soc.* **2024**, *107*, 105440. [CrossRef]
8. Maiullari, D.; Nageli, C.; Rudena, A.; Isacson, Å.; Dokter, G.; Ellenbroek, I.; Wallbaum, H.; Thuvander, L. Digital twin for supporting decision-making and stakeholder collaboration in urban decarbonization processes. A participatory development in Gothenburg. *Environ. Plan. B Urban Anal. City Sci.* **2024**. [CrossRef]
9. Zhou, S.; Ng, S.T.; Huang, G.; Dao, J.; Li, D. Extracting interrelated information from road-related social media data. *Adv. Eng. Inform.* **2022**, *54*, 101780. [CrossRef]
10. Hua, Y.; Kang, M.; Yao, H.; Fu, Y. How to Foster Project Organization Resilience: The Mechanism of Data Governance Capabilities. *Buildings* **2025**, *15*, 1219. [CrossRef]

buildings

MDPI

Article

A Deep Learning Framework for Corrosion Assessment of Steel Structures Using Inception v3 Model

Xinghong Huang [1], Zhen Duan [1], Shaojin Hao [1], Jia Hou [2,3,*], Wei Chen [2,3] and Lixiong Cai [2,3]

1 CSCEC 7th Division International Engineering Co., Ltd., Guangzhou 510080, China; 349228@whut.edu.cn (X.H.); 274419@whut.edu.cn (Z.D.); lh2691121083@126.com (S.H.)
2 School of Civil Engineering and Architecture, Wuhan University of Technology, Wuhan 430062, China; iamhappychen@163.com (W.C.); cailixiong@whut.edu.cn (L.C.)
3 Sanya Science and Education Innovation Park, Wuhan University of Technology, Sanya 572024, China
* Correspondence: houjia@whut.edu.cn

Abstract: Corrosion detection plays a crucial role in the effective lifecycle management of steel structures, significantly impacting maintenance strategies and operational performance. This study presents a machine vision-based approach for classifying corrosion levels in Q235 steel, providing valuable insights for lifecycle assessment and decision-making. Accelerated salt spray tests were performed to simulate corrosion progression over multiple cycles, resulting in a comprehensive dataset comprising surface images and corresponding eight loss measurements. A comparative evaluation with other architectures, namely, AlexNet, ResNet, and VggNet, demonstrated that the Inception v3 model achieved superior classification accuracy, exceeding 95%. This method offers an effective and precise solution for corrosion evaluation, supporting proactive maintenance planning and optimal resource allocation throughout the lifecycle of steel structures. By leveraging advanced deep learning techniques, the approach provides a scalable and efficient framework for enhancing the sustainability and safety of steel infrastructure.

Keywords: steel structure; corrosion level; machine vision; convolutional neural network; Inception V3

1. Introduction

Steel structures have gained widespread adoption due to their advantages, including lightweight design, high industrialization, environmental sustainability, energy efficiency, and recyclability [1]. Although steel structures offer numerous advantages, their susceptibility to corrosion in environments with high salt spray concentrations has multifaceted implications for long-term infrastructure planning [1,2]. Economically, corrosion significantly increases the full lifecycle cost of projects. During the construction phase, additional anti-corrosion measures must be implemented, raising the initial cost. In the operational phase, continuous capital investment is required for frequent inspections, maintenance, and repairs. In terms of safety, corrosion threatens the strength and stability of steel structures. A safety incident resulting from corrosion could lead to severe casualties, property damage, and social disruption. Therefore, in-depth research into the corrosion issues of steel structures and the exploration of effective solutions are crucial for ensuring the safe and stable operation of long-term infrastructure. Currently, manual visual inspection is the most commonly used method in engineering. This approach depends on the inspector's experience, leading to low efficiency, high subjectivity, elevated costs, and considerable safety risks [1–3]. Although non-destructive testing techniques such as ultrasonic testing, eddy current testing, and infrared testing offer accurate results and are frequently used for factory testing of parts and components, they often incur prohibitive time and cost implications when applied to large spans and complex structures in engineering [3–6]. With the continuous advancements in image processing and neural network technology, machine vision detection has emerged as a new inspection method [7–10]. The advent of Convolutional Neural Networks (CNNs) has greatly enhanced the accuracy of image classification and object detection. They not only improve the accuracy and efficiency of inspections but also provide a more intelligent solution for the maintenance of steel structures [11–13].

Convolutional Neural Networks (CNNs) are capable of automatically extracting important features from images through multi-layer convolution operations and performing complex pattern recognition layer by layer during corrosion detection tasks [14–18]. This capability for automated feature extraction and learning enables CNNs to handle corrosion images under different scales and environmental conditions, thereby improving detection accuracy and consistency. Additionally, the deep learning structure of CNNs demonstrates better generalization ability when processing large volumes of data, allowing adaptation to various steel surface issues such as cracks and rust [9,13,19,20]. Faster R-CNN enhances detection accuracy by precisely locating corrosion regions using Region Proposal Networks (RPNs) and further improving it with a classification network [21–23]. U-Net and SegNet focus on image segmentation tasks; through encoder–decoder architectures and feature decoders, they can accurately segment corrosion areas and handle corrosion images with high complexity and noise [24–26]. AlexNet, as an early deep learning model, although having fewer layers, effectively extracts high-level features from images, laying the foundation for steel structure corrosion detection [27–29]. VGGNet, with its deep convolutional network structure, can identify complex corrosion patterns and perform fine analysis of subtle corrosion phenomena. ResNet addresses the gradient vanishing problem in deep network training by introducing residual blocks, resulting in higher precision and stability when processing complex corrosion images [30].

The Inception module in Convolutional Neural Networks (CNNs) enhances network expressiveness and efficiency by applying various convolution and pooling operations in parallel, enabling the extraction of features at multiple scales [31,32]. The introduction of depthwise separable convolutions and 1 × 1 convolutions reduces computational complexity, enabling the model to operate efficiently in environments with limited computational resources [33,34]. Inception v3 further optimizes this design by combining convolutional kernels of different sizes, improving its ability to recognize complex images. With its multi-scale feature extraction capabilities and efficient computational performance, Inception v3 provides accurate and real-time detection results for steel structure corrosion images. These characteristics make it particularly effective in addressing complex corrosion patterns [35–37].

In this study, weight loss and surface images of Q235 steel at various corrosion cycles were obtained through indoor accelerated salt spray corrosion experiments. A corrosion level standard was established to determine the approximate corrosion levels of the images, which were used as the dataset for the classification model. To validate the Inception v3 model's effectiveness for corrosion classification, three classic CNN architectures—AlexNet, ResNet, and VGGNet—were selected as control models. These models were compared with the lightweight Inception v3 model using the same dataset and accuracy (Acc) as the primary evaluation metric while keeping model training hyperparameters consistent. This comparison assessed the applicability and accuracy of the Inception v3 model for corrosion classification, offering a new technological approach for the efficient operation and maintenance of steel structure engineering.

2. Related Works

2.1. Inception

The Inception module in CNNs is an innovative architecture known for its modular design and multi-path parallel processing [38,39]. It combines convolutional kernels of different sizes with pooling layers, allowing the model to extract features at various scales while reducing computational complexity through 1×1 convolutions [40]. The core advantage of the Inception module lies in its flexible structure and efficient feature extraction capabilities, making it highly effective across multiple visual tasks and achieving a good balance in resource utilization [41].

The most direct way to enhance network performance is by increasing the depth of the convolutional network. However, as the number of layers increases, so do the parameters, which can lead to issues such as overfitting and gradient vanishing. Additionally, the explosion in parameter numbers causes a significant increase in memory usage during network training, leading to longer training times and suboptimal training results [42]. As shown in Figure 1, the core idea of the Inception module is to combine different convolutional layers in parallel. The result matrices processed by these convolutional layers are concatenated along the depth dimension to form a deeper matrix. The Inception module can be stacked repeatedly to form larger networks, allowing for efficient expansion in both depth and width. This approach enhances the accuracy of deep learning networks while preventing overfitting [43,44]. An advantage of the Inception module is its ability to perform dimensionality reduction on larger matrices while aggregating visual information at different sizes, facilitating feature extraction from various scales [45–47].

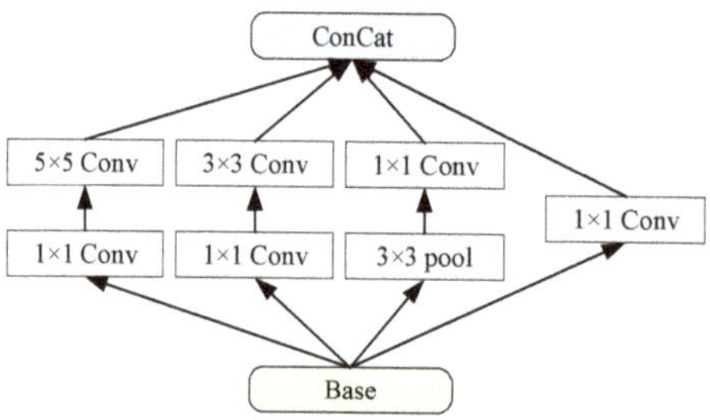

Figure 1. Inception module.

The Inception module is capable of flexibly selecting appropriate computational paths while maintaining high performance. The introduction of depthwise separable convolutions and 1×1 convolutions reduces computational complexity and parameter count, enabling the model to operate efficiently in resource-constrained environments [38–40]. Furthermore, parameter sharing and path optimization within the module further decrease computational overhead, making the Inception module particularly well suited for lightweight applications. It simultaneously extracts multi-scale and multi-level features, enhancing the network's expressive power [41–43]. The Inception module overcomes the computational bottlenecks associated with deep networks in traditional convolutional architectures and can effectively adapt to various complex tasks, as demonstrated by its widespread use in several CNN variants, including Inception v3 [48–50].

The Inception v3 model is an efficient architecture within CNNs, incorporating a "modular" design that combines convolutional kernels and pooling operations of different sizes to extract rich features. This approach allows the model to capture features at various scales while reducing computational burden [51,52]. By introducing combinations of multiple kernel sizes, the network can extract features across different scales, thereby improving its ability to recognize complex images. The Inception v3 model offers significant advantages in the construction industry. Its high accuracy excels in classifying corrosion levels, where subtle differences are evident, and its deep network structure optimizes computational efficiency through branching mechanisms. This reduces the demand for computational resources, enabling real-time monitoring of large-scale building structures [53–55].

Inception v3 effectively extracts features at different scales, enhancing the ability to identify various building surface issues such as cracks and corrosion. The model performs stably under different conditions and adapts well to actual building inspection environments [48–50]. These advantages make Inception v3 highly promising for structural health monitoring and maintenance in construction. Through efficient image analysis, it can assist in identifying building types, detecting damage, monitoring construction quality, and aiding design optimization and automated measurement. These applications significantly enhance the management and maintenance efficiency of construction projects.

2.2. AlexNet

AlexNet marked the beginning of the era of deep convolutional networks. With its powerful image processing capabilities, AlexNet can accurately identify different types of buildings, monitor construction progress in real time, detect potential issues in building structures, and assist in optimizing architectural design, thereby enhancing management efficiency and quality control in construction projects [56–58]. Compared to LeNet, AlexNet utilizes ImageNet as the training dataset, resulting in a substantial increase in both the number of image categories and the amount of data. The incorporation of dropout, a regularization technique, mitigates the issue of network overfitting. Additionally, the use of the ReLU activation function, which has a faster convergence rate, improves model training efficiency [59–61].

The application of AlexNet in steel structure corrosion detection primarily leverages its powerful feature extraction capabilities provided by its deep convolutional neural network. As a classic convolutional neural network, AlexNet effectively extracts complex features from images of steel structure surfaces, capturing subtle corrosion details through its multi-layered network structure. Its robust image classification performance enables AlexNet to identify and classify varying degrees of corrosion, facilitating efficient and accurate detection of steel structure corrosion and providing a scientific basis for structural maintenance [62–66].

2.3. ResNet

ResNet builds upon the VGG architecture by introducing residual connections (skip connections), addressing the gradient vanishing problem in deep networks. This innovation enables the network to train deeper layers, improving both model performance and accuracy. In summary, ResNet enhances the depth and learning capability of the network based on VGG's design principles through its residual structure [67–69]. ResNet introduces a residual module named "residual", as shown in Figure 2. In a typical network module, the desired nonlinear mapping is F(x). With the introduction of an identity shortcut path, the desired mapping becomes H(x) = F(x) + x. The purpose of this adjustment is to combine the learned F(x) with the original x to better capture features. Compared to F(x), H(x) is easier to optimize. In extreme cases, if F(x) = 0, then H(x) = x indicates that the module has not learned any features and has only performed an identity mapping, which ensures that network performance is not adversely affected. In practice, the residual F(x) will not be zero, allowing the stacked layers to learn new features based on the input features, thereby achieving better performance.

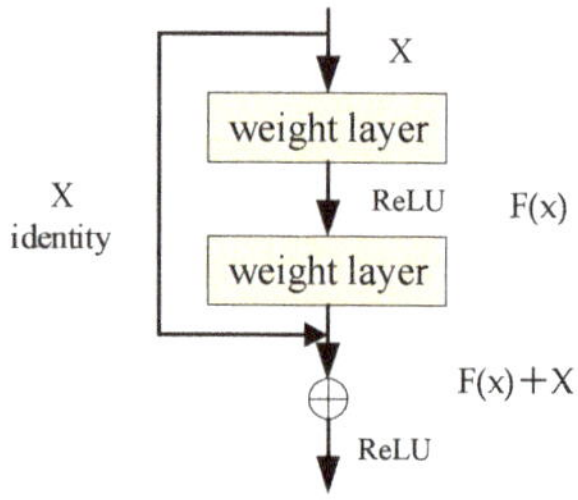

Figure 2. Residual module.

The results indicate that, although ResNet uses convolutional kernels of the same scale as VGG-19, its resolution of the network degradation problem allows it to construct deeper networks [70–72]. Compared to VGG, ResNet achieves lower training errors and higher accuracy on test sets. In the construction industry, ResNet's applications include building image classification, construction progress monitoring, and structural defect detection. Its innovative residual learning mechanism enables deep networks to effectively capture complex features in building images, accurately identify different types of buildings, detect issues such as cracks and corrosion in structural elements, and monitor construction quality and progress in real time. This enhances management efficiency and maintenance effectiveness for construction projects [71–75].

2.4. SegNet

ResNet has further advanced the exploration of network depth by addressing challenges in training deep networks [76,77]. SegNet improves upon the FCN's transmission logic by modifying the VGG-16 network as its backbone structure. During encoding, the model extracts features through convolution operations, with SegNet employing "same" convolutions to maintain the same dimensions as the original image after convolution. During decoding, the model also uses "same" convolutions, but these convolutions serve to enrich the information in the upsampled image, allowing information lost during pooling to be recovered through learning in the decoding process. SegNet's ability to recover information lost during pooling is particularly beneficial for construction applications. By maintaining spatial integrity and enhancing boundary detection, SegNet provides accurate and reliable segmentation masks, enabling precise structural detail extraction. This capability is crucial for tasks such as damage assessment, condition monitoring, and maintenance planning, ultimately contributing to safer and more efficient construction practices. In the decoding stage, the downsampled feature maps are upsampled, followed by convolutional processing of the upsampled images to refine the geometric shapes of objects in the image. This process restores the features obtained during encoding to the specific pixel locations in the original image [78–80].

SegNet is primarily used in the construction industry for image segmentation tasks, such as automated building outline recognition, structural detail extraction, and construction area marking. Its efficient encoder–decoder network structure allows SegNet to accurately segment different regions in building images, identify and label building components or damaged areas, and enhance the accuracy of design plans and the efficiency of construction process monitoring [81–83].

3. Methodology

3.1. Datasets

As the primary model for corrosion level assessment in this study, the classification model must ensure that the training dataset is large and representative. Q235 steel is a commonly used carbon structural steel, known for its good machinability and weldability, and is widely applied in construction, bridges, and machinery. Its chemical composition and physical properties make it representative of the general corrosion behavior of steel, making it ideal for corrosion experiments. Using Q235 steel ensures that the experimental results are both representative and practical, providing valuable insights for corrosion protection in steel structures. Therefore, this study selects Q235 specimens, as shown in Figure 3, to investigate the corrosion behavior of Q235 steel in a salt fog environment. Physical data, including surface changes and mass loss across different corrosion cycles, are collected, and the corrosion level for each cycle is determined according to relevant standards [23]. Additionally, photographs are taken of each specimen for every cycle to capture images corresponding to the corrosion state, and these images, along with the corrosion levels, constitute the dataset for steel corrosion levels.

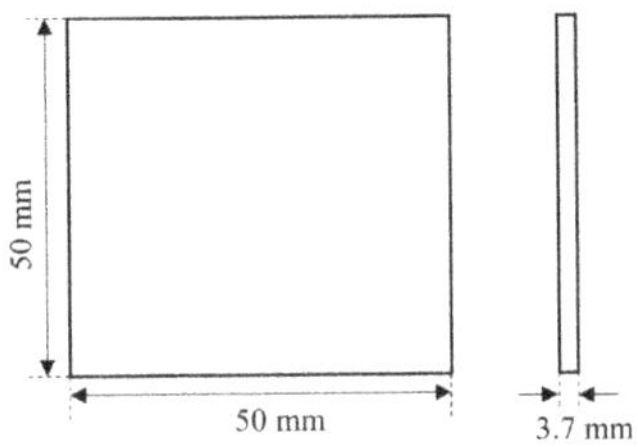

Figure 3. Specifications of experimental specimens.

3.1.1. Salt Fog Test

As shown in Figure 3, the specimens used in this experiment are 50 mm × 50 mm × 3.7 mm in size and made of Q235B steel. Figure 4 shows the most pristine polished steel sheet. The specific types of materials and the number of specimens used in the experiment are detailed in Table 1. Table 2 presents the list of experimental equipment.

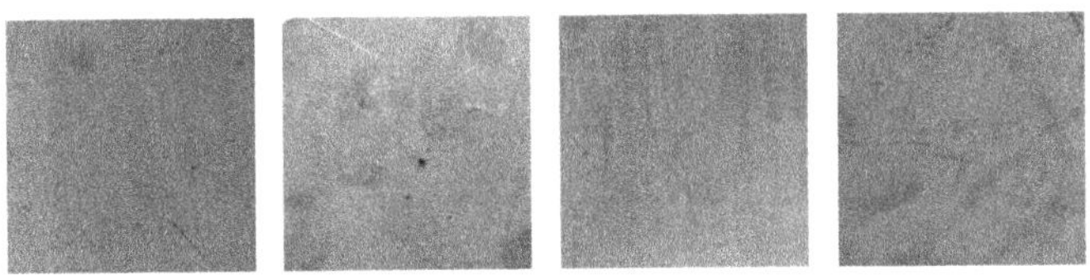

Figure 4. Experimental specimen setup.

Table 1. Experimental material setup.

Experiment Number	Specimen Material	Quantity (Pieces)
1	Q235	40
2	Q235	40
3	Q235	40
4	Q235	40
5	Q235	40

The salt fog test was conducted over 8 cycles, with each cycle lasting 4 h. A 5% NaCl (w/v) solution was used to generate the salt fog, and the experiment followed the American Society for Testing and Materials (ASTM) standard ASTM B117-19 [84]. The experimental steps are as follows:

(1) Prepare the reagents and add a sufficient amount to the salt fog chamber.
(2) Open the salt fog chamber, place the specimens inside as shown in Figure 5, and adjust the working environment (temperature, air pressure, etc.).

(3) After the first experimental cycle, remove the fog from the salt fog chamber, take out the specimens, and rinse the surface of the specimens with distilled water to remove salt.
(4) Place all specimens in the constant temperature drying oven to dry.
(5) Take photographs to obtain images of the corrosion on the specimens' surfaces.
(6) Remove the specimens (do not return them to the salt fog chamber), use an ultrasonic acid cleaning machine to remove surface corrosion, and dry them.
(7) Weigh the cleaned specimens, record the data, and determine the corrosion level based on mass loss and the proportion of the corrosion area.
(8) Repeat steps (2)–(7) until the full cycle experiment is completed.

Table 2. Experimental equipment list.

Equipment Name	Model	Purpose
Electronic Salt Fog Chamber (JITAI Corporation, Beijing, China)	JiTai 90	Indoor accelerated salt fog laboratory
Silent Air Compressor (AOTUSI Corporation, Taizhou, China)	ZhiPu 1490W-30L	Supplying air pressure to the salt fog chamber
Constant Temperature Drying Oven (SUBO Corporation, Shaoxing, China)	SuBo 101-00B	Drying specimens
Ultrasonic Cleaning Machine (SHANGPU Corporation, Shanghai, China)	ShangYi SN-QX-08E	Cleaning corrosion sites
Precision Electronic Scale (YOUKEWEITE Corporation, Kunshan, China)	LeQi 0.001 g	Weighing specimens

Figure 5. Placement of specimens in the salt fog chamber.

A preliminary experiment is first conducted to observe and determine the corrosion cycles and characteristics of the material in the salt fog environment, preventing potential data gaps in the formal experiments due to an overly large cycle span. The preliminary experiment on Q235 steel revealed the following patterns: initially, the corrosion rate of the specimens is high, and surface changes are clearly noticeable. However, once extensive corrosion coverage is reached, the corrosion rate decreases, and surface changes become less observable to the naked eye. In response to this, non-uniform intervals for the experimental cycles were established based on the preliminary experiment. Table 3 and Figure 6 illustrate the time intervals and corresponding corrosion conditions for each cycle, using Q235 steel specimens as an example. The formal experiment was then carried out in accordance with the corrosion patterns outlined in Table 3 to collect image and mass loss data for each specimen.

Table 3. Time corrosion cycles and surface conditions.

Title 1	Title 2	Title 3	Title 4
1	4 h	4 h	Pitting corrosion observed
2	8 h	4 h	Pitting corrosion intensifies and spreads further
3	12 h	4 h	Spot corrosion appears
4	24 h	12 h	Extensive corrosion observed
5	48 h	24 h	Corrosion almost fully covered
6	72 h	24 h	Corrosion fully covered with deepened color
7	96 h	24 h	Deepened corrosion color and beginning to flake off

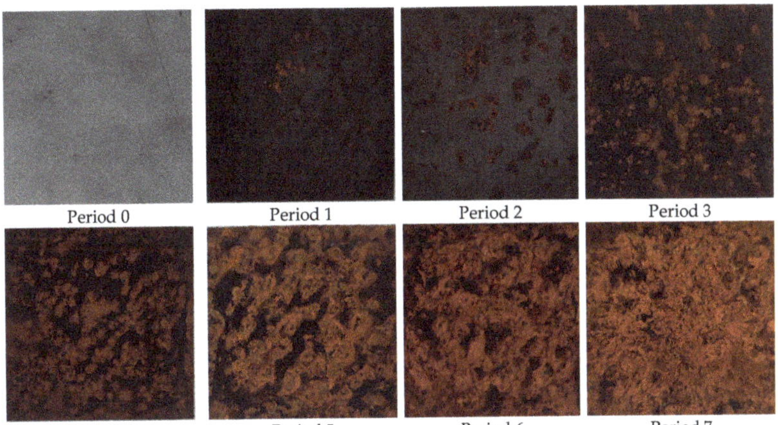

Figure 6. The surface of each corrosion period of the specimen.

3.1.2. Classification Dataset

1. Judgment Criteria

Corrosion level refers to the extent of corrosion on metal specimens and is another key aspect of this experiment. It has been found that surface observation alone is insufficient for accurately determining the corrosion level, especially when the specimen is fully covered by corrosion in the later stages, making surface changes less noticeable. Therefore, it was decided to use the weight loss method in conjunction with surface observation to determine the corrosion level. To ensure consistency and synchronization, we strictly control the time intervals between surface imaging and weight loss measurements in each experiment. Standardized sample handling procedures are employed to minimize error sources. During preliminary experiments, it was observed that the mass of the specimens might initially increase before decreasing over time due to corrosion loss. This makes weight gain calculations inadequate for accurately reflecting the corrosion extent, particularly in the later stages of the experiment. Consequently, in the formal experiment, the weight loss method (mass loss method) was adopted to calculate the mass lost due to corrosion oxidation. According to the American Society for Testing and Materials (ASTM) standard ASTM G1-03 [85] for the preparation, cleaning, and evaluation of corrosion specimens, the formula for calculating the corrosion rate using weight loss is given in Equation (1).

$$\eta = (W_0 - W_1)/W_0 \tag{1}$$

In the formula, η represents the corrosion rate, W_0 is the initial mass of the specimen, and W_1 is the mass of the specimen measured during the experiment.

The final corrosion levels, corresponding corrosion cycles, mass loss, and surface corrosion conditions are summarized in Table 4.

Table 4. Corrosion level system.

Corrosion Level	Corrosion Cycle	Mass Loss	Surface Corrosion Condition
0	Initial	0%	Smooth surface, no corrosion
1	1	0–0.5%	Pitting, corrosion at scratches
2	2, 3	0.5–2%	Patches of corrosion, strip corrosion
3	3, 4, 5	2–5%	Extensive corrosion, full coverage
4	5, 6, 7	>5%	Complete coverage, uneven surface, deepened color

2. Dataset

The corrosion level dataset consists of surface images of specimens at various corrosion cycles obtained from the experiments, along with the corresponding corrosion levels determined based on mass loss and surface corrosion conditions. The samples are organized into folders according to their corrosion levels, with images of the same corrosion level stored in the same folder and named accordingly. These images are then stored in the training files for the steel corrosion level classification model. Due to time constraints, a total of 651 valid corrosion images were collected and classified. The classification was based on image quality, clarity of corrosion features, consistency across experimental cycles, and relevance to the experimental conditions. Each experimental cycle provided approximately 150 valid images. To mitigate the issue of limited data, data augmentation techniques were employed, as shown in Figure 7. These techniques, including rotation and flipping, were applied to generate additional images within each category, thus expanding the dataset. After augmentation, a total of 2000 images were randomly selected from each category, with 400 images per category. The dataset was then split into training, validation, and testing sets, as shown in Table 5.

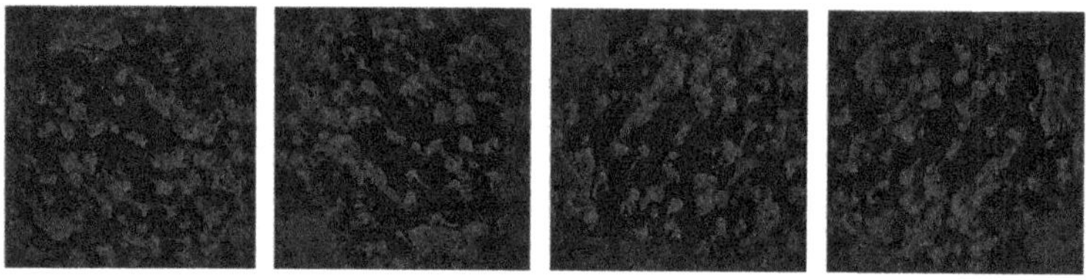

Figure 7. Data enhancement.

Table 5. Dataset partitioning for classification training.

Dataset	Level 0	Level 1	Level 2	Level 3	Level 4	Total
Training Set	320	320	320	320	320	1600
Validation Set	40	40	40	40	40	200
Testing Set	40	40	40	40	40	200

3.2. Inception Module Lightweight

The Inception module's lightweight design significantly reduces computational complexity and model size, thereby enhancing computational efficiency and real-time performance. This is achieved through several key optimizations: reducing convolutional kernel sizes, incorporating depthwise separable convolutions, and employing parameter-sharing techniques. These strategies enable substantial reductions in computational demands while preserving robust feature extraction capabilities [40–43]. As a result, the model becomes more suitable for deployment in devices with limited computational resources. Larger convolutional kernels, such as 5×5 or 7×7, incur disproportionately higher memory usage and computational costs. For example, a 5×5 convolution operation requires 2.78 times more computation than a 3×3 convolution with an equal number of filters. Since Inception networks are fully convolutional, each weight corresponds to a multiplication for each activation. Therefore, any reduction in computational cost directly translates to a decrease in the number of parameters. Through appropriate decomposition, more disentangled parameters can be obtained, which accelerates training speed and improves overall model efficiency [48–50]. The modified Inception module structure is illustrated in Figures 8 and 9. These modules will be directly applied in the subsequent Inception v3 model.

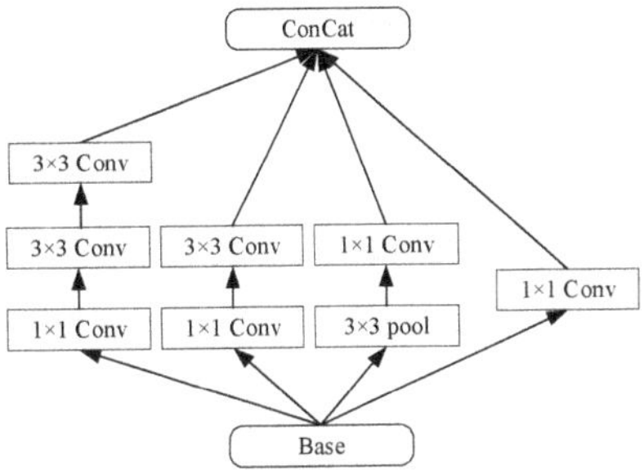

Figure 8. Replacement of 5 × 5 convolution with 3 × 3 convolutions.

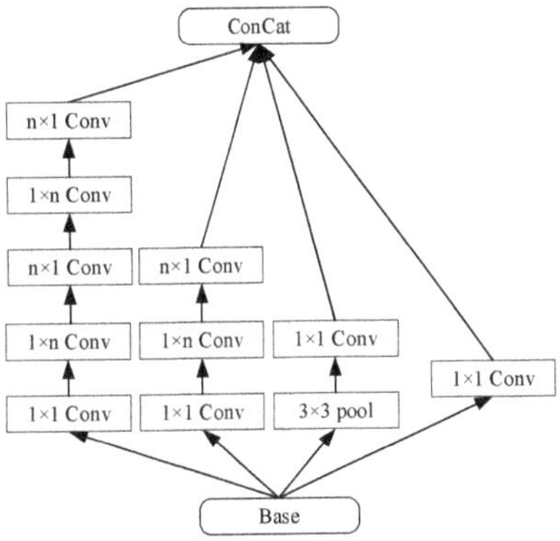

Figure 9. Replacement of high-dimensional convolutions with 1 × n and n × 1 convolutions.

3.3. *Inception v3*

The 3 × 3 convolution module, as shown in Figure 8, is utilized as the initial Inception module. The 1 × n module, depicted in Figure 9 with n set to 7, is applied to the intermediate Inception layers. Convolutions through this module effectively simulate a 17 × 17 convolution operation, as illustrated in Figure 10. Additionally, a high-dimensional separable Inception module is implemented, as shown in Figure 11, before the final output classification layer. This module expands feature dimensions to generate high-dimensional sparse features, thereby increasing the number of features. The specific structure of the Inception v3 model is detailed in Table 6.

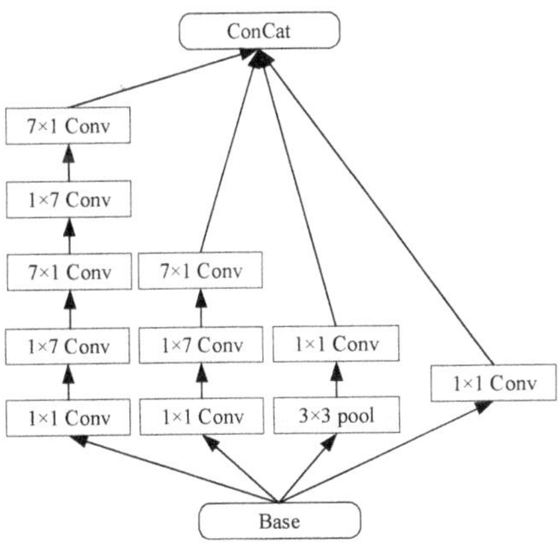

Figure 10. Replacing high-dimensional convolutions with 1 × 7 and 7 × 1 convolutions.

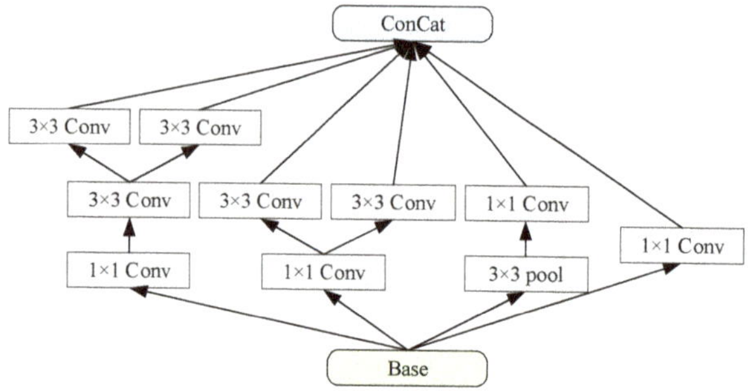

Figure 11. High-dimensional separable Inception module.

Table 6. Inception v3 architecture.

Type	Receptive Field/Stride	Input Size
conv	3 × 3/2	299 × 299 × 3
conv	3 × 3/1	149 × 149 × 32
conv padded	3 × 3/1	147 × 147 × 32
pool	3 × 3/2	147 × 147 × 64
conv	3 × 3/1	73 × 73 × 64
conv	3 × 3/2	71 × 71 × 80
conv	3 × 3/1	53 × 53 × 192
3 × Inception	Figure 8	35 × 35 × 288
5 × Inception	Figure 10	17 × 17 × 768
2 × Inception	Figure 11	8 × 8 × 1280
pool	8 × 8	8 × 8 × 2048
linear	logits	1 × 1 × 2048
softmax	classifier	1 × 1 × 1000

4. Experiments and Results

4.1. *Training and Finetuning*

Although the model has undergone lightweight improvements, issues such as a high number of training parameters and significant data memory usage still persist. To ensure smooth model training, computing devices with higher configurations that meet the training requirements were selected. The software and hardware configurations used for model training are outlined in Table 7.

Table 7. Inception v3 training environment.

Component		Title 3
Hardware	CPU	Intel(R) Core(TM) i7-10700 CPU @290 GHz (Intel Corporation, Santa Clara, CA, USA)
	GPU	NVIDIA GeForce RTX 3060 (NVIDIA Corporation, Santa Clara, CA, USA)
	RAM	DDR4 16 GB (Samsung Electronics, Suwon, Republic of Korea)
Software	Cuda	Cuda 10.0
	IDE	PyCharm Community Edition 2017.2.2
	Programming Language	Python 3.8
	Framework	PyTorch 1.2.0

4.2. *Results and Analysis*

4.2.1. Accuracy Analysis

Accuracy (Acc), also known as the accuracy rate, is a crucial metric for evaluating the performance of a neural network. It measures the proportion of correctly predicted samples to the total number of predicted samples, regardless of whether the samples are positive or negative. If only positive samples are considered in the model, the negative samples can be omitted from the numerator in the formula. Accuracy is one of the most commonly used metrics in neural network evaluations, with values ranging from 0 to 1. A value of 0 indicates that the model has classified all samples incorrectly, while a value of 1 signifies that all samples have been classified correctly. The formula for accuracy is given by Equation (2):

$$\mathrm{Acc} = \frac{TP + TN}{TP + TN + FP + FN} \tag{2}$$

In the formula, TP represents the number of true positive samples (correctly classified positive samples); TN represents the number of true negative samples (correctly classified negative samples); FP denotes the number of false positive samples (incorrectly classified positive samples); and FN indicates the number of false negative samples (incorrectly classified negative samples).

To demonstrate the superior performance of the lightweight Inception v3 model for Q235 steel corrosion classification, comparative experiments were conducted using the same dataset to train AlexNet, ResNet, and VggNet as control models. All models were trained with consistent hyperparameter settings. The classification performance of each model was evaluated based on accuracy (Acc), as shown in Figure 12. It is evident that all four classification models were well trained, with accuracies exceeding 85% after 200 iterations. Among them, AlexNet exhibited the lowest classification performance, with an accuracy of only 87%. Both ResNet and VggNet showed comparable results, achieving accuracies of over 90%. The Inception v3 model demonstrated the highest performance for corrosion classification tasks, with an accuracy of 95.5%. Therefore, it can be concluded that the lightweight Inception v3 model is the most suitable for classifying the corrosion levels of steel.

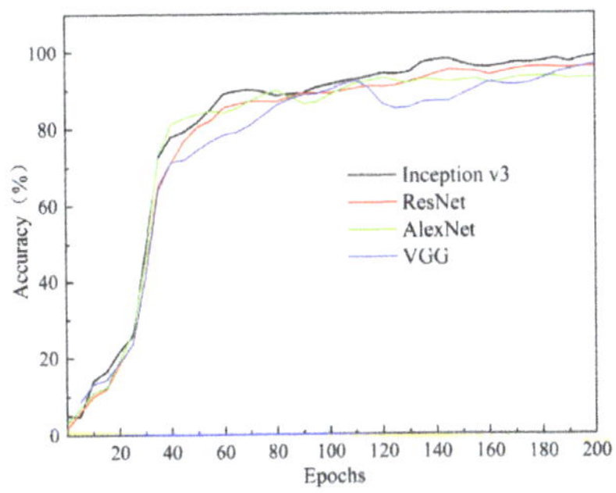

Figure 12. Classification accuracy of various models.

4.2.2. Error Analysis

A confusion matrix is a table used in machine learning and statistics to evaluate the performance of a classification model. It is commonly used to assess the quality of a classifier by providing a visual representation of classification results through matrix data. Its basic structure is illustrated in Table 8.

Table 8. Confusion matrix.

	Predicted Positive	Predicted Positive
Actual Positive	True Positive (TP)	False Negative (FN)
Actual Negative	False Positive (FP)	True Negative (TN)

The classification results of the lightweight Inception v3 network for a total of 200 test samples were exported into a confusion matrix, as shown in Figure 13. The confusion matrix indicates that the model accurately classifies corrosion levels overall, with only a few misclassified images. Notably, for images with a corrosion level of 0, all classifications were correct. This high accuracy is due to the fact that level 0 images have not undergone corrosion, resulting in smooth, uniformly colored surfaces with fewer and more distinct features.

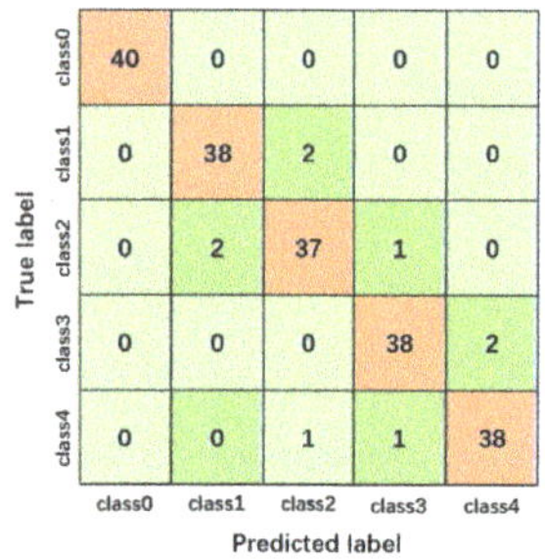

Figure 13. Confusion matrix for Inception v3.

The remaining misclassified test samples across the four corrosion levels were selected and analyzed to investigate the reasons behind these classification errors. Both the original images and the corresponding classification weights were examined, as shown in Figure 14. For Level 1, two test samples—sample "a" and its rotated version—were misclassified as Level 2. The primary reason was that sample "a" had minor surface scratches before the experiment, which, after corrosion, resulted in a yellowed appearance resembling patchy corrosion. This visual similarity likely led the model to classify it as Level 2. Among the 40 test samples for Level 2, three samples—b, c, and d—were misclassified. Sample "b" was incorrectly predicted as Level 1 with an 81.4% probability, possibly because it was captured at an intermediate stage between pitting and patchy corrosion, making it difficult to differentiate, even for human evaluators. Sample "c" was also misclassified as Level 1 due to its extensive corrosion coverage, which blended with the original surface color, preventing the model from effectively capturing corrosion edge features. Sample "d" was erroneously classified as Level 3, likely because the model perceived the corrosion as more severe or misinterpreted the darker corrosion color as a more advanced stage.

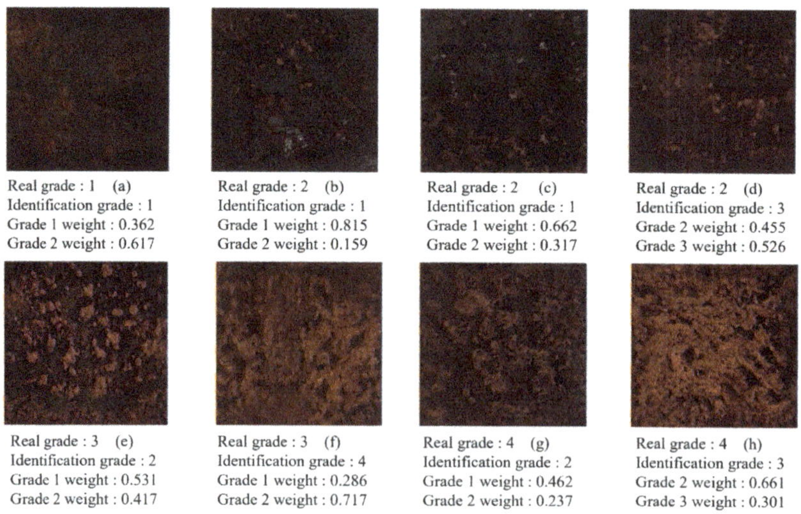

Figure 14. Classification weights for misclassified samples.

For Levels 3 and 4, four samples were misclassified. Sample "e" was misclassified due to pre-cleaning before imaging and poor lighting conditions during capture, which hindered the model's ability to extract corrosion features accurately. Sample "f" was predicted with over a 70% probability as the highest corrosion level (Level 4), while sample "h" was classified as Level 3 with a 66% probability. The challenge in distinguishing these samples arose because, as corrosion advanced, it accumulated and thickened, leading to minimal apparent changes that complicated classification. Sample "g" was misclassified as Level 2, likely because the corrosion had progressed to a stage where peeling occurred, darkening the surface. This visual shift caused the model to misinterpret the darkened regions as non-corroded areas, leading to an incorrect classification.

Overall, the errors encountered by the model are within a controllable range. Most of the misclassified samples result from issues during dataset creation, such as specimen wear, ambiguous classification, and poor lighting conditions during image capture. This underscores the importance of constructing a high-quality dataset before model training. For future classification model research, it is crucial to ensure that the dataset is clearly categorized and features are distinct. A well-constructed dataset is essential for training an effective neural network model.

5. Conclusions

In conclusion, this study demonstrates the effectiveness of a machine vision-based approach for corrosion level classification in Q235 steel, offering a robust tool for lifecycle management and maintenance optimization of steel structures. By integrating accelerated salt spray tests with advanced deep learning techniques, we successfully generated a comprehensive dataset of corrosion progression and achieved a classification accuracy exceeding 95% using the Inception v3 model. This approach not only outperforms traditional methods and other architectures such as AlexNet, ResNet, and VggNet but also provides a scalable and efficient framework for real-world applications. The results highlight the potential of this method to enhance the sustainability, safety, and cost-effectiveness of steel infrastructure through proactive maintenance planning and data-driven decision-making.

While machine vision detection provides significant advantages over traditional inspection methods, large-scale implementation faces challenges such as deployment costs, adaptation to varying environmental conditions, and ensuring reliability across diverse corrosion patterns. Future research should focus on improving model generalization and developing cost-effective deployment strategies to enhance practical applicability. By addressing these challenges, machine vision based corrosion detection can contribute to more efficient and intelligent maintenance strategies for steel structures, particularly in high-risk environments.

Author Contributions: Conceptualization, X.H.; methodology, X.H.; software, X.H.; validation, S.H.; formal analysis, Z.D.; investigation, Z.D.; resources, L.C.; data curation, J.H.; writing—original draft preparation, J.H.; writing—review and editing, S.H.; visualization, W.C.; supervision, W.C.; project administration, W.C.; funding acquisition, L.C. All authors have read and agreed to the published version of the manuscript.

Funding: This research was funded by the Special research and development plan of China Construction Seventh Engineering Division Co., Ltd. (Grant No. CSCEC7b-2023-Z-19).

Data Availability Statement: The original contributions presented in this study are included in the article, further inquiries can be directed to the corresponding author.

Conflicts of Interest: Authors Zhen Duan, Xinghong Huang, and Shaojin Hao are working for the company CSCEC 7th Division International Engineering Co., Ltd. The remaining authors declare that the research was conducted in the absence of any commercial or financial relationships that could be construed as a potential conflict of interest.

References

1. Khayatazad, M.; Honhon, M.; De Waele, W. Detection of corrosion on steel structures using an artificial neural network. *Struct. Infrastruct. Eng.* **2023**, *19*, 1860–1871. [CrossRef]
2. Khayatazad, M.; De Pue, L.; De Waele, W. Detection of corrosion on steel structures using automated image processing. *Dev. Built Environ.* **2020**, *3*, 12. [CrossRef]
3. Han, Q.H.; Liu, X.; Xu, J. Detection and Location of Steel Structure Surface Cracks Based on Unmanned Aerial Vehicle Images. *J. Build. Eng.* **2022**, *50*, 14. [CrossRef]
4. Imran, M.M.H.; Jamaludin, S.; Ayob, A.F.M.; Ali, A.; Ahmad, S.; Akhbar, M.F.A.; Suhrab, M.I.R.; Zainal, N.; Norzeli, S.M.; Mohamed, S.B. Application of Artificial Intelligence in Marine Corrosion Prediction and Detection. *J. Mar. Sci. Eng.* **2023**, *11*, 25. [CrossRef]
5. Han, Q.H.; Zhao, N.; Xu, J. Recognition and location of steel structure surface corrosion based on unmanned aerial vehicle images. *J. Civ. Struct. Health Monit.* **2021**, *11*, 1375–1392. [CrossRef]
6. Han, S.X.; Li, B.D.; Li, W.; Zhang, Y.; Liu, P.Y. Intelligent analysis of corrosion characteristics of steel pipe piles of offshore construction wharfs based on computer vision. *Heliyon* **2024**, *10*, 14. [CrossRef]
7. Xu, Z.C.; Cai, B.Y.; Yan, L.C.; Pang, X.L.; Gao, K.W. Statistical analysis of metastable pitting behavior of 2024 aluminum alloy based on deep learning. *Corros. Sci.* **2024**, *233*, 14. [CrossRef]
8. Yang, L.Y.; Huang, X.B.; Ren, Y.C.; Han, Q.; Huang, Y.C. Steel plate surface defect classification technology based on image enhancement and combination feature extraction. *Eng. Comput.* **2023**, *40*, 1305–1329. [CrossRef]
9. Yu, Q.F.; Han, Y.D.; Lin, W.G.; Gao, X.J. Detection and Analysis of Corrosion on Coated Metal Surfaces Using Enhanced YOLO v5 Algorithm for Anti-Corrosion Performance Evaluation. *J. Mar. Sci. Eng.* **2024**, *12*, 19. [CrossRef]
10. Liu, T.T.; Kang, K.; Zhang, F.; Ni, J.L.; Wang, T.Y. A Corrosion Detection Algorithm Via the Random Forest Model. In Proceedings of the 17th International Conference on Optical Communications and Networks (ICOCN), Zhuhai, China, 16–19 November 2018.
11. Atha, D.J.; Jahanshahi, M.R. Evaluation of deep learning approaches based on convolutional neural networks for corrosion detection. *Struct. Health Monit.* **2018**, *17*, 1110–1128. [CrossRef]
12. Chen, S.K.; Huang, I.F.; Chen, P.H. Applying fully convolutional neural networks for corrosion semantic segmentation for steel bridges: The use of U-Net. In Proceedings of the 10th International Conference on Bridge Maintenance, Safety and Management (IABMAS), Sapporo, Japan, 11–18 April 2021; pp. 341–346.
13. Xu, J.; Gui, C.Q.; Han, Q.H. Recognition of rust grade and rust ratio of steel structures based on ensembled convolutional neural network. *Comput.-Aided Civil Infrastruct. Eng.* **2020**, *35*, 1160–1174. [CrossRef]
14. Barakbayeva, T.; Demirci, F.M. Fully automatic CNN design with inception and ResNet blocks. *Neural Comput. Appl.* **2023**, *35*, 1569–1580. [CrossRef]
15. Li, X.; Hao, T.X.; Li, F.; Zhao, L.Z.; Wang, Z.H. Faster R-CNN-LSTM Construction Site Unsafe Behavior Recognition Model. *Appl. Sci.* **2023**, *13*, 16. [CrossRef]
16. Mannem, K.R.; Mengiste, E.; Hasan, S.; de Soto, B.G.; Sacks, R. Smart audio signal classification for tracking of construction tasks. *Autom. Constr.* **2024**, *165*, 13. [CrossRef]
17. Meng, Q.H.; Zhu, S.Y. Construction Activity Classification Based on Vibration Monitoring Data: A Supervised Deep-Learning Approach with Time Series RandAugment. *J. Constr. Eng. Manag.* **2022**, *148*, 11. [CrossRef]
18. Zou, G.F.; Fu, G.X.; Gao, M.L.; Shen, J.; Yin, L.J.; Ben, X.Y. A novel construction method of convolutional neural network model based on data-driven. *Multimed. Tools Appl.* **2019**, *78*, 6969–6987. [CrossRef]
19. Jiang, X.; Qi, H.; Qiang, X.H.; Zhao, B.S.; Dong, H. A Convolutional Neural Network-Based Corrosion Damage Determination Method for Localized Random Pitting Steel Columns. *Appl. Sci.* **2023**, *13*, 21. [CrossRef]
20. Li, Z.J.; Shao, P.; Zhao, M.H.; Yan, K.; Liu, G.X.; Wan, L.; Xu, X.L.; Li, K.L. Optimized deep learning for steel bridge bolt corrosion detection and classification. *J. Constr. Steel Res.* **2024**, *215*, 13. [CrossRef]
21. Cui, Y.; Lei, D.F. Optimizing Internet of Things-Based Intelligent Transportation System's Information Acquisition Using Deep Learning. *IEEE Access* **2023**, *11*, 11804–11810. [CrossRef]
22. Panmatharit, A.; Jiraraksopakun, Y.; Siripanichgorn, A.; Siricharoen, P. Bolt Looseness Identification using Faster R-CNN and Grid Mask Augmentation. In Proceedings of the 14th Annual Summit and Conference of the Asia-Pacific-Signal-and-Information-Processing-Association (APSIPA ASC), Chiang Mai, Thailand, 7–10 November 2022; pp. 1632–1637.
23. Zhao, D.B.; Li, H. Forward Vehicle Detection Based on Deep Convolution Neural Network. In Proceedings of the 3rd International Conference on Advances in Materials, Machinery, Electronics (AMME), Wuhan, China, 19–20 January 2019.

24. Wang, H.; Cao, G.M.; Liu, J.J.; Wu, S.W.; Li, Z.F.; Liu, Z.Y. Development and application of automatic identification methods based on deep learning for oxide scale structures of iron and steel materials. *J. Mater. Sci.* **2023**, *58*, 17675–17690. [CrossRef]
25. Zhang, M.Y.; Wang, W.L. Deep learning-based extraction and quantification of features in XCT images of steel corrosion in concrete. *Case Stud. Constr. Mater.* **2024**, *20*, 18. [CrossRef]
26. Zhou, G.; Sun, H.Y. Defect Detection Method for Steel Based on Semantic Segmentation. In Proceedings of the IEEE 5th Information Technology and Mechatronics Engineering Conference (ITOEC), Chongqing, China, 12–14 June 2020; pp. 975–979.
27. Jafari, F.; Dorafshan, S.; Kaabouch, N. Segmentation of fatigue cracks in ancillary steel structures using deep learning convolutional neural networks. In Proceedings of the IEEE/ASME International Conference on Advanced Intelligent Mechatronics (AIM), Seattle, WA, USA, 28–30 June 2023; pp. 872–877.
28. Santos, R.; Ribeiro, D.; Lopes, P.; Cabral, R.; Calcada, R. Detection of exposed steel rebars based on deep-learning techniques and unmanned aerial vehicles. *Autom. Constr.* **2022**, *139*, 17. [CrossRef]
29. Zhu, Z.L.; Liang, Y.L. Prediction of Residual Stress of Carburized Steel Based on Machine Learning. *Appl. Sci.* **2020**, *10*, 16. [CrossRef]
30. Bao, N.X.; Zhang, T.; Huang, R.Z.; Biswal, S.; Su, J.Y.; Wang, Y. A Deep Transfer Learning Network for Structural Condition Identification with Limited Real-World Training Data. *Struct. Control Health Monit.* **2023**, *2023*, 8899806. [CrossRef]
31. Ali, R.; Cha, Y.J. Subsurface damage detection of a steel bridge using deep learning and uncooled micro-bolometer. *Constr. Build. Mater.* **2019**, *226*, 376–387. [CrossRef]
32. Chi, Y.L.; Cai, C.Z.; Ren, J.H.; Xue, Y.F.; Zhang, N. Damage location diagnosis of frame structure based on wavelet denoising and convolution neural network implanted with Inception module and LSTM. *Struct. Health Monit.* **2024**, *23*, 57–76. [CrossRef]
33. Liu, Z.; Wang, X.S.; Chen, X. Inception Dual Network for steel strip defect detection. In Proceedings of the 16th IEEE International Conference on Networking, Sensing and Control (ICNSC), Banff, AB, Canada, 9–11 May 2019; pp. 409–414.
34. Ren, J.H.; Cai, C.Z.; Chi, Y.L.; Xue, Y.F. Integrated Damage Location Diagnosis of Frame Structure Based on Convolutional Neural Network with Inception Module. *Sensors* **2023**, *23*, 418. [CrossRef]
35. Bouguettaya, A.; Zarzour, H. CNN-based hot-rolled steel strip surface defects classification: A comparative study between different pre-trained CNN models. *Int. J. Adv. Manuf. Technol.* **2024**, *132*, 399–419. [CrossRef]
36. Ivo, R.F.; Rodrigues, D.D.; Bezerra, G.M.; Freitas, F.N.C.; de Abreu, H.F.G.; Rebouc, P.P. Non-grain oriented electrical steel photomicrograph classification using transfer learning. *J. Mater. Res. Technol-JMRT* **2020**, *9*, 8580–8591. [CrossRef]
37. Sundarrajan, K.; Rajendran, B.K. Explainable efficient and optimized feature fusion network for surface defect detection. *Int. J. Adv. Manuf. Technol.* **2023**, *126*, 1–18. [CrossRef]
38. Ahmed, S.; Cho, S.H. Hand Gesture Recognition Using an IR-UWB Radar with an Inception Module-Based Classifier. *Sensors* **2020**, *20*, 18. [CrossRef] [PubMed]
39. Baixo, S.; Ribeiro, T.; Lopes, G.; Ribeiro, A.F. 3D Face Recognition using Inception Networks for Service Robots. In Proceedings of the IEEE International Conference on Autonomous Robot Systems and Competitions (ICARSC), Santa Maria da Feira, Portugal, 29–30 April 2022; pp. 47–52.
40. Chavan, S.; Nair, L.; Nimbalkar, N.; Solkar, S. Karyotyping of human chromosomes in metaphase images using faster R-CNN and inception models. *Int. J. Imaging Syst. Technol.* **2024**, *34*, 24. [CrossRef]
41. Huang, Q.Q.; Cai, Q.; Chen, Y.; Huang, J.B. Single image dehazing network based on inception module. In Proceedings of the 3rd International Conference on Electronics and Communication; Network and Computer Technology (ECNCT), Xiamen, China, 3–5 December 2021.
42. Liu, F.L.; Qin, D.B.; Yang, S.; Du, R.Y. WS-ICNN algorithm for robust adaptive beamforming. *Wirel. Netw.* **2024**, *30*, 5201–5210. [CrossRef]
43. Wang, J.K.; He, X.H.; Faming, S.; Lu, G.L.; Cong, H.; Jiang, Q.Y. A Real-Time Bridge Crack Detection Method Based on an Improved Inception-Resnet-v2 Structure. *IEEE Access* **2021**, *9*, 93209–93223. [CrossRef]
44. Yong, L.; Bo, Z. An Intrusion Detection Model Based on Multi-scale CNN. In Proceedings of the IEEE 3rd Information Technology, Networking, Electronic and Automation Control Conference (ITNEC), Chengdu, China, 15–17 March 2019; pp. 214–218.
45. Zhai, Z.L.; Feng, S.; Yao, L.Y.; Li, P.H. Retinal vessel image segmentation algorithm based on encoder-decoder structure. *Multimed. Tools Appl.* **2022**, *81*, 33361–33373. [CrossRef]
46. Zhang, C.Z.; Zhang, Y.Z.; Huang, Z.Y.; Lv, C.; Hao, D.; Liang, C.; Deng, C.H.; Chen, J.R. Real-Time Optimization of Energy Management Strategy for Fuel Cell Vehicles Using Inflated 3D Inception Long Short-Term Memory Network-Based Speed Prediction. *IEEE Trans. Veh. Technol.* **2021**, *70*, 1190–1199. [CrossRef]
47. Zheng, G.Y.; Han, G.H.; Soomro, N.Q. An Inception Module CNN Classifiers Fusion Method on Pulmonary Nodule Diagnosis by Signs. *Tsinghua Sci. Technol.* **2020**, *25*, 368–383. [CrossRef]
48. Chandankhede, C.; Sachdeo, R. Offline MODI script character recognition using deep learning techniques. *Multimed. Tools Appl.* **2023**, *82*, 21045–21056. [CrossRef]

49. Jeyakumar, J.P.; Jude, A.; Henry, A.G.P.; Hemanth, J. Comparative Analysis of Melanoma Classification Using Deep Learning Techniques on Dermoscopy Images. *Electronics* **2022**, *11*, 11. [CrossRef]
50. Liu, K.; Yu, S.T.; Liu, S.D. An Improved InceptionV3 Network for Obscured Ship Classification in Remote Sensing Images. *IEEE J. Sel. Top. Appl. Earth Observ. Remote Sens.* **2020**, *13*, 4738–4747. [CrossRef]
51. Mputu, H.S.; Abdel-Mawgood, A.; Shimada, A.; Sayed, M.S. Tomato Quality Classification Based on Transfer Learning Feature Extraction and Machine Learning Algorithm Classifiers. *IEEE Access* **2024**, *12*, 8283–8295. [CrossRef]
52. Pravin, S.C.; Rohith, G.; Kiruthika, V.; Manikandan, E.; Methelesh, S.; Manoj, A. Underwater Animal Identification and Classification Using a Hybrid Classical-Quantum Algorithm. *IEEE Access* **2023**, *11*, 141902–141914. [CrossRef]
53. Singh, N.; Tripathi, P. An efficient model for detecting real-time facemask based on different Classification Algorithms. *Multimed. Tools Appl.* **2023**, *83*, 55175–55198. [CrossRef]
54. Sriraam, N.; Srinivasulu, A. Performance evaluation of convolution neural network models for detection of abnormal and ventricular ectopic beat cardiac episodes. *Multimed. Tools Appl.* **2024**, *83*, 65149–65188. [CrossRef]
55. Thentu, S.; Cordeiro, R.; Park, Y.; Karimian, N. ECG biometric using 2D Deep Convolutional Neural Network. In Proceedings of the IEEE International Conference on Consumer Electronics (ICCE), Las Vegas, NV, USA, 10–12 January 2021.
56. Abd Almisreb, A.; Jamil, N.; Din, N.M. Utilizing AlexNet Deep Transfer Learning for Ear Recognition. In Proceedings of the 4th International Conference on Information Retrieval and Knowledge Management (CAMP), Sabah, Malaysia, 26–28 March 2018; pp. 8–12.
57. Revathi, M.; Raghuraman, G. Kidney Stone Detection from CT Images Using ALEXNET and Hybrid ALEXNET-RF Models. *J. Circuits Syst. Comput.* **2024**, *33*, 16. [CrossRef]
58. Toliupa, S.; Tereikovskyi, I.; Tereikovskyi, O.; Tereikovska, L.; Nakonechnyi, V.; Akov, Y.K. Keyboard Dynamic Analysis by Alexnet Type Neural Network. In Proceedings of the 15th International Conference on Advanced Trends in Radioelectronics, Telecommunications and Computer Engineering (TCSET), Lviv, Ukraine, 25–29 February 2020; pp. 416–420.
59. Xiao, L.S.; Yan, Q.; Deng, S.Y. Scene Classification with Improved AlexNet Model. In Proceedings of the 12th International Conference on Intelligent Systems and Knowledge Engineering (IEEE ISKE), Nanjing, China, 24–26 November 2017.
60. Yang, M.Y.; Xie, K.; Li, T.; Ye, Y.H.; Yang, Z.P. Color Constancy Using AlexNet Convolutional Neural Network. In Proceedings of the 6th International Workshop on Pattern Recognition (IWPR), Beijing, China, 25–27 June 2021.
61. Zhang, Y.M.; Chang, F.L.; Li, N.J.; Liu, H.B.; Gai, Z.D. Modified AlexNet for Dense Crowd Counting. In Proceedings of the 2nd International Conference on Computer Engineering, Information Science and Internet Technology (CII), Sanya, China, 11–12 November 2017; pp. 351–357.
62. Ge, Y.F.; Liu, G.; Tang, H.M.; Zhao, B.B.; Xiong, C.R. Comparative analysis of five convolutional neural networks for landslide susceptibility assessment. *Bull. Eng. Geol. Environ.* **2023**, *82*, 26. [CrossRef]
63. Lu, G.Y.; Cao, B.; Zhu, X.D.; Lin, Z.S.; Bai, D.X.; Tao, C.Y.; Li, Y.N. Identification of rock mass discontinuity from 3D point clouds using improved fuzzy C-means and convolutional neural network. *Bull. Eng. Geol. Environ.* **2024**, *83*, 18. [CrossRef]
64. Wang, Z.J.; Zhao, W.L.; Du, W.H.; Li, N.P.; Wang, J.Y. Data-driven fault diagnosis method based on the conversion of erosion operation signals into images and convolutional neural network. Process Saf. *Environ. Protect.* **2021**, *149*, 591–601. [CrossRef]
65. Xie, G.B.; Shi, B.H.; Su, Y.X.; Wu, X.R.; Zhou, G.; Shi, J.F. Research on the Vanishing Point Detection Method Based on an Improved Lightweight AlexNet Network for Narrow Waterway Scenarios. *J. Mar. Sci. Eng.* **2024**, *12*, 17. [CrossRef]
66. Yuan, X.; Ren, J.W.; Cheng, G.F.; Xu, J. Toward Alleviating the Data Sparsity Problem of Deep Learning Based Underwater Target Classification. In Proceedings of the OCEANS Conference, San Diego, CA, USA, 20–23 September 2021.
67. Yang, S.; Xue, L.Z.; Hong, X.; Zeng, X.Y. A Lightweight Network Model Based on an Attention Mechanism for Ship-Radiated Noise Classification. *J. Mar. Sci. Eng.* **2023**, *11*, 17. [CrossRef]
68. Zecchetto, S.; Zanchetta, A. Validation of high resolution SAR winds fields obtained by Deep Learning. In Proceedings of the IEEE International Workshop on Metrology for the Sea Learning to Measure Sea Health Parameters (MetroSea), Milazzo, Italy, 3–5 October 2022; pp. 501–505.
69. Zhang, P.Y.; Jiang, W.L.; Zheng, Y.F.; Zhang, S.Q.; Zhang, S.; Liu, S.Y. Hydraulic-Pump Fault-Diagnosis Method Based on Mean Spectrogram Bar Graph of Voiceprint and ResNet-50 Model Transfer. *J. Mar. Sci. Eng.* **2023**, *11*, 31. [CrossRef]
70. Dai, Z.Z.; Liang, H.; Duan, T. Small-Sample Sonar Image Classification Based on Deep Learning. *J. Mar. Sci. Eng.* **2022**, *10*, 22. [CrossRef]
71. Luo, X.W.; Zhang, M.H.; Liu, T.; Huang, M.; Xu, X.G. An Underwater Acoustic Target Recognition Method Based on Spectrograms with Different Resolutions. *J. Mar. Sci. Eng.* **2021**, *9*, 20. [CrossRef]
72. Xie, J.L.; Shi, W.F.; Xue, T.; Liu, Y.H. High-Resistance Connection Fault Diagnosis in Ship Electric Propulsion System Using Res-CBDNN. J. *Mar. Sci. Eng.* **2024**, *12*, 17. [CrossRef]
73. Chen, Y.N.; Tian, Z.G.; Wei, H.T.; Dong, S.H. Reliability analysis of corroded pipes using MFL signals and Residual Neural Networks. Process Saf. *Environ. Protect.* **2024**, *184*, 1131–1142. [CrossRef]

74. Lin, K.S.; Zhao, Y.C.; Wang, L.A.; Shi, W.J.; Cui, F.F.; Zhou, T. MSWNet: A visual deep machine learning method adopting transfer learning based upon ResNet 50 for municipal solid waste sorting. *Front. Env. Sci. Eng.* **2023**, *17*, 12. [CrossRef]
75. Peng, L.G.; Zhang, J.C.; Lu, S.Q.; Li, Y.Q.; Du, G.F. One-dimensional residual convolutional neural network and percussion-based method for pipeline leakage and water deposit detection. Process Saf. *Environ. Protect.* **2023**, *177*, 1142–1153. [CrossRef]
76. Chen, L.S.; Peng, H.Y.; Yang, D.D.; Wang, T.Z. An attachment recognition method based on semi-supervised video segmentation for tidal stream turbines. *Ocean Eng.* **2024**, *293*, 17. [CrossRef]
77. Dong, X.R.; Li, J.S.; Li, B.; Jin, Y.Q.; Miao, S.F. Marine Oil Spill Detection from Low-Quality SAR Remote Sensing Images. *J. Mar. Sci. Eng.* **2023**, *11*, 20. [CrossRef]
78. Kim, K.; Kim, J. Semantic Segmentation of Marine Radar Images using Convolutional Neural Networks. In Proceedings of the OCEANS—Marseille Conference, Marseille, France, 17–20 June 2019.
79. O'Byrne, M.; Pakrashi, V.; Schoefs, F.; Ghosh, B. Semantic Segmentation of Underwater Imagery Using Deep Networks Trained on Synthetic Imagery. *J. Mar. Sci. Eng.* **2018**, *6*, 15. [CrossRef]
80. Yu, F.; He, B.; Li, K.G.; Yan, T.H.; Shen, Y.; Wang, Q.; Wu, M.H. Side-scan sonar images segmentation for AUV with recurrent residual convolutional neural network module and self-guidance module. *Appl. Ocean Res.* **2021**, *113*, 14. [CrossRef]
81. Nunes, A.; Gaspar, A.R.; Matos, A. Comparative Study of Semantic Segmentation Methods in Harbour Infrastructures. In Proceedings of the OCEANS Conference, Limerick, Ireland, 5–8 June 2023.
82. Nunes, A.; Matos, A. Improving Semantic Segmentation Performance in Underwater Images. *J. Mar. Sci. Eng.* **2023**, *11*, 26. [CrossRef]
83. Zhou, H.X.; Tao, G.X.; Nie, Y.X.; Yan, X.Y.; Sun, J. Outdoor thermal environment on road and its influencing factors in hot, humid weather: A case study in Xuzhou City, China. *Build. Environ.* **2022**, *207*, 15. [CrossRef]
84. *ASTM B117-19*; Standard Practice for Operating Salt Spray (Fog) Apparatus. ASTM International: West Conshohocken, PA, USA, 2019.
85. *ASTM G1-03*; Standard Practice for Preparing, Cleaning, and Evaluating Corrosion Test Specimens. ASTM International: West Conshohocken, PA, USA, 2017.

buildings

MDPI

Article

Design and Validation of a Real-Time Maintenance Monitoring System Using BIM and Digital Twin Integration

Seung-Won Yang [1], Yuki Lee [2] and Sung-Ah Kim [3,*]

[1] Department of Global Smart City, Sungkyunkwan University, 2066 Seobu-ro, Suwon 16419, Republic of Korea; dirotab@skku.edu
[2] Department of Computer Science and Engineering, Sungkyunkwan University, 2066 Seobu-ro, Suwon 16419, Republic of Korea; yusun91317@skku.edu
[3] Department of Architecture, Sungkyunkwan University, 2066 Seobu-ro, Suwon 16419, Republic of Korea
* Correspondence: sakim@skku.edu

Abstract: This study presents a real-time monitoring system integrating Building Information Modeling (BIM) and digital twin technology to enhance maintenance efficiency and safety in urban infrastructure. Unlike conventional periodic inspections, which miss dynamic changes and increase costs, this system uses a BIM model at LOD 400 for a solar-powered noise barrier tunnel integrated with the Wansan Tunnel in South Korea. It incorporates IoT sensor data, including vibration, tilt, light, air quality, and water detection, which are synchronized via the Autodesk Forge API, and WebSockets and visualized on a web-based dashboard. A demonstration from 22 October to 7 November 2024 confirmed that this system had stable data transmission, with light sensor rates exceeding 90%, and enabled the detection of anomalies such as irregular illuminance and structural shifts, thereby supporting informed maintenance decisions. While it is proven that BIM–digital twin integration improves NBT management, partial air quality data gaps highlight areas for refinement. This framework lays the groundwork for predictive maintenance through advanced analytics.

Keywords: real-time monitoring; building information modeling (BIM); digital twin; noise barrier tunnel (NBT); maintenance efficiency; urban infrastructure management

1. Introduction

1.1. Research Background and Context

Smart cities are increasingly recognized as a model capable of addressing future climate challenges driven by rapid urbanization, prompting the development of related research and business models [1]. Concurrently, the Fourth Industrial Revolution has introduced digital twin technology—leveraging IoT, AI, and big data—to drive transformative innovations in construction and infrastructure management [2]. The concept of the digital twin was first proposed by Grieves in a 2002 lecture [3] and was later refined into a more advanced framework by NASA in 2010 [4]. Within smart cities, digital twins are being increasingly applied to fields such as urban planning, traffic management, and environmental monitoring [2]. Notably, ongoing research actively explores the integration of real-time data with Building Information Modeling (BIM) for infrastructure management [5,6]. BIM serves as an effective data format for managing project information across its entire lifecycle, enabling facility condition monitoring through real-time data synchronization [7]. Numerous efforts are underway to enhance construction and maintenance efficiency through this integration [8,9], with BIM improving data interoperability and

https://doi.org/10.3390/buildings15081312

optimizing maintenance processes for architectural asset performance [7]. Furthermore, BIM is evolving into a technology that bridges the physical and digital realms by merging real-time data with static models [10]. Recent studies suggest that integrating BIM with digital twins can significantly enhance Life Cycle Sustainability Assessments (LCSAs) and maintenance efficiency [11,12]. Against this backdrop, there is a pressing need for innovative approaches to overcome the limitations of traditional static maintenance methods and enable real-time monitoring and predictive maintenance in urban infrastructure.

1.2. Motivation and Objectives

Noise Barrier Tunnels (NBTs) are essential urban infrastructures that mitigate traffic noise, requiring advanced technologies for effective long-term maintenance and performance [13,14]. However, the current methods, which rely on periodic visual inspections and passive maintenance, fail to capture real-time conditions, reducing efficiency and compromising safety [7]. The 2022 fire at the North Uiwang Interchange NBT on South Korea's Second Gyeongin Expressway, which caused five deaths and over 40 injuries, highlights the urgent need for systematic maintenance and real-time monitoring [15].

Despite increasing interest, the integrated use of BIM and digital twins in infrastructure like NBTs remains limited. While prior studies have explored BIM and IoT integration in construction and maintenance workflows, most remain theoretical with few real-world applications. For NBTs, which require continuous oversight, research on digital twin applications is particularly needed [13]. Most existing work focuses on conceptual frameworks for data integration using BIM and IoT, yet practical validation in infrastructure contexts remains inadequate [7]. In contrast, combining BIM and digital twins offers clear advantages, such as intuitive 3D-based maintenance, real-time condition monitoring via IoT, and predictive maintenance support [16]. However, applying these integrated technologies to NBTs is still in its nascent stages [17]. The utilization of real-time sensor data within a BIM-based digital twin facilitates the comparison of actual structural or environmental conditions with predefined performance thresholds in the model. In the event of the detection of anomalies, such as abnormal vibration, pitch variation, or irregular illuminance, the system provides immediate feedback to enable preventive action.

Therefore, this study proposes the development of a BIM-based digital twin monitoring system for solar-powered NBTs in South Korea. The system aims to overcome the limitations of traditional periodic inspections, namely their inability to capture dynamic changes and to improve maintenance efficiency by supporting early anomaly detection through data analysis.

1.3. Research Scope and Methodology

To surmount the limitations of conventional periodic inspections and enhance the efficiency of maintenance, this study developed and implemented a BIM-based digital twin monitoring system for solar-powered NBTs in South Korea. The research commenced with a review of extant studies on digital twins and BIM to define the project's objectives and requirements. A 3D BIM model of the tunnel and a design automation tool were created using Autodesk Revit and Dynamo. An IoT sensor network was then deployed to collect real-time data, including but not limited to vibration, tilt, illuminance, air quality, and water detection, from both the interior and exterior of the tunnel.

The collected data were stored in CSV and JSON formats and synchronized with the BIM model to dynamically reflect the tunnel's condition within the digital twin environment. The sensor readings were interpreted in two distinct ways—firstly, as direct measurements, and secondly, as values compared against predefined baselines or thresholds according to the sensor type. For instance, initial tilt angles were used to detect abnormal pitch changes, and vibration intensity was assessed relative to safe operating limits. While real-time alerts are delivered through the dashboard based on detected deviations, the BIM model provides a spatial reference to contextualize sensor locations and conditions. Therefore, anomalies are interpreted through the data stream itself, rather than being directly visualized within the 3D model.

The overall methodology is outlined in Figure 1. It began with a literature review, followed by the development of the BIM model and the sensor-integrated digital twin system. After deploying the system on a solar-powered NBT, data were collected and partially visualized in real time via a web-based interface using the Autodesk Forge API. While basic alerts based on device-reported flags were displayed on the dashboard, a more detailed evaluation of anomaly patterns, such as irregular time intervals or abnormal sensor readings, was performed offline using raw data analysis. This two-tiered approach allowed for verification of data integrity and sensor reliability under field conditions. The final phase of the methodology evaluated the system's integration performance and its ability to support real-time monitoring and maintenance decision-making, providing insight for future enhancements.

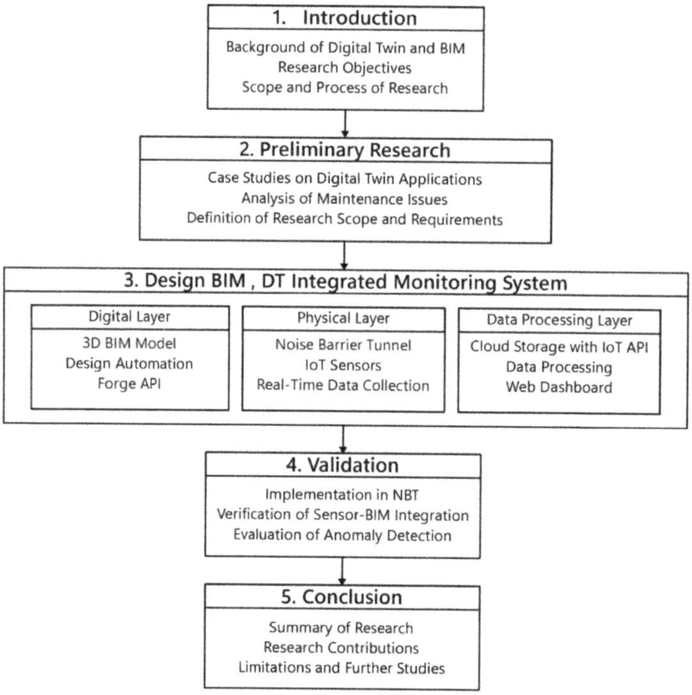

Figure 1. Research methodology and process.

2. Literature Review

2.1. Integration of BIM and Digital Twin Concepts

A digital twin is an advanced technology that integrates real-time data with a virtual representation of a physical object or system to enable continuous monitoring and predictive analysis. Originally introduced by Grieves in 2002, and further developed by NASA in 2010, the digital twin concept has evolved into a dynamic system for prediction and operational optimization [3,4,18]. Initially defined as a triad consisting of the physical entity, its digital counterpart, and the data interface, the digital twin has matured through the adoption of IoT, Big Data, and AI technologies [19,20]. These advances have extended its application beyond manufacturing to infrastructure domains, such as tunnels, NBTs, and underground facilities, where it supports accurate structural assessment and proactive maintenance strategies [21,22].

Building Information Modeling (BIM) is a digital framework that captures the three-dimensional geometry, material properties, and construction sequence of buildings and road infrastructure, enabling lifecycle data management. Although BIM is inherently static, its integration with digital twins transforms it into a dynamic environment capable of reflecting real-time conditions via sensor and operational data [13,23]. Recent studies have shown that BIM provides the foundational layer for digital twin environments, where the integration of IoT-based sensing and AI-driven predictive algorithms significantly improves operational monitoring and safety management, particularly in tunnel and NBT applications [24–27]. Continuous condition updates in this integrated environment improve maintenance decision-making and infrastructure resilience.

In addition, Kritzinger et al. (2018) categorized digital twins in manufacturing and affirmed their potential in predicting and maintaining complex systems [28]. However, their focus on manufacturing highlights the need for further exploration in the context of construction infrastructure. Kim and Kim (2020) applied digital twins to predict the lifetime of NBT components, demonstrating how they address the limitations of periodic inspections and validate the integration of BIM with IoT-based monitoring [13]. Similarly, Xu et al. (2021) demonstrated that real-time prediction and maintenance models based on digital twins can enhance the management of complex systems, providing valuable insights for the system design in this study [20].

Interdisciplinary advances are essential for the development of BIM-DT frameworks, especially for NBT monitoring. Alshammari et al. (2021) highlighted the importance of cybersecurity, such as encryption and authentication, to protect real-time sensor data, which is critical for NBT reliability [29]. Lopez and Akundi (2022) proposed model-based systems engineering (MBSE) to integrate IoT, analytics, and BIM within a unified architecture to optimize system performance [30]. Soman et al. (2025) highlighted the role of human–machine interfaces in improving operator decision-making through actionable insights [31]. Lampropoulos et al. (2024) emphasized the contribution of digital twins to NBT resilience and safety through predictive maintenance [32], while Noroozinejad et al. (2024) extended the scope by incorporating digital twins into immersive metaverse environments to support enhanced design review, training, and lifecycle management [33].

In summary, the integration of BIM and digital twins enables dynamic infrastructure representation that goes beyond the static limitations of BIM alone. Previous studies and interdisciplinary developments confirm the value of this approach in improving maintenance efficiency and safety. Building on these findings, the present study demonstrates the feasibility and practical utility of this approach.

2.2. Limitations of Traditional Maintenance Approaches

Traditional infrastructure maintenance relies primarily on scheduled visual inspections and manual assessments to evaluate asset conditions and determine necessary interventions. However, this approach has several limitations. First, fixed inspection cycles make it difficult to detect sudden structural changes or rapidly developing damage in a timely manner [20,22]. Second, inspections rely heavily on human judgment, making the process labor-intensive, error-prone, and inconsistent, which in turn increases maintenance costs and reduces emergency response [23,34]. Third, critical infrastructure, such as tunnels, underground spaces, and bridges, are particularly vulnerable to fatigue and degradation caused by environmental and traffic variations that may occur between inspections but remain unmonitored [21]. Fourth, the lack of real-time data and digital integration results in poor coordination and slow decision-making [35].

Previous studies have highlighted these shortcomings. For example, Zhao et al. (2022) reported that traditional periodic inspections are insufficient for tunnel safety because they cannot capture dynamic structural changes [25]. Wang et al. (2024) also found that traditional maintenance of large underground facilities fails to integrate dispersed information in real time, leading to inefficient and reactive management [26]. These limitations are particularly critical in infrastructure environments, such as tunnels, NBTs, and underground spaces, where internal and external conditions are highly variable and complex. Recent studies have addressed these challenges by introducing dynamic maintenance frameworks that incorporate digital twin and IoT technologies [21,27]. Such approaches are expected to play a key role in overcoming the limitations of traditional methods, especially for NBTs, which require customized maintenance strategies due to their unique structural and operational characteristics [7]. The 2022 fire at the North Uiwang Interchange NBT on South Korea's Second Gyeongin Expressway, which resulted in five deaths and over 40 injuries, highlights the urgent need for real-time monitoring and systematic maintenance [15].

The BIM-based digital twin monitoring system developed in this study responds to these needs by combining static BIM design data with real-time sensor inputs to continuously update the condition of solar-powered NBTs. This integration enables rapid anomaly detection and response, while reducing reliance on manual labor and minimizing data inconsistencies. Ultimately, it supports a more systematic and proactive maintenance and safety management framework.

2.3. Review of Prior Research

Prior studies have shown that the integration of digital twins with BIM goes beyond the replication of 3D models, enabling real-time monitoring and dynamic maintenance through sensor and IoT technologies. Grieves and Vickers laid the theoretical foundation for digital twins by introducing the concept of data-driven connectivity between physical and virtual systems [18]. Building on this, Tao et al. (2018) explored the evolution of digital twins with IoT and AI integration, highlighting their potential for real-time condition prediction and proactive maintenance [19].

In the infrastructure domain, Zhao et al. (2022) highlighted the role of sensor data in detecting sudden structural changes during tunnel safety assessments [25], while Wang et al. (2024) demonstrated that integrating distributed data streams improves consistency and responsiveness in managing large underground facilities [26]. Focusing on NBTs, Kim and Kim (2020) proposed fatigue-based lifetime prediction models for tunnel components, offering insights into structural performance but limiting their scope to component-level analysis without comprehensive real-time monitoring [13]. Mohammadi et al. (2023) demonstrated the integration of BIM with real-time vibration monitoring for bridge asset management, reducing inspection frequency and enhancing operational efficiency [36]. Similarly, Kaewunruen et al. (2020) applied digital twins to subway stations, achieving cost savings and sustainability improvements through real-time audits [37]. Zhong et al. (2023) provided an overview of predictive maintenance using digital twins, emphasizing algorithmic advancements for fault prediction and decision-making across industries, including infrastructure [38]. Wang et al. (2024) further reviewed the adoption of digital twins in construction projects, identifying opportunities for optimizing maintenance workflows and addressing implementation challenges [39]. While these studies offer robust frameworks and practical applications, empirical validation of fully integrated BIM–digital twin systems in real-world infrastructure settings remains limited. To address this gap, the present study evaluates the effectiveness of a BIM-based digital twin monitoring system implemented in a solar-powered NBT to support real-time maintenance and safety management in the field.

Additionally, Kritzinger et al. (2018) proposed a classification framework for digital twins in manufacturing, highlighting their predictive maintenance potential, though its focus on manufacturing limits direct applicability to infrastructure [28]. Xu et al. (2021) introduced a digital twin-based model for performance optimization and predictive maintenance in aviation systems, demonstrating real-time applicability to complex infrastructure [20]. Yu et al. (2021) presented a digital twin framework for tunnel maintenance decision-making, achieving notable efficiency improvements through integrated real-time data and predictive analytics [40]. While these studies showcase the potential of real-time, IoT-based digital twin solutions, most remain confined to specific components or simulation environments, lacking comprehensive on-site validation. This study addressed these gaps by implementing a BIM-based digital twin monitoring system in an operational NBT context, tailored to site-specific conditions, thus laying the groundwork for practical field adoption and future integration of advanced analytics.

Table 1 summarizes key prior studies, which suggest that the integration of digital twins with BIM supports maintenance management by leveraging real-time sensor data and advanced analytics. While some research remains industry-specific—focused on manufacturing or aviation—recent studies have begun addressing civil infrastructure, including bridges, subway stations, and construction projects. However, empirical validation in fully integrated, complex infrastructure contexts, such as tunnels and NBTs, remains limited. To address this gap, the present study implements a BIM-based digital twin monitoring system in an actual NBT setting, drawing on the findings and limitations identified in Table 1. This empirical application demonstrates the practical feasibility of BIM-DT integration for real-time maintenance and safety management in urban infrastructure, while laying the foundation for future expansion into predictive algorithms and advanced analytics.

Table 1. Summary of prior research.

Author	Research Focus	Application Areas and Strengths	Shortcomings	Implications for This Study
Kim and Kim (2020) [13]	Fatigue-based life prediction	NBT components; structural assessment	Limited to fatigue analysis	Basis for real-time NBT monitoring
Yu et al. (2021) [40]	Decision support framework	Tunnel O&M; data integration	Scalability unverified	Validates system integration
Wang, H. et al. (2024) [26]	DT for underground spaces	O&M management; real-life validation	Specific to underground spaces	Empirical O&M applicability
Mohammadi et al. (2023) [36]	BIM-DT for bridge management	Bridge O&M; reduced inspection frequency	Focused on bridges, not tunnels	Real-time monitoring validation
Kaewunruen et al. (2020) [37]	DT for subway sustainability	Subway stations; cost efficiency	Specific to subway infrastructure	Supports cost-effective maintenance
Zhong et al. (2023) [38]	Predictive maintenance with DT	Multi-industry; algorithmic advancements	General overview, not NBT-specific	Framework for predictive analytics
Wang, M. et al. (2024) [39]	DT in construction projects	Construction O&M; workflow optimization	Broad review, lacks specific cases	Supports practical DT adoption
Kritzinger et al. (2018) [28]	DT classification	Manufacturing; maturity assessment	Manufacturing-focused	Foundation for construction O&M
Tao et al. (2019) [19]	DT application review	Cyber-physical integration; analytics	Construction specificity lacking	Identifies core DT technologies
Xu et al. (2021) [20]	DT optimization	Aviation; real-time feedback	Limited to aviation	Real-time optimization potential

3. Design of a Noise Barrier Tunnel Maintenance Monitoring System

Traditional maintenance practices for NBTs rely primarily on periodic visual inspections, which are inadequate for detecting real-time structural or environmental changes. These limitations can lead to delayed responses, increased costs, and safety risks. While BIM has been widely adopted during design and construction, its integration with real-time data during operational phases remains limited. To address this gap, this study introduced a digital twin-based monitoring system that integrates BIM and IoT sensor data to improve real-time condition monitoring and predictive maintenance in NBTs. This section presents the system architecture, its component layers, and the implementation strategy adopted to support practical applicability in field conditions.

3.1. System Overview and Objectives

This study developed a BIM-based digital twin monitoring system and demonstrated its potential to improve maintenance efficiency and safety in NBTs. By integrating BIM, DT technologies, and IoT sensors, the system addresses the limitations of conventional periodic inspections and enables real-time condition tracking and predictive maintenance. The BIM model provides static design data, such as geometric configuration and material specifications, while the DT integrates this with real-time environmental and structural data (e.g., temperature, vibration) collected by IoT sensors. This integration enables dynamic monitoring of tunnel conditions in a virtual environment, facilitating early anomaly detection and data-driven decision-making.

The system architecture consists of three interrelated layers, as shown in Figure 2. The digital layer establishes a maintenance-oriented environment using a 3D BIM model developed in Autodesk Revit at LOD 400, supported by design automation workflows. The physical layer includes the actual NBT and a sensor network that measures key parameters, including temperature, humidity, vibration, and light levels. The data-processing layer transmits and stores sensor data via a commercial IoT API, pre-processes them, and synchronizes them with the BIM model in real time. The data are transferred using WebSocket protocols, and visualization is enabled through the Autodesk Forge API and a web-based dashboard. This layered framework supports continuous monitoring and system responsiveness, positioning BIM-DT integration as a practical solution for predictive maintenance in urban infrastructure.

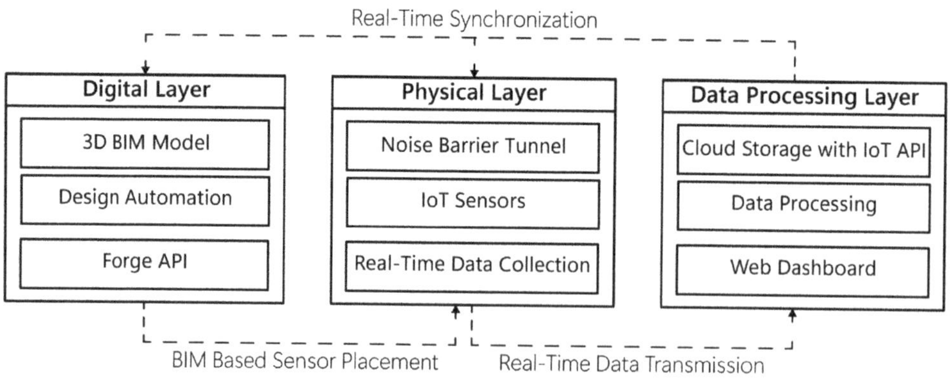

Figure 2. Main layer composition of the NBT maintenance monitoring system.

Technically, the system uses Autodesk Revit and Dynamo to develop a detailed BIM model at LOD 400, while design automation increases model flexibility and responsiveness. Real-time data transfer is enabled through WebSockets, and synchronization with the BIM model is maintained through the Autodesk Forge API. This infrastructure supports a web-based dashboard that provides real-time visualization, anomaly detection, and decision support for maintenance operations.

This study helps overcome the limitations of traditional inspection-based maintenance through three key contributions: enhanced real-time monitoring, predictive maintenance capabilities, and improved maintenance efficiency and safety. By combining BIM and IoT within a unified platform, the system provides a structured framework for real-time, data-driven tunnel management. It drives the shift from reactive to predictive maintenance practices, providing practical value to infrastructure operators.

3.2. Digital Environment and BIM Model Development

Traditional BIM models have prioritized static data management during the design and construction phases, limiting their ability to reflect dynamic changes in infrastructure, such as NBTs. This lack of real-time integration hinders the timely detection of structural or environmental changes, reducing maintenance effectiveness and increasing operational risk. To overcome these limitations, this study established a digital environment that combines BIM with digital twin technology, enabling real-time synchronization, continuous condition monitoring, and predictive maintenance for NBTs.

A BIM model at LOD 400 was developed using Autodesk Revit, supported by a set of parametric modeling scripts created in Dynamo. As shown in Figure 3, the modeling process begins with the import of roadway geometry (Road.dwg) and predefined Revit family components. The scripting pipeline automatically generates the tunnel floor, structural members, panel systems, connectors, and rafter beam cover assemblies, systematically constructing the entire geometry with precision and consistency. The resulting model is saved in RVT format for integration and visualization and converted into SVF format for web-based interaction. The LOD 400 model includes detailed attributes, such as photovoltaic module dimensions, bolted joints, and structural framing, to support sensor placement and long-term asset management.

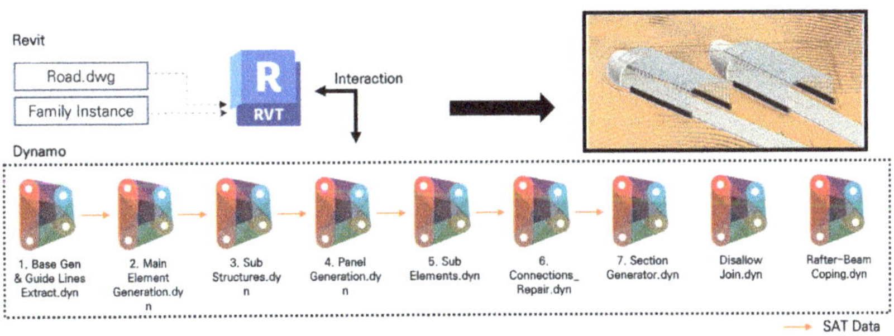

Figure 3. BIM model with LOD 400 level created using Autodesk Revit and Dynamo.

Real-time sensor data are collected from vibration sensors (mm/s^2), tilt detectors (degrees), air quality monitors (PM1, PM2.5, PM10 in μg/m^3), light sensors (lux), and water detection devices (Boolean). These values are transmitted to the cloud via a commercial IoT aggregation API, refined and validated through a sensor API, and synchronized with the BIM model using the Autodesk Forge API and WebSockets. This real-time streaming supports the visualization of current conditions and alerts on a web-based dashboard. As illustrated in Figure 4, the system architecture comprises three key stages. First, sensor data are acquired in real time from the physical tunnel environment. Second, the data are transmitted to a central repository via authenticated WebSocket communication, undergoing format conversion for compatibility. Third, the monitoring client integrates the LOD

BIM model and sensor data via the Forge API, rendering the model in SVF format for interactive visualization. This integration enables anomaly detection, facilitates intuitive monitoring, and supports data-driven maintenance decisions.

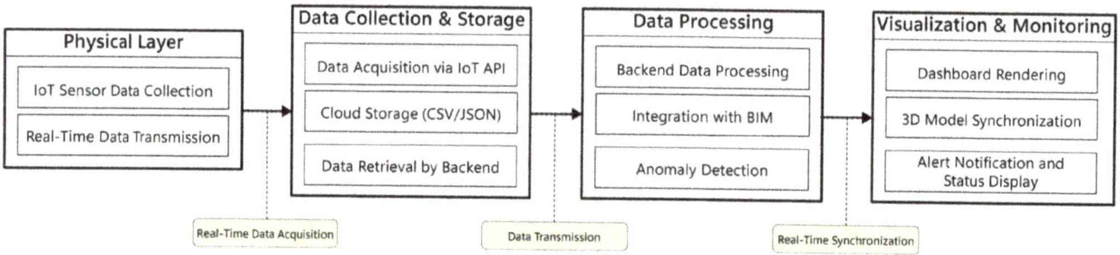

Figure 4. Process of transmitting and processing the acquired data.

This approach moves beyond traditional periodic inspections by enabling real-time responsiveness, reducing manual intervention, and enhancing safety and operational efficiency for urban infrastructure maintenance. Although the project did not follow a rigid linear development model, initial sensor testing and connectivity verification were conducted in a lab-scale prototype environment. This informal preliminary phase helped validate data acquisition, synchronization, and alert thresholds before full-scale deployment in the field. Future implementations may benefit from a more formalized testing framework and a clearly defined development timeline.

3.3. *Physical Environment Setup and IoT Sensor Network*

To establish a reliable real-time monitoring system for the NBT, this study implemented an integrated IoT sensor network within the physical tunnel environment. This network serves as the primary interface between the tunnel's real-world conditions and the digital twin system, enabling continuous monitoring and timely anomaly detection. The sensors collect structural and environmental data, which are then synchronized with the BIM model to reflect the dynamic state of the tunnel.

Five types of sensors were selected based on their relevance to tunnel safety, operational efficiency, and maintenance requirements: vibration sensors, tilt sensors, light sensors, air quality sensors, and water detection sensors. Each type of sensor captures a specific variable and transmits it in a standardized digital format suitable for real-time analysis and visualization. Table 2 summarizes the measured variables, data formats, and maintenance-related functions of each sensor.

Table 2. Overview of sensors.

Sensor Type	Measured Variable	Unit	Sampled Input Format	Maintenance Purpose in NBT	Twin Application
Vibration	Acceleration/intensity	mm/s^2, m/s^2	Float (e.g., 0.62)	Detects abnormal stress/crack	Structural health model
Tilt	Pitch angle (θ)	Degrees (°)	Float (e.g., 1.24°)	Detects foundation or frameshift	Deformation visualization
Light (Illuminance)	Illuminance	Lux	Integer (e.g., 1540 lux)	Checks visibility and solar panel output	Solar efficiency mapping
Air Quality	PM1, PM2.5, PM10	$\mu g/m^3$	Float (e.g., 42.1)	Assesses tunnel air safety	Environmental safety index
Water Detection	Flood presence (binary)	Boolean	True/false	Detects road flooding and blockage	Accessibility alert

To provide a clearer understanding of the sensor deployment, Figure 5 shows two views (frontal and isometric) of a single NBT. In the front view, vibration meters and light meters are installed at the top truss level (Z: +11,072) to monitor structural vibration and light exposure from external sources. Tilt sensors, temperature sensors, air quality meters, and water detection sensors are distributed along vertical positions from Z: +5035 to Z: +1095 on the steel columns and floor to comprehensively monitor structural tilt, environmental conditions, and potential flooding at the tunnel floor. This placement strategy ensures effective coverage of critical zones, including the roof structure, access level ventilation, and flood-prone areas. All sensors are configured to collect data at 10-min intervals, balancing energy efficiency, transmission reliability, and maintenance responsiveness. The collected data are transmitted via WebSocket protocol to a cloud server, enabling near real-time synchronization with the digital twin platform. This supports dynamic visualization, automated anomaly detection, and maintenance decision-making through a unified monitoring interface. The same sensor layout and vertical positioning strategy is mirrored on the opposite road, where another NBT structure is installed. This ensures consistent monitoring in both tunnel directions.

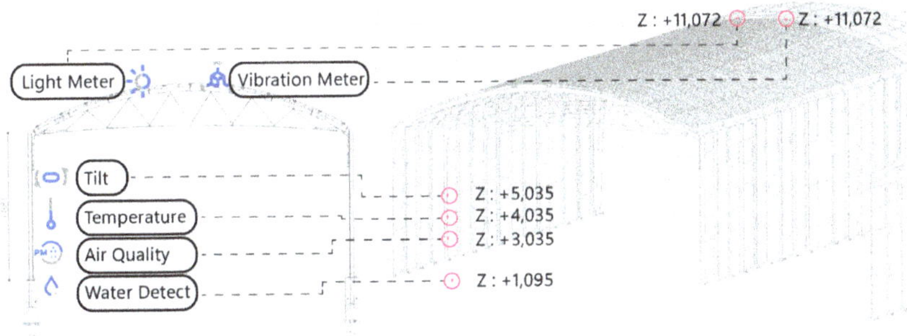

Figure 5. Overview of the installation of sensors in NBT (Tunnel 1).

The sensor network is designed to transmit collected data to a centralized cloud repository using a commercial IoT aggregation API. Its low-power architecture and robust communication protocols ensure long-term autonomous operation. Sensor readings are processed and delivered to the digital twin via a structured WebSocket stream, enabling near real-time updates on conditions such as vibration intensity, pitch angle variations, and the presence of surface water. Unlike traditional inspections that rely on infrequent site visits and may miss transient anomalies, this configuration supports continuous tracking and rapid anomaly detection based on predefined thresholds. This real-time responsiveness helps reduce maintenance costs and improve structural safety.

Figure 6 illustrates the data pipeline from sensor acquisition to cloud transmission and synchronization with the BIM-based monitoring interface, including integration via the Autodesk Forge API. Although the system does not employ equation-based numerical modeling, each sensor input is evaluated through logic-based rules implemented via API integration and condition monitoring functions. These rules use threshold comparisons and hardware status flags (e.g., Sensor State = 2) to determine real-time alert levels and system responses. Although the system does not use explicit numerical equations for state prediction, each sensor reading is interpreted through threshold-based logic, anomaly rules, and comparison to baseline values to represent behavior in the digital twin environment.

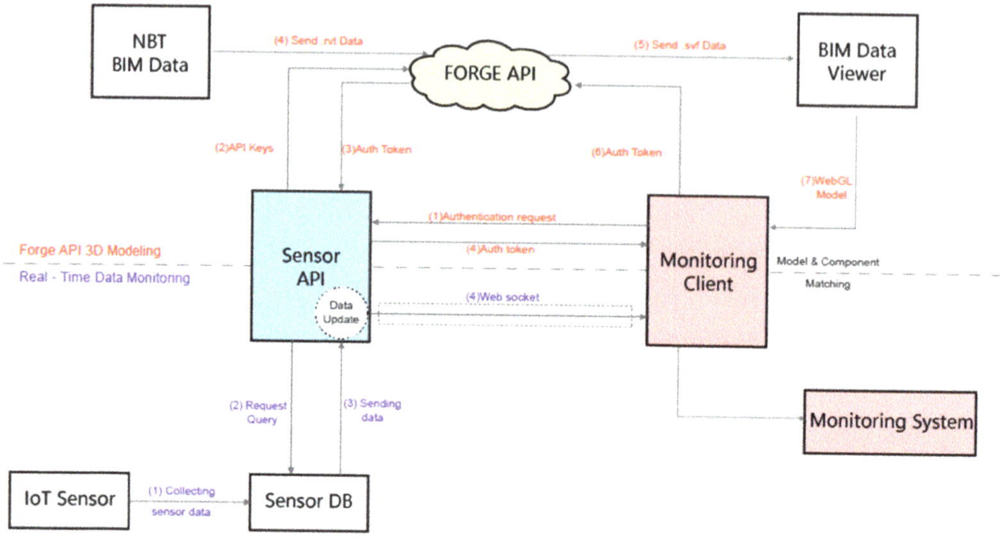

Figure 6. Composition and data flow of the sensor network.

Each sensor measures a specific physical variable, but several measurements are functionally related. For example, the light levels indirectly reflect solar irradiance, which is tied to photovoltaic performance. Similarly, the tilt and vibration sensors complement each other in detecting structural instability; tilt sensors are sensitive to gradual positional shifts, while vibration sensors capture dynamic impacts. By analyzing these inputs together, the system enhances the contextual understanding of the tunnel's condition within the digital twin.

3.4. Real-Time Monitoring System Design

The real-time monitoring system developed in this study enables continuous tracking of the physical condition of the NBT and supports timely maintenance decisions. As shown in Figure 6, sensor data are transmitted from the tunnel site to a central cloud repository, where they are processed and synchronized with the BIM model. This end-to-end data pipeline—from acquisition to visualization—is the backbone of the monitoring framework discussed in this section. Sensor readings are transmitted via a commercial IoT API and delivered to the client in real time using WebSocket communication. These data streams are parsed and structured in the cloud to support integration with the NBT's LOD 400 BIM model. The model is rendered using the Autodesk Forge API and converted into Simple Vector Format (SVF) for fast and interactive access via web browsers.

On the client side, a web-based dashboard provides intuitive monitoring of the system status, structural conditions, and environmental variables. The dashboard displays real-time sensor values, synchronized 3D views of the NBT, and visual alerts when anomalies occur. As shown in Figure 7, this interface integrates tabular data, live graphics, and interaction with the BIM model, allowing maintenance personnel to efficiently assess the condition of the tunnel. To ensure secure transmission and controlled access, all API communications are authenticated and encrypted using platform-level protocols, such as token-based access and TLS encryption. While this provides a basic level of security, more robust system-level features, including user role access control and data-level encryption, are not yet fully implemented. These enhancements have been identified as key priorities for future releases of the platform. By eliminating reliance on periodic inspections and enabling continuous data-driven monitoring, the system provides a proactive framework for infrastructure maintenance and security management.

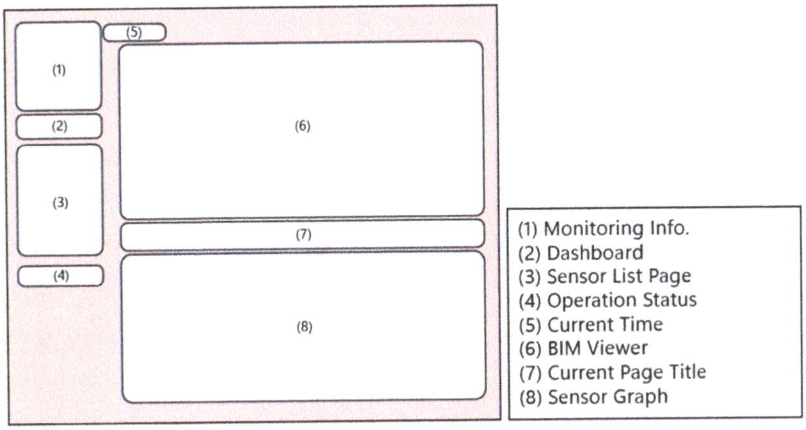

Figure 7. Dashboard configuration of the monitoring system.

4. Demonstration and Performance Assessment

4.1. Demonstration Targets and Site Conditions

This study focused on a solar-powered NBT integrated with the Wansan Tunnel in Jeonju, Republic of Korea. Designed to mitigate traffic noise and enhance safety in urban transportation infrastructure, the NBT serves as the demonstration target, as illustrated in Figure 8. The tunnel measures approximately 7~8 m in height and 14~17 m in width. Its geometric structure and component layout formed the basis for developing a detailed BIM model. A range of IoT sensors, including water detection, motion detection, tilt, temperature, air quality, and vibration sensors, were deployed throughout the tunnel to monitor structural and environmental conditions in real time (see Figure 5). Each sensor is supported by an optimized power supply and communication module, transmitting data to a centralized cloud repository via a commercial IoT data aggregation API.

Figure 8. BIM-based digital model and physical implementation of the NBT on site. (Location: N 35°48′46.44″, E 127°11′29.4″).

The BIM model, developed at LOD 400 using Autodesk Revit, contains detailed geometric and operational data. It is synchronized with the digital twin environment using the Autodesk Forge API. This integration provides a robust foundation for real-time structural monitoring, anomaly detection, and maintenance support. The site's physical conditions and sensor deployment details serve as critical baseline data for system validation and performance evaluation, demonstrating the practical applicability of BIM–digital twin integration in urban infrastructure management.

4.2. Prototype Integration and Functional Validation

Figure 9 illustrates the comprehensive process from the design and construction of the NBT to the implementation of the digital twin monitoring system. The top section shows the digital development pipeline, from element modeling and design automation, using Autodesk Revit and Dynamo, to BIM file conversion, using the Forge platform. The lower section depicts the physical implementation, showing the field installation of sensors and the real-time dashboard interface. This illustration shows the system's integration path from virtual modeling to field deployment and web-based visualization.

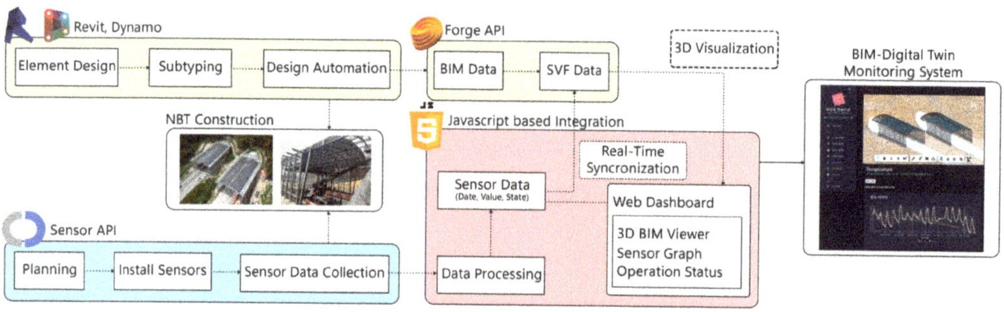

Figure 9. BIM and digital twin integrated monitoring system configuration process.

To ensure real-time monitoring, an array of IoT sensors was deployed inside and outside the NBT structure, including vibration, tilt, light, air quality, and water detection units. Figure 10a,b presents the hardware configuration of these sensors prior to deployment, with modular components prepared for various structural and environmental monitoring needs. These sensors serve as primary input nodes to the digital twin platform, providing the data backbone for anomaly detection and condition assessment. Figure 10c–e details the on-site sensor installation process, capturing how the sensors were mounted on the tunnel's interior steel columns and exterior surfaces. The images show practical considerations, such as exposure to sunlight for light sensors and ground-level placement for water detection. These visual records validate the system's field readiness and the integration of sensor networks with the tunnel's architectural elements.

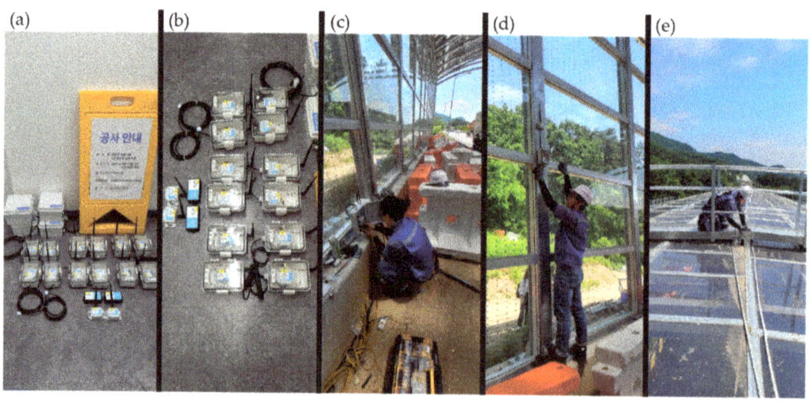

Figure 10. (**a**,**b**) Sensors prepared before installation. (**c**–**e**) Field installation of sensors in the NBT.

Sensor data are transmitted to a centralized cloud repository via a commercial IoT data aggregation API and synchronized with the BIM model via WebSocket communication, enabling real-time visualization and alerts. Initial post-installation testing confirmed stable sensor communication and real-time responsiveness, with no observable latency or data loss, supporting the system's potential for proactive maintenance management.

4.3. Results and Performance Analysis

To assess the reliability and performance of the deployed monitoring system, this section analyzes the sensor data collected during the demonstration period. While the light sensors initially provided the most consistent data stream, additional analysis confirmed that other sensors—vibration, tilt, air quality, and water detection—also transmitted data successfully and were visualized through the digital twin dashboard (Figure 11). This confirmed that the platform infrastructure and integration logic were operational across the sensor types.

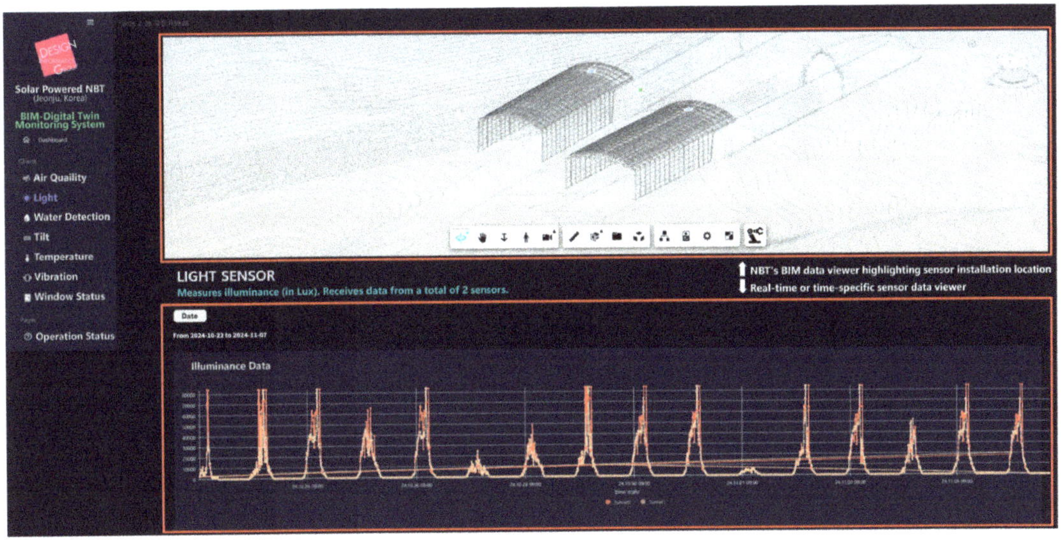

Figure 11. Visualization of the sensor data-based monitoring system.

Anomaly detection was performed using a complementary two-part approach. The first approach used system-reported status flags, such as Sensor State = 2, Alert Sent = True, and Battery = 0, which were monitored in real time via a dedicated status page on the dashboard. This mechanism enabled immediate operator notification in response to deviations from expected sensor ranges, such as sudden spikes or drops in illuminance, as shown in Figure 12. The second approach involved post hoc analysis of raw sensor data to identify indirect anomalies not flagged by device logic, including irregular sampling intervals, flatline trends, and unexpected environmental values. Although not embedded in the live monitoring platform, this analysis provided valuable diagnostic insights into data reliability and sensor behavior. Together, these two approaches strengthened the overall robustness of the system and provided direction for future enhancements.

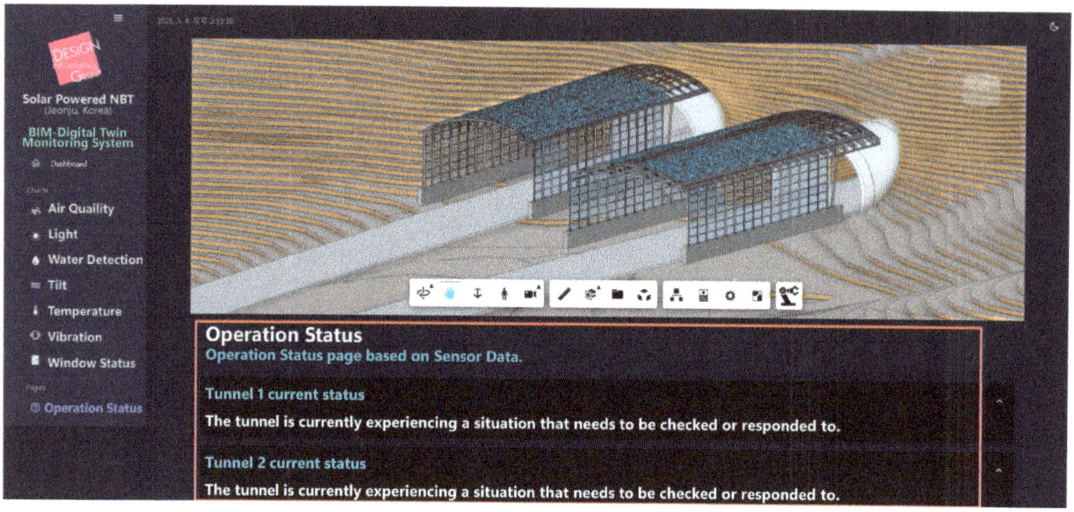

Figure 12. Sensor transmission data status: Data-based anomaly notification. (Translated from Korean to English).

Vibration sensors 01 and 02 exhibited stable performance throughout the demonstration period. No system-level alarms (e.g., Sensor State = 2) were triggered, and acceleration values remained within the expected limits. Data logging intervals were consistent, confirming both sensor integrity and communication reliability. Tilt sensors 01 and 02 recorded tilt values within a narrow band of $\pm 1°$, well below the alert threshold of $\pm 3°$. No abrupt changes were detected, indicating structural stability during the monitoring period. These patterns are illustrated in Figures 13 and 14, which show the vibration time series and pitch angle trends, respectively, with threshold values superimposed.

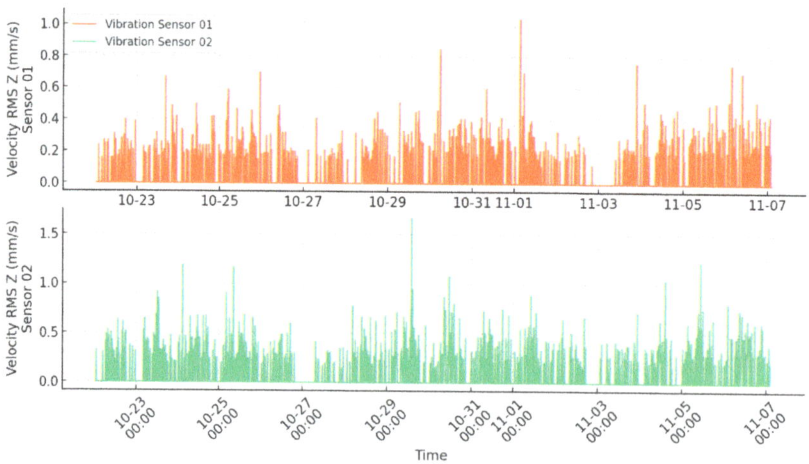

Figure 13. Vibration sensor: Velocity RMS Z.

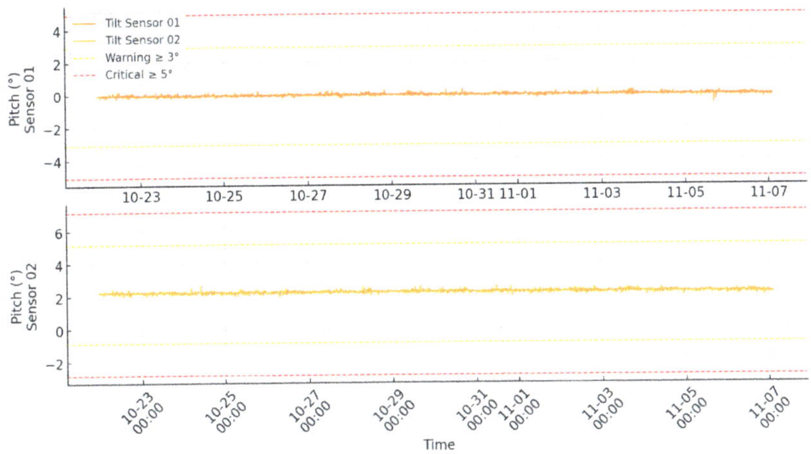

Figure 14. Tilt sensor: Pitch angle variation.

Light sensors 01 and 02 achieved high data transmission rates of 92.1% and 97.0%, respectively, confirming regular synchronization with the cloud repository. However, offline analysis revealed anomalies such as 1-min recording intervals, repeated maximum illuminances (e.g., 83,865.6 lux), and zero lux values during daylight hours. These may be due to transient occlusions or processing delays at the sensor level. These anomalies are visualized in Figure 15, with shading indicating irregular intervals and abnormal values. Air Quality Sensor 01 transmitted 647 out of 933 expected records between 29 October and 5 November, resulting in 30.65% data loss. However, 99.07% of the data received was within the 10-min recording cycle, reflecting high temporal accuracy when active. Sensor 02 did not transmit any data. Figure 16 shows the PM1, PM2.5, and PM10 values over time, with shaded areas indicating gaps caused by irregular logging. Water Detection Sensors 01 and 02 recorded binary outputs with no flood detection events. Sensor 01 transmitted data consistently; sensor 02 did not transmit at all, indicating a persistent error. Although no actual flooding events occurred, the system's ability to record "false" baseline values verified its readiness to detect critical events.

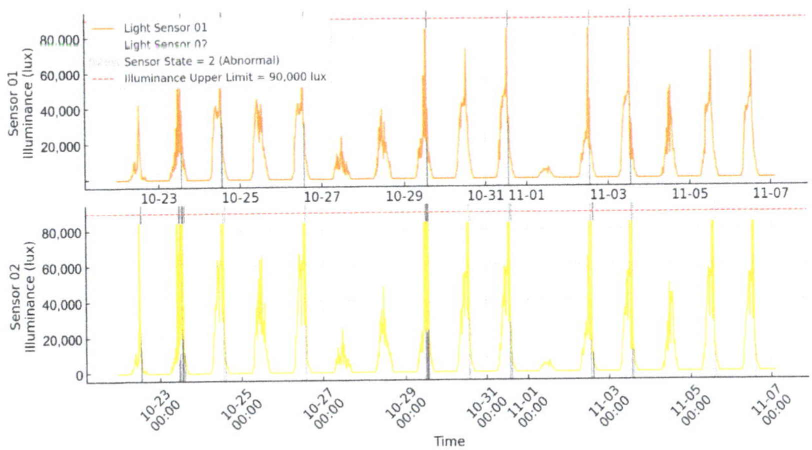

Figure 15. Light sensor: Illuminance with abnormality.

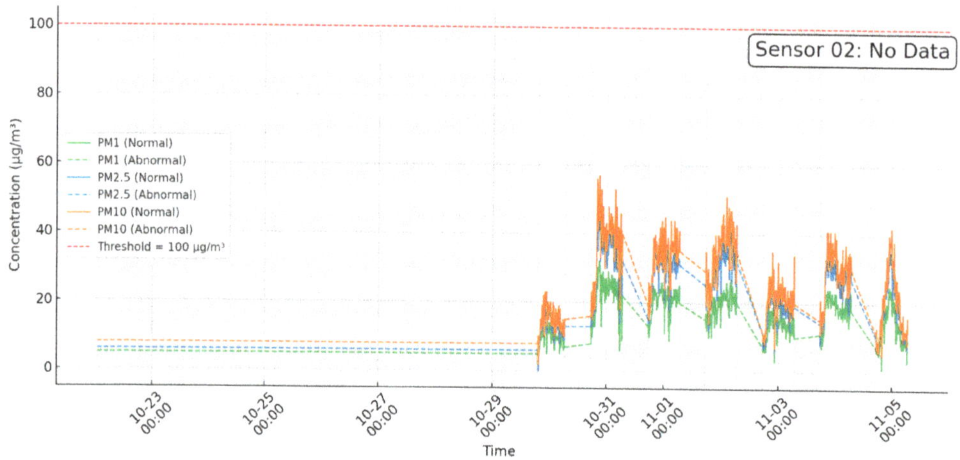

Figure 16. Air quality: PM trends with data gaps.

Anomaly assessment highlights the complementary value of hardware-level alerts and offline interpretation of raw data. In several cases, such as anomalous illuminance values or subtle tilt angle shifts, no system alert was triggered, yet post-analysis revealed clear anomalies. This underscores the importance of external validation to detect anomalies that may go unnoticed by device-level logic alone. Conversely, communication failures, such as those seen in Water Sensor 02, were immediately apparent from the lack of data and could be visualized directly on the monitoring platform. These findings support the integration of both alert-based and data-driven diagnostics for increased reliability.

Although the offline analysis in this study was performed retrospectively, the findings show clear potential for real-time application. Visual elements, such as threshold overlays, shaded data gaps, and highlighted anomalies, as illustrated in Figures 15 and 16, can be adapted for real-time anomaly detection and alert generation. Future iterations of the system may incorporate these techniques into the dashboard to support predictive maintenance workflows and enable proactive responses in tunnel operations scenarios.

4.4. Discussion and Future Enhancements

The demonstration confirmed that the proposed BIM-based digital twin monitoring system effectively integrates real-time sensor data with a high-resolution BIM model of the NBT site. The system employs a dual anomaly detection strategy, integrating direct alerts based on sensor-reported flags with offline analysis grounded in data-driven criteria. This integrated approach enhances the identification of hardware defects, environmental irregularities, and structural anomalies, providing significant advantages over conventional periodic inspections. Among the deployed sensors, light sensors exhibited the highest data coverage and demonstrated stable latency. The system's capacity to detect subtle anomalies beyond the thresholds of built-in sensor alerts is validated by additional insights from the vibration, tilt, air quality, and water detection sensors, including pitch deviations, illuminance anomalies, and irregular transmission intervals.

These outcomes align with prior research findings. For instance, Mohammadi et al. (2023) demonstrated that integrating BIM with real-time vibration monitoring reduced bridge inspection frequency and operational downtime [36]. Similarly, Kaewunruen et al. (2020) reported cost savings and efficiency gains in subway stations through BIM–digital twin platforms [38]. This study lends further support to the field applicability of digital twins for NBT maintenance.

Notwithstanding these strengths, this study has several limitations. While the dashboard successfully visualizes real-time sensor values and status flags (e.g., sensor state, battery status), more nuanced detections, such as gradual pitch drift or irregular illuminance cycles, require offline post-processing. This limitation restricts the system's responsiveness to emergent or low-severity conditions that do not trigger predefined thresholds. As Zhong et al. (2023) emphasized, effective predictive maintenance requires algorithmic intelligence beyond threshold-based alerts [38]. Additionally, while the system performed reliably in the NBT use case, its broader application to bridges, tunnels, and other asset types will necessitate calibration and tuning.

Future improvements should prioritize the incorporation of sophisticated analytics, including real-time pitch deviation detection, illumination pattern recognition, and inter-valve gap monitoring, directly into the monitoring platform. These enhancements would enable proactive maintenance through automated alerts based on contextual sensor trends, as suggested by Wang et al. (2024) [39]. User interface enhancements based on operator feedback, reduced data transmission latency, and enhanced cybersecurity protocols are also essential for long-term deployment. Given its demonstrated flexibility and effectiveness, the proposed framework is well-positioned for extension to a wider range of urban infrastructure applications. With these enhancements, the BIM–digital twin system can evolve into a core platform for data-driven, predictive asset management.

5. Conclusions

5.1. Research Summary and Key Contributions

This study developed and validated a real-time infrastructure monitoring system by integrating BIM and digital twin technologies at a solar-powered NBT site in Jeonju, Republic of Korea. The proposed system combined a high-resolution BIM model with real-time IoT sensor data, including vibration, inclination, illumination, air quality, and water presence. These data were synchronized through cloud-based architecture using the Autodesk Forge API and WebSocket communication protocols.

The system enabled continuous visualization and analysis of structural and environmental conditions and supported anomaly detection through a two-tiered approach: (1) direct alerts triggered by device-reported status flags, and (2) post-processed identification of anomalous trends, such as tilt angle drift, illuminance spikes, and irregular data intervals. These results demonstrate the system's ability to detect subtle irregularities, even when hardware-based thresholds were not exceeded. Real-time data transmission from the light sensors achieved rates of more than 90%, while offline analysis confirmed the stable performance of the tilt and vibration modules. Although air quality data coverage was partial, high interval fidelity underscored reliable timing and communication when active. This dual-insight strategy enhances situational awareness of emerging anomalies and complements traditional hardware-based monitoring.

These results are consistent with recent research highlighting the benefits of integrating sensor data into BIM-based systems. Mohammadi et al. (2023) demonstrated that integrating vibration monitoring into BIM models reduced inspection frequency and downtime in bridge infrastructure [36]. Similarly, Kaewunruen et al. (2020) demonstrated cost savings and efficiency gains in subway stations through digital twin applications [37], while Wang et al. (2024) emphasized the practical utility of digital twins in optimizing maintenance workflows in construction projects [39]. In addition, Zhong et al. (2023) highlighted the role of digital twin-based predictive maintenance in leveraging algorithmic intelligence to overcome the limitations of threshold-based approaches and provide a robust framework for real-time failure prediction and decision-making [38].

The key contributions of this study are as follows:

First, it implemented and validated a BIM–digital twin infrastructure monitoring system in a real-world context, integrating high-resolution BIM models with real-time environmental and structural sensor data.

Second, it demonstrated stable data collection, cloud-based synchronization, and integrated visualization via a web-based dashboard, confirming the feasibility of the system for real-time condition monitoring and maintenance support in complex infrastructure environments.

Third, sensor interoperability and data reliability were empirically evaluated under field conditions, confirming the robustness of the system architecture and its adaptability to infrastructure environments.

Taken together, these results highlight the pragmatic benefits of BIM–digital twin integration in enhancing situational awareness, enabling preventive maintenance, and improving operational efficiency in infrastructure management.

5.2. Study Limitations

The proposed system successfully demonstrated real-time integration of BIM and IoT sensor data in an operational NBT. However, several limitations emerged that constrain its full practical application. Most notably, intermittent power supply issues during the demonstration led to data loss and delayed sensor activation, particularly for the air quality and tilt modules. For instance, Air Quality Sensor 01 did not commence data collection until the first week of the demonstration and exhibited approximately 69.3% coverage during its active period. Despite the implementation of proper synchronization mechanisms, sporadic network instability also contributed to temporal discontinuities in the data stream.

Another key constraint lies in the limited real-time diagnostic capability of the current dashboard implementation. While the system does support immediate alerts based on sensor-reported flags (e.g., battery failure or sensor state anomalies), more advanced anomaly patterns (e.g., gradual pitch angle drift or irregular illuminance cycles) require offline post-analysis. This architectural separation, therefore, limits the system's capacity to provide immediate, data-driven maintenance suggestions. Furthermore, the demonstration was conducted on a single infrastructure type and site, which limits the generalizability of the findings. To ensure the effectiveness of the system in real-world scenarios, further validation is necessary to ascertain the compatibility of sensors with diverse infrastructure types, such as bridges or metro corridors, as well as to determine environment-specific thresholds and the system's long-term operational resilience.

Finally, the current system is deficient in the realm of real-time recovery and buffering mechanisms, which are imperative for the management of transient data outages resulting from power or network disruptions. Without such mechanisms, the reliability and precision of monitoring may be jeopardized under field conditions.

5.3. Direction for Future Research

To enhance the scalability and utility of the proposed BIM–digital twin system, future research should prioritize the integration of real-time analytics and predictive models. While this study applied threshold-based and interval analyses offline, the incorporation of such logic into the live platform will enable automated anomaly detection and early intervention. As Kaewunruen et al. (2021) observed, predictive maintenance necessitates dynamic analysis that extends beyond basic alerts. Secondly, the validation of the system across diverse infrastructure types, including bridges and tunnels, will assess its scalability and adaptability [37]. It is acknowledged that each context may necessitate customized thresholds and sensor configurations, and the broader implementation will contribute to the generalization of BIM–digital twin frameworks. Thirdly, enhancing the user experience and system resilience is imperative. Operator feedback has underscored the need for more intuitive dashboards, customized alerts, and real-time visualization. Future systems should incorporate secure data transmission, buffering, and failover mechanisms to ensure continuity during power or network disruptions. Finally, the establishment of interoperable data-sharing frameworks could facilitate collaboration among cities, operators, and researchers. In accordance with Wang et al. (2024), the implementation of standardized platforms would expedite the adoption of digital twin technologies and facilitate the development of more comprehensive predictive maintenance strategies [39].

In summary, advancing analytics, expanding deployment, and improving usability and security are key to realizing the full potential of BIM–digital twin systems as robust infrastructure management platforms.

Author Contributions: S.-W.Y. conducted research planning and field demonstrations, wrote the manuscript, planned the initial monitoring system, and developed and executed the NBT BIM data creation script. Y.L. developed the API interpretation and monitoring system, and processed and visualized the collected sensor data. S.-A.K. supervised the research. All authors have read and agreed to the published version of the manuscript.

Funding: This work was financially supported by the Korean Ministry of Land, Infrastructure, and Transport (MOLIT) through the Innovative Talent Education Program for Smart Cities. This work was also supported by a grant from the Korea Institute of Energy Technology Evaluation and Planning (KETEP), funded by the Korean government (MOTIE) (20213030010280, Development and field demonstration of PV modules for soundproofing facilities with secured safety and maintainability).

Data Availability Statement: Due to national research funding policies and infrastructure security regulations, the full BIM model and real-time monitoring dashboard cannot be publicly released. These models contain structural and proprietary information that was developed in collaboration with public infrastructure authorities and private companies in South Korea. However, anonymized sample sensor data collected during the demonstration period may be made available upon reasonable request to the author.

Conflicts of Interest: The authors declare that they have no known competing financial interests or personal relationships that could have influenced the work reported in this study.

References

1. Cho, Y. Criteria for Smart Cities, Smart City Indexes. *Sci. Technol. Innov. Policy* **2016**, *26*, 38–43. Available online: https://www.stepi.re.kr/site/stepiko/PeriodicReportList.do?cateCont=A0504&pageIndex=2&reIdx=0&searchYear=&searchCondition=1&searchKeyword=&searchSort=PUBLIC_DT#modal1 (accessed on 30 December 2024).
2. El-Agamy, R.F.; Sayed, H.A.; ALAkhatatneh, A.M.; Aljohani, M.; Elhosseini, M. Comprehensive Analysis of Digital Twins in Smart Cities: A 4200-Paper Bibliometric Study. *Artif. Intell. Rev.* **2024**, *57*, 154. [CrossRef]
3. Grieves, M.W. Product Lifecycle Management: The New Paradigm for Enterprises. *Int. J. Prod. Dev.* **2005**, *2*, 71–84. [CrossRef]
4. National Aeronautics and Space Administration (NASA). *Encyclopedia of Geography*; National Aeronautics and Space Administration (NASA): Washington, DC, USA, 2010. [CrossRef]
5. Dave, B.; Kubler, S.; Främling, K.; Koskela, L. Opportunities for Enhanced Lean Construction Management Using Internet of Things Standards. *Autom. Constr.* **2016**, *61*, 86–97. [CrossRef]
6. Dave, B.; Buda, A.; Nurminen, A.; Främling, K. A Framework for Integrating BIM and IoT through Open Standards. *Autom. Constr.* **2018**, *95*, 35–45. [CrossRef]
7. Nour El-Din, M.; Pereira, P.F.; Poças Martins, J.; Ramos, N.M.M. Digital Twins for Construction Assets Using BIM Standard Specifications. *Buildings* **2022**, *12*, 2155. [CrossRef]
8. Cheng, J.C.; Chen, W.; Chen, K.; Wang, Q. Data-Driven Predictive Maintenance Planning Framework for MEP Components Based on BIM and IoT Using Machine Learning Algorithms. *Autom. Constr.* **2020**, *112*, 103087. [CrossRef]
9. Teizer, J.; Wolf, M.; Golovina, O.; Perschewski, M.; Propach, M.; Neges, M.; König, M. Internet of Things (IoT) for Integrating Environmental and Localization Data in Building Information Modeling (BIM). In Proceedings of the International Symposium on Automation and Robotics in Construction, Taipei, Taiwan, 28 June–1 July 2017; Volume 34. [CrossRef]
10. UK BIM Alliance. *BIM and Digital Twins [Positioning Statement]*; UK BIM Alliance: London, UK, 2021.

11. Boje, C.; Guerriero, A.; Kubicki, S.; Rezgui, Y. A Framework Using BIM and Digital Twins in Facilitating LCSA for Buildings. *J. Build. Eng.* **2023**, *76*, 107232. [CrossRef]
12. Opoku, D.G.J.; Perera, S.; Osei-Kyei, R.; Rashidi, M. Digital Twin Application in the Construction Industry: A Literature Review. *J. Build. Eng.* **2021**, *40*, 102726. [CrossRef]
13. Kim, J.; Kim, S.-A. Lifespan Prediction Technique for Digital Twin-Based Noise Barrier Tunnels. *Sustainability* **2020**, *12*, 2940. [CrossRef]
14. Son, J. Case Study on the Planning Procedure of Traffic Soundproof Facilities and Alternative Decision Making. Master's Thesis, Bukyeong University, Busan, Republic of Korea, 25 February 2009. Available online: https://repository.pknu.ac.kr:8443/handle/2021.oak/10689 (accessed on 31 December 2024).
15. Kim, J.-M.; Son, S.-B.; Kim, H.-B.; Cho, J.-W. Apocalyptic Highway Fire Exposes Dangers of Plastic Tunnels. Available online: https://koreajoongangdaily.joins.com/2022/12/30/national/socialAffairs/korea-fire-gyeongin-expressway/20221230154034712.html (accessed on 28 December 2024).
16. Biagini, C.; Bongini, A. BIM and Data Integration: A Workflow for the Implementation of Digital Twins. In *Beyond Digital Representation. Digital Innovations in Architecture, Engineering and Construction*; Giordano, A., Russo, M., Spallone, R., Eds.; Springer: Cham, Switzerland, 2024; Volume 1, p. 822. [CrossRef]
17. Deren, L.; Wenbo, Y.; Zhenfeng, S. Smart City Based on Digital Twins. *Comput. Urban Sci.* **2021**, *1*, 4. [CrossRef]
18. Grieves, M.; Vickers, J. Digital twin: Mitigating unpredictable, undesirable emergent behavior in complex systems. In *Transdisciplinary Perspectives on Complex Systems. New Findings and Approaches*; Springer: Cham, Switzerland, 2017; pp. 85–113. [CrossRef]
19. Tao, F.; Zhang, H.; Liu, A.; Nee, A. Digital twin in industry: State-of-the-art. *IEEE Trans. Ind. Inform.* **2019**, *15*, 2405–2415. [CrossRef]
20. Xu, Z.; Ji, F.; Ding, S.; Zhao, Y.; Zhou, Y.; Zhang, Q.; Du, F. Digital Twin-Driven Optimization of Gas Exchange System of 2-Stroke Heavy Fuel Aircraft Engine. *J. Manuf. Syst.* **2021**, *58*, 132–145. [CrossRef]
21. Sychev, S.; Batako, A.; Nguyen, T.; Xavior, A.; Antosz, K.; Machado, J. An Investigation into Connection between BIM and Digital Twins Technologies. *IFAC-Pap.* **2024**, *58*, 234–239. [CrossRef]
22. Honghong, S.; Gang, Y.; Haijiang, L.; Tian, Z.; Annan, J. Digital twin enhanced BIM to shape full life cycle digital transformation for bridge engineering. *Autom. Constr.* **2023**, *147*, 104736. [CrossRef]
23. Boje, C.; Guerriero, A.; Kubicki, S.; Rezgui, Y. Towards a Semantic Construction Digital Twin: Directions for Future Research. *Autom. Constr.* **2020**, *114*, 103179. [CrossRef]
24. Pan, Y.; Zhang, L. A BIM-Data Mining Integrated Digital Twin Framework for Advanced Project Management. *Autom. Constr.* **2021**, *124*, 103564. [CrossRef]
25. Zhao, Y.; Cao, C.; Liu, Z. A Framework for Prefabricated Component Hoisting Management Systems Based on Digital Twin Technology. *Buildings* **2022**, *12*, 276. [CrossRef]
26. Wang, H.; Yu, C.; Zheng, J.; Jia, Y.; Liu, Z.; Yang, K. Digital-Twin-Based Operation and Maintenance Management Method for Large Underground Spaces. *Buildings* **2024**, *14*, 2577. [CrossRef]
27. Negri, E.; Fumagalli, L.; Macchi, M. A review of the roles of digital twin in CPS-based production systems. *Procedia Manuf.* **2017**, *11*, 939–948. [CrossRef]
28. Kritzinger, W.; Karner, M.; Traar, G.; Henjes, J.; Sihn, W. Digital Twin in manufacturing: A categorical literature review and classification. *IFAC-Pap.* **2018**, *51*, 1016–1022. [CrossRef]
29. Alshammari, K.; Beach, T.; Rezgui, Y. Cybersecurity for Digital Twins in the Built Environment: Current Research and Future Directions. *J. Inf. Technol. Constr.* **2021**, *26*, 159–173. [CrossRef]
30. Lopez, V.; Akundi, A. A Conceptual Model-based Systems Engineering (MBSE) Approach to Develop Digital Twins. In Proceedings of the 2022 IEEE International Systems Conference (SysCon), Montreal, QC, Canada, 25–28 April 2022; IEEE: New York, NY, USA, 2022; pp. 1–5. [CrossRef]
31. Soman, R.K.; Farghaly, K.; Mills, G.; Whyte, J. Digital Twin Construction with a Focus on Human Twin Interfaces. *Autom. Constr.* **2025**, *170*, 105924. [CrossRef]
32. Lampropoulos, G.; Larrucea, X.; Colomo-Palacios, R. Digital Twins in Critical Infrastructure. *Information* **2024**, *15*, 454. [CrossRef]
33. Noroozinejad Farsangi, E.; Shehata, A.O.; Rashidi, M.; Ghassempour, N.; Mirjalili, S. Transitioning from BIM to Digital Twin to Metaverse. *Front. Built Environ.* **2024**, *10*, 1486423. [CrossRef]
34. Liu, S.; Ren, S.; Jiang, H. Predictive Maintenance of Wind Turbines Based on Digital Twin Technology. *Energy Rep.* **2023**, *9*, 1344–1352. [CrossRef]
35. Lee, J.; Bagheri, B.; Kao, H. A Cyber-Physical Systems architecture for Industry 4.0-based manufacturing systems. *Manuf. Lett.* **2015**, *3*, 18–23. [CrossRef]

36. Mohammadi, M.; Rashidi, M.; Yu, Y.; Samali, B. Integration of TLS-derived bridge information modeling (BrIM) with a decision support system (DSS) for digital twinning and asset management of bridge infrastructures. *Comput. Ind.* **2023**, *147*, 103881. [CrossRef]
37. Kaewunruen, S.; Peng, S.; Phil-Ebosie, O. Digital twin aided sustainability and vulnerability audit for subway stations. *Sustainability* **2020**, *12*, 7873. [CrossRef]
38. Zhong, D.; Xia, Z.; Zhu, Y.; Duan, J. Overview of predictive maintenance based on digital twin technology. *Heliyon* **2023**, *9*, e14534. [CrossRef]
39. Wang, M.; Ashour, M.; Mahdiyar, A.; Sabri, S. Opportunities and threats of adopting digital twin in construction projects: A review. *Buildings* **2024**, *14*, 2349. [CrossRef]
40. Yu, G.; Wang, Y.; Mao, Z.; Hu, M.; Sugumaran, V.; Wang, Y.K. A Digital Twin-Based Decision Analysis Framework for Operation and Maintenance of Tunnels. *Tunn. Undergr. Space Technol.* **2021**, *116*, 104125. [CrossRef]

buildings

MDPI

Article

Bridge Geometric Shape Measurement Using LiDAR–Camera Fusion Mapping and Learning-Based Segmentation Method

Shang Jiang [1,*], Yifan Yang [1], Siyang Gu [2], Jiahui Li [1] and Yingyan Hou [3]

[1] School of Transportation and Civil Engineering, Nantong University, Nantong 226019, China; yangyifan@stmail.ntu.edu.cn (Y.Y.); 2233110034@stmail.ntu.edu.cn (J.L.)
[2] School of Civil Engineering, Southeast University, Nanjing 211189, China; gusiyang@seu.edu.cn
[3] Target Key Laboratory of Cognition and Application Technology, Aerospace Information Research Institute, Beijing 100190, China; houyy@aircas.ac.cn
* Correspondence: shangjiang@ntu.edu.cn

Abstract: The rapid measurement of three-dimensional bridge geometric shapes is crucial for assessing construction quality and in-service structural conditions. Existing geometric shape measurement methods predominantly rely on traditional surveying instruments, which suffer from low efficiency and are limited to sparse point sampling. This study proposes a novel framework that utilizes an airborne LiDAR–camera fusion system for data acquisition, reconstructs high-precision 3D bridge models through real-time mapping, and automatically extracts structural geometric shapes using deep learning. The main contributions include the following: (1) A synchronized LiDAR–camera fusion system integrated with an unmanned aerial vehicle (UAV) and a microprocessor was developed, enabling the flexible and large-scale acquisition of bridge images and point clouds; (2) A multi-sensor fusion mapping method coupling visual-inertial odometry (VIO) and Li-DAR-inertial odometry (LIO) was implemented to construct 3D bridge point clouds in real time robustly; and (3) An instance segmentation network-based approach was proposed to detect key structural components in images, with detected geometric shapes projected from image coordinates to 3D space using LiDAR–camera calibration parameters, addressing challenges in automated large-scale point cloud analysis. The proposed method was validated through geometric shape measurements on a concrete arch bridge. The results demonstrate that compared to the oblique photogrammetry method, the proposed approach reduces errors by 77.13%, while its detection time accounts for 4.18% of that required by a stationary laser scanner and 0.29% of that needed for oblique photogrammetry.

Keywords: bridge geometric shape; LiDAR; point cloud; 3D reconstruction; deep learning

1. Introduction

The measurement of bridge geometric shape is one of the key contents throughout the entire life cycle of a bridge, including the construction phase, the load test process, and the service phase. However, bridge structures are generally characterized by large spans and high heights, which pose a serious challenge to the fast and accurate measurement of their geometric shape. Existing geometric shape measurement methods rely on the manual operation of a total station to measure point by point. This kind of method is simple and reliable, and the error can be controlled within centimeters when measuring kilometer-level bridges. However, it has the disadvantage of relying on the manual pre-laying of reflecting prisms, which is dangerous and time-consuming when measuring the geometric shape of large-span bridges. Additionally, this point-by-point method can only measure a small

https://doi.org/10.3390/buildings15091458

number of points and, therefore, may fail to monitor the deformation of some locations where no measurement points have been arranged.

Non-contact measurement methods, such as vision-based methods and radar-based methods, have been rapidly developed in recent years, providing efficiency and accuracy improvements for bridge inspection, and these methods have been extensively applied to the surface defect detection of bridges [1], the deformation measurement of bridges [2], the tension detection of cables [3], and the 3D model reconstruction of bridges [4]. However, as revealed by algorithmic principles, such deformation monitoring methods yield temporal variations in bridge morphology rather than quantifying absolute geometric states. The prerequisite for measuring 3D bridge geometric shapes lies in acquiring high-fidelity 3D point cloud representations of the structure. Commonly used techniques for acquiring 3D point clouds include stereo vision-based methods [5], structured light-based methods [6], laser scanning-based methods [7], and millimeter-wave radar-based methods [8]. The equipment for acquiring 3D information includes UAVs equipped with mapping cameras, stereo vision scanners, structured light scanners, 3D Light Laser Detection and Ranging (LiDAR), and millimeter-wave radar systems [9,10]. Among them, the point cloud density generally acquired by millimeter-wave radar is quite sparse, so it is usually not used in the dense 3D reconstruction of large scenes. Three-dimensional scanners based on the stereo vision principle or based on the structured light principle are generally characterized by a short baseline, high accuracy, and a high frame rate, but they can generally only be used for high-precision small-volume 3D scanning in indoor environments [11]. Multi-view image 3D reconstruction based on an unmanned airborne camera is a 3D reconstruction method widely used in structural inspection. The cost of this method is much lower than that of other methods. Still, the reconstruction accuracy of this method is affected by a variety of factors, such as the quality of the camera, the lighting situation, the shadow occlusion, and the overlap rate of the image. The accuracy of the large-scale 3D reconstruction can hardly reach the centimeter level, and the reconstruction algorithms can be extremely time-consuming [4].

Three-dimensional laser scanning has the characteristics of rapidity, non-contact, active point cloud acquisition, high density, and high accuracy of point cloud quality. The two main types of scanning are static Terrestrial Laser Scanning (TLS) and mobile dynamic scanning. For a static TLS implementation, multiple scanning stations must be strategically installed around the bridge peripheries and deck surfaces. Each station conducts localized scanning, subsequently registering all station-specific datasets into a unified point cloud. This methodology necessitates over 30 scanning stations for long-span bridges, resulting in prolonged field operations. Post-acquisition workflows demand labor-intensive or automated point cloud registration processes, significantly increasing temporal costs. For instance, Lin et al. [12] implemented TLS for bridge scanning, reporting 3 h of scanning for the bridge, and the point cloud density of the bridge reached 7700 points per square meter. For large-scale bridge point cloud data acquisition, two primary scanning methods exist due to the classification of 3D laser scanning into stationary static scanning and mobile dynamic scanning. In static scanning, multiple measurement stations are established around and on the bridge deck, with scanning performed at each station, followed by data stitching. Long-span linear bridges typically require setting up over 30 stations, resulting in a time-consuming process. Lu et al. proposed a top-down scanning strategy tested on over 10 highway bridges, showing an average scanning time of 2.8 h across 17 scanning sessions [13]. Mohammadi et al. utilized TLS to scan Werrington Bridge, obtaining 525 million points requiring approximately 15 GB of storage space [14]. Post-scanning, manual or automated point cloud registration is needed to align point clouds from all stations, a process that remains labor-intensive. While automated registration methods have emerged, such as Shao et al. proposed a marker-free approach using structured light scanners (SLSs) and TLS point clouds, which replaces marker-based coarse registration with an iterative closest point (ICP) algorithm to refine alignment, reducing per-station registration time to 2 min [15]. However, existing methods often depend on additional sensors or targets and face limitations in processing massive point clouds [16]. In recent years, growing applications of vehicle-mounted [17], UAV-borne [18], and handheld [19] laser scanners have been seen in dynamic scanning. These mobile platforms equipped with LiDAR can reconstruct bridges by flying predefined routes over the structure, using inertial navigation systems (INS) and GPS data to calculate trajectory and stitch temporal point clouds. For instance, Zhu et al. investigated UAV-based LiDAR-RGB fusion for bridge reconstruction, demonstrating its capacity to provide precise spatial information, though it requires post-processing for point cloud alignment [20]. Li et al. achieved a high-precision bridge model (overall accuracy: 1.70 cm) by fusing UAV imagery with TLS point clouds [21]. While dynamic scanning significantly reduces field time, post-processing alignment remains crucial, particularly for INS-based systems where positioning errors degrade reconstruction accuracy. For example, INS-estimated UAV positions typically accumulate substantial errors, leading to reduced model fidelity. Therefore, developing efficient and robust laser scanning equipment specifically for bridges, along with advanced point cloud registration methodologies, constitutes an essential research direction.

After obtaining the 3D point cloud of bridges, how to rapidly and automatically extract the linear alignment of key bridge components from massive 3D point clouds becomes a critical focus. Huang et al. proposed a power line detection method utilizing UAV LiDAR edge computing and 4G real-time transmission. This approach first employs semantic feature constraints to classify laser points in power line corridors into ground, vegetation, towers, cables, and building types, followed by trained deep learning models to further identify tower types for strain span separation [22]. The second challenge involves segmenting component-specific point clouds from bridge point clouds. For instance, extracting the point cloud of cable clamps and deriving their central coordinates enables the acquisition of cable linear alignment. Xu et al. developed a deep learning pre-segmentation method for bridge 3D reconstruction to detect cable alignment. Their methodology first applies deep learning-based object detection combined with Euclidean clustering to generate cropped sub-image sets of cable clamps automatically, then employs a 3D-to-2D grid projection-based image mask generation technique to retain valid pixels for camera pose estimation [23]. Kim et al. proposed a similar approach for segmenting bridge piers and girders, initially performing deep learning segmentation on 3D reconstruction images before projecting the segmentation results onto image-based reconstructed 3D point clouds [24]. Regarding point cloud segmentation from LiDAR data, Xia et al. presented a method combining local descriptors with machine learning for automatic detection of bridge structural components. Based on bridge geometric features, they designed a multi-scale local descriptor for training a deep classification neural network, achieving 97.26% accuracy in concrete girder bridge tests [25]. Yang et al. developed a weighted superpoint graph (WSPG) method for direct component recognition from large-scale bridge point clouds. The bridge point cloud is first clustered into hundreds of semantically homogeneous superpoints, then classified into different bridge components using PointNet and graph neural networks, attaining 99.43% accuracy in both real and synthetic datasets [26]. Jing et al. proposed BridgeNet, a 3D deep learning neural network for automatic segmentation of masonry arch bridges, achieving a maximum mean intersection-over-union of 94.2% in synthetic dataset training and testing [27].

While these studies demonstrate that deep learning-based point cloud segmentation methods can achieve satisfactory accuracy, current research encounters two significant limitations. First, most studies rely on synthetic datasets due to data scarcity, with each dataset containing only thousands of bridge point clouds, significantly fewer than the hundreds of millions of points in real LiDAR-acquired bridge point clouds. Second, the computational efficiency of algorithms remains insufficiently addressed. Practical applications reveal that processing bridge point clouds containing tens of millions of points using 3D deep learning methods imposes enormous demands on Graphics Processing Unit (GPU) memory. Even processing bridge point clouds with less than 1.4 million points requires over 30 min. Therefore, developing algorithms with lower computational demands and higher efficiency for processing large-scale real bridge point clouds remains another critical research challenge.

Based on the above review, this study proposes an integrated camera–LiDAR fusion framework for rapid 3D bridge reconstruction and a point cloud segmentation method for bridge components to efficiently measure bridge geometric linearity. The methodology centers on three contributions: (1) A multi-sensor acquisition system integrating LiDAR, cameras, and an inertial measurement unit (IMU) is developed, featuring an embedded real-time reconstruction processor designed for UAV deployment to perform dynamic large-scale scanning. (2) A coupled visual-LiDAR odometry Simultaneous Localization and Mapping (SLAM) algorithm is implemented to synchronously process point cloud and image data during dynamic scanning, enabling real-time point cloud stitching and visualization. (3) A deep learning segmentation method based on image-point cloud feature matching is proposed to automatically extract bridge components from reconstructed point clouds, allowing for the rapid geometric measurement of structural elements.

The remainder of this study is structured as follows: The general framework of the proposed LiDAR camera fusion-based geometric shape measurement method for bridges is presented in Section 2. The designed system is presented in Section 3. Section 4 introduces the 3D reconstruction methods based on LiDAR and camera fusion, and the proposed method is verified by field-testing a bridge in Section 5. Section 6 describes the conclusions of this work.

2. Framework of the Proposed Method

The core of the proposed method is to develop a LIDAR, camera, and IMU fusion system to rapidly and dynamically reconstruct a 3D point cloud of the bridge. A learning-based image segmentation network is proposed to automatically extract key components of the bridge from the point cloud based on the matching relationship between the camera and LiDAR. The 3D line shape of the components can be extracted, as shown in Figure 1. Based on this, the proposed method can be divided into three parts: establishing a multi-sensor fusion data acquisition system, the mapping method based on image and laser point cloud fusion, and the point cloud extraction method for bridge components based on image segmentation.

The first part addresses two fundamental technical challenges: temporal synchronization within multi-sensor fusion systems and comprehensive system calibration. Temporal synchronization is achieved through a GNSS (Global Navigation Satellite System)-synchronized processing unit that timestamps camera and LiDAR data streams, followed by the temporal alignment of multi-sensor measurements based on their respective sampling rates. The calibration process encompasses three essential procedures: (1) intrinsic camera calibration for lens distortion correction, (2) LiDAR-IMU calibration to compensate for motion-induced point cloud distortions during dynamic operations, and (3) cross-sensor calibration to determine precise extrinsic parameters between the camera and LiDAR systems. Only after completing these synchronization and calibration protocols can the acquired multi-modal data be effectively utilized for subsequent sensor fusion and high-precision mapping operations.

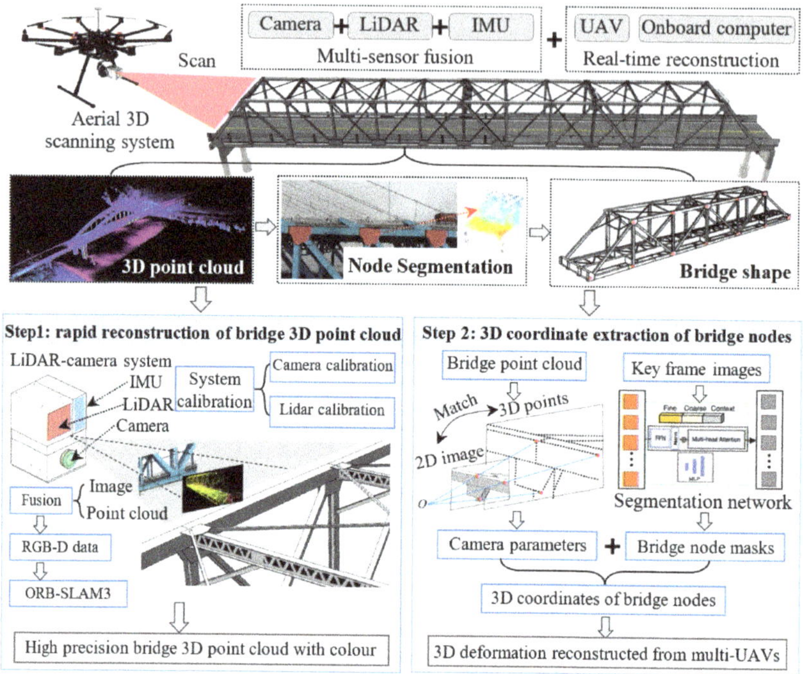

Figure 1. Framework of the proposed method.

The second part focuses on reconstructing bridge 3D point clouds using preprocessed sensor data. This approach distinguishes itself from conventional LiDAR-based mapping methodologies through the implementation of a dual-odometry framework that synergistically combines VIO and LIO. A tightly coupled fusion framework is developed to integrate pose estimation outcomes from both modalities, effectively capitalizing on the complementary advantages of visual data's dense feature representation in proximate ranges and LiDAR point clouds' superior spatial constraint accuracy at extended distances. This hybrid architecture significantly enhances system robustness in 6-Degrees of Freedom (DoF) pose estimation and enables the generation of globally consistent 3D point cloud models.

The third part focuses on automatically identifying bridge components from collected bridge images, segmenting these elements, and restoring their 2D coordinates to the global 3D coordinate system of the bridge through image–point cloud alignment. Specifically, a deep learning-based image segmentation technique is first proposed to automatically segment bridge joints from the massive images acquired in the initial phase. Subsequently, the spatial correspondence between each image and its associated laser point cloud data are utilized to derive the 3D point cloud representation of each segmented part. The 3D coordinates of the components are then extracted from their corresponding point clouds. By integrating all detected component coordinates through this workflow, the geometric shape configuration of the entire bridge structure is reconstructed.

3. Establishment of LiDAR–Camera Fusion Measurement System

As a commonly used measurement device, cameras have undergone significant development and reached maturity. Therefore, this study employs an industrial camera as the image acquisition sensor. Meanwhile, LiDAR, as a precision measurement sensor, has traditionally been characterized by high costs; some stationary high-precision LiDAR scanners exceed hundreds of thousands of US dollars, while handheld dynamic laser scanners typically cost tens of thousands of dollars. With the rapid development of autonomous driving technology in recent years, low-cost sensor-grade LiDARs have been developed, making the widespread application of LiDAR feasible. Notably, the Livox series LiDARs are priced at only a few hundred dollars, yet their non-repetitive scanning pattern enables point cloud density comparable to more expensive scanners during low-speed mobile scanning. Based on these advantages, this research adopted the Livox LiDAR as the scanning sensor in a low-cost scanning system.

Specifically, the industrial camera employed is the HIKROBOT MV-CE120-10UC equipped with a 5 mm focal-length lens, capable of capturing images at 4000 × 3036 pixel resolution. The LiDAR unit selected is the Livox Avia, featuring a scanning range exceeding 190 m, point cloud density of 240,000 points/second, and an integrated BMI088 IMU with 0.09 mg resolution. The system also incorporates a processing unit for point cloud and image storage and processing. The processor used in this study is the Intel NUC12, configured with an i5-12500H CPU, 16 GB of Random Access Memory (RAM), and 256 GB of storage.

To minimize the weight of the system and make it easy to mount on the UAV, the aforementioned LiDAR and camera were connected together by a 3D-printed frame. The processor used for data analysis is placed in the lower part of the UAV to balance the center of gravity of the UAV. The UAV equipped with the complete system is the M300 RTK UAV from DJI Company, Shen Zhen, China, which can provide about 35 min of flight time. The complete data acquisition system is shown in Figure 2.

Figure 2. The developed unmanned airborne LiDAR camera fusion system.

Sensor calibration is the basic work before measurement, and the proposed method includes three kinds of sensors: LiDAR, camera, and IMU, so each sensor needs to be calibrated separately first. For digital cameras, a variety of calibration methods can be used to calibrate the camera, depending on different camera models and different calibrators. Among them, Zhang's calibration method is one of the most commonly used calibration methods, and in this part, Zhang's calibration method is adopted to calibrate the intrinsic parameter matrix of the camera.

Regarding LiDAR calibration, since the 3D reconstruction in this study involves dynamic scenarios where the UAV maintains continuous motion during laser scanning, the following challenge arises: As the LiDAR collects each segment of point cloud data, it must accumulate laser returns over a specific time interval. During this interval, the UAV's movement causes shifts in the coordinate system of points within each point cloud frame, resulting in distortion. To address this, motion-induced distortion in the LiDAR data must be eliminated through calibration between the LiDAR and the IMU prior to measurement.

Unlike conventional rotating mechanical LiDAR systems, the Livox LiDAR adopted in this research utilizes a semi-solid-state design incorporating a rotating prism mechanism. This configuration projects laser beams through precisely coordinated rotations of multiple prisms, generating a predetermined scanning trajectory. While significantly more cost-effective than traditional rotating mechanical LiDARs, its scanning mechanism fundamentally differs through the implementation of non-repetitive scanning patterns. The non-repetitive scanning achieves progressive field-of-view (FOV) coverage proportional to integration duration. At an integration time of 100 ms, the point cloud coverage approximates that of a conventional 32-beam LiDAR. Furthermore, the Livox Avia scanner exhibits a distinctive flower-petal-shaped scanning trajectory during operation, as illustrated in Figure 3a. Figure 3b demonstrates point cloud distortions induced by rotational and translational motions. During operation, the LiDAR operates at a standard sampling frequency of 10 Hz, corresponding to a 100 ms interval for laser emission and reception. When divided into three temporal segments (T_1, T_2, T_3), the following effects are observed: During the first temporal segment (T_1) with static LiDAR conditions, point cloud data are acquired at positions a_1–c_1. In the second segment (T_2) under rotational motion, sampling positions shift to a_2–c_2. During the third segment (T_3) involving translational motion, sampling positions further shift to a_3–c_3.

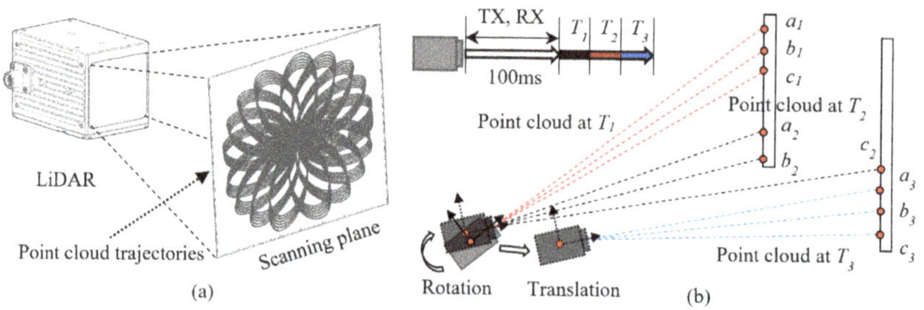

Figure 3. (**a**) Trajectory of the point cloud and (**b**) the onset of motion blur in point clouds.

Motion distortion correction requires transforming the coordinates of each point within a LiDAR frame to their corresponding LiDAR origin at respective timestamps. Under minimal motion conditions, adjacent points demonstrate sufficient spatial continuity and overlap, enabling the application of point cloud registration-based estimation methods. These methods typically utilize the Iterative Closest Point (ICP) algorithm or its derivatives to compute inter-frame transformations for point cloud alignment [28]. However, such approaches inherently assume constant velocity motion and yield approximate solutions, leading to cumulative errors during extended scanning sessions. An alternative methodology integrates IMU data to estimate LiDAR motion states. By fusing high-frequency local pose estimations with kinematic state calculations, this approach reconstructs the LiDAR's instantaneous orientation and velocity during data acquisition, enabling precise point cloud reprojection.

We assume the motion state of the LiDAR during scanning is represented as $S = \left[R^T, P^T, V^T, b^T\right]^T$, where R denotes the rotation matrix, P is the position vector, V represents the velocity, and b corresponds to the accelerometer bias. Let the world coordinate system be denoted as W and the LiDAR coordinate system as B. To analyze the LiDAR's motion using IMU data, the objective reduces to solving R, P, and V. In the implemented hardware configuration, the IMU operates at a data acquisition frequency of 200 Hz, while the LiDAR operates at 10 Hz. To correct LiDAR point cloud distortion using

IMU data, the IMU measurements must first undergo integration. The angular velocity and acceleration from the IMU are defined as

$$\begin{cases} \hat{\omega}_t = \omega_t + b_t^{\omega} + n_t^{w} \\ \hat{a}_t = R_t^{BW}(a_t - g) + b_t^a + n_t^a \end{cases} \tag{1}$$

where $\hat{\omega}_t$ and $\hat{a}_t$ represent the angular velocity and acceleration measured by the IMU at time t, ω_t denotes the true angular velocity, and b_t^{ω}, b_t^a, n_t^w, and n_t^a correspond to the biases and noise in the measurements. The rotation matrix R_t^{BW} transforms coordinates from the world frame W to the LiDAR frame B, and g is the gravitational acceleration vector. Based on kinematic relationships among displacement, velocity, and acceleration, the position vector, velocity, and accelerometer bias from time t to $t + \Delta t$ can be expressed as

$$\begin{cases} V_{t+\Delta t} = V_t + g\Delta t + R_t(\hat{a}_t - b_t^a - n_t^a)\Delta t \\ P_{t+\Delta t} = P_t + V_t\Delta t + \frac{1}{2}g\Delta t^2 + \frac{1}{2}R_t(\hat{a}_t - b_t^a - n_t^a)\Delta t^2 \\ R_{t+\Delta t} = R_t exp((\hat{\omega}_t - b_t^{\omega} - n_t^{w}))\Delta t \end{cases} \tag{2}$$

Applying the inertial pre-integration method, the change in velocity, position, and rotation matrices from time i to time j can be expressed as

$$\begin{cases} \Delta V_{ij} = R_i^T(V_j - V_i - g\Delta t_{ij}) \\ \Delta P_{ij} = R_i^T\left(P_j - P_i - V_i\Delta t_{ij} - \frac{1}{2}g\Delta t_{ij}{}^2\right) \\ \Delta R_{ij} = R_i^T R_j \end{cases} \tag{3}$$

After obtaining the position and attitude transformation matrices of the system using the above equations, the point clouds during an acquisition time period (T_1–T_3) can be transformed to a coordinate system.

Following the calibration of the LiDAR-IMU system, the extrinsic matrix between the LiDAR and camera must be determined to establish spatiotemporal correspondence between image frames and 3D point clouds. Temporal synchronization across multi-sensor systems forms the foundation of sensor fusion methodologies. In this study, the LiDAR's integrated IMU inherently ensures synchronization between the IMU and LiDAR, enabling simultaneous data acquisition through the direct invocation of the LiDAR driver. However, the camera and LiDAR operate as independent acquisition units, necessitating explicit temporal alignment between image frames and point cloud sequences prior to calibration and fusion.

The synchronization challenge between LiDAR and camera encompasses two aspects: clock synchronization and data acquisition synchronization. For clock synchronization, a common solution involves integrating a GPS into the acquisition system, where the GPS timing signal is fed to the central processing unit to unify timestamps. Regarding data acquisition synchronization, multiple approaches exist depending on synchronization precision requirements. A simplified method employs software-based triggering by initiating both sensors through a unified software command. For instance, if both LiDAR and camera operate at 10 Hz, their acquisition cycles theoretically align. However, this software-triggered approach suffers from inherent limitations: neither sensor strictly adheres to the nominal 10 Hz frequency due to timing jitter, causing progressive temporal misalignment during prolonged operation. While negligible for short-duration acquisitions, cumulative errors become significant over extended periods. A more robust solution utilizes external hardware triggering. This method applies to sensors supporting external trigger modes, where a dedicated device generates synchronization signals. The LiDAR and camera initiate acquisition upon receiving these signals, reducing inter-sensor synchronization errors to approximately 10 ms. In this study, the Livox Avia LiDAR supports Pulse Per Second (PPS) synchronization, while the HIKROBOT camera accommodates hardware-triggered acquisition. Thus, an STM32 microcontroller generates trigger signals: a 1 Hz Pulse Width Modulation (PWM) square wave for LiDAR and a 10 Hz PWM signal for the camera. The architecture of this temporal synchronization and triggering system is illustrated in Figure 4.

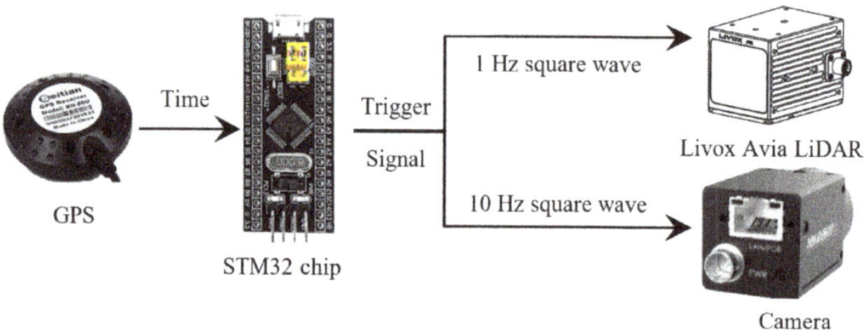

Figure 4. Synchronized triggering of camera and LiDAR.

4. 3D Reconstruction Using LiDAR–Camera Fusion

4.1. Calibration and Matching of Laser Point Clouds to Images

The subsequent step involves calibrating the camera and LiDAR to establish 2D-3D correspondences, thereby achieving spatial synchronization between the two sensors. This process fundamentally relies on constructing matched point pairs between 2D images and 3D point clouds using a geometric calibration target to estimate the relative rotation and translation. The calibration of multi-sensor systems, such as cameras and LiDARs, constitutes the foundation for integrated measurement platforms. While this field has been extensively researched in recent years, most existing calibration methods target 360° rotating mechanical LiDARs. When applied to solid-state LiDARs or non-repetitive scanning LiDARs, challenges arise due to inhomogeneous point cloud density across scanning directions and inconsistent ranging error distributions. To address these challenges, specialized calibration methods for solid-state LiDARs have been proposed. For instance, the ACSC method [29] employs a spatiotemporal geometric feature extraction approach combined with intensity distribution analysis to estimate 3D corner points in point clouds. This technique enables precise corner detection under non-repetitive scanning conditions, even with uneven point density and reflectivity variations. Once corner points are detected in both point clouds and images, a system of equations is constructed to compute the extrinsic matrix.

The ACSC calibration method comprises three sequential phases: refined point cloud processing, corner detection, and extrinsic matrix estimation. The first phase involves denoising and enhancing point cloud features to mitigate inhomogeneity caused by non-repetitive scanning. Initially, temporal integration of the point cloud is performed. Compared to single-frame processing, continuous scanning over an extended duration generates denser point clouds for 3D objects, leveraging the inherent advantages of non-repetitive scanning. In this study, a checkerboard panel is selected as the calibration target. When visualizing the point cloud via intensity values, distinct patterns emerge between white and black checkerboard squares, analogous to image representations. However, the direct accumulation of raw point clouds introduces noise aggregation across frames, leading to blurred results. To counteract this, statistical outlier removal is applied to each input frame by analyzing neighborhood point density in the temporal domain. Post-integration, the surface normal distribution exhibits enhanced continuity compared to single-scan data. Subsequent segmentation based on normal vector differences and target similarity metrics enables cluster ranking, with the minimal-variation cluster retained for checkerboard localization. Furthermore, point cloud feature enhancement techniques are implemented to refine the checkerboard point cloud, ensuring noise-free and geometrically precise data. The specific procedure begins with detecting the boundary lines of the checkerboard panel. Subsequently, background point clouds outside these boundaries are filtered out, yielding a refined point cloud P_c exclusively containing the checkerboard. An iterative optimization method is then applied to align P_c with an idealized plane. The optimization process is formulated as follows:

$$P_c' = \left\{ (x_0, y_0, z_0) \in P_c \middle| \frac{|Ax_0 + By_0 + Cz_0 + D|}{\sqrt{A^2 + B^2 + C^2}} < \sigma \right\} \quad (4)$$

where A, B, C, and D denote the estimated parameters of the fitted plane, and σ represents the standard deviation of distances from points in P_c to the fitted plane. To address the inhomogeneous point cloud density, a voxel grid-based stochastic sampling method is applied to the refined point cloud P_c'. This sampling strategy ensures that the post-sampling point density does not exceed the maximum density observed in the original voxel grid, thereby homogenizing spatial distribution while preserving geometric fidelity.

After optimizing the point cloud, the next step is to detect the coordinates of corner points in the point cloud. The ACSC calibration method employs a nonlinear optimization approach to detect corner points. The core of this method lies in designing a similarity function to estimate the pose difference between the measured values P of checkerboard corners solved from reflectance and the mathematical model S of checkerboard corners. Based on this difference, the L-BFGS algorithm [30] is used to optimize the measured values of the corner points.

Once the corner coordinates are obtained, the Perspective-n-Point (PnP) algorithm can be directly applied to solve the extrinsic matrix. This study adopted the subpixel corner detection algorithm for extracting checkerboard corners from images. When using the algorithms mentioned above to calibrate the extrinsic parameters between the camera and LiDAR, the collected point cloud and final matching results are shown in Figure 5. In this section, the LiDAR and camera calibration toolbox in MATLAB (version 2022b) is used to calibrate the image and point cloud. After importing the calibrated image and the preprocessed point cloud, the program will compute the rotation and translation matrices for transforming the point cloud to the camera coordinates. After the calibration is completed, to verify the accuracy of the calibration, the point cloud is projected onto the image using the transformation matrix obtained from the calibration, and the translation error and reprojection error are obtained to be 2.61 mm and 2.48 pixels, respectively, as shown in Figure 5b.

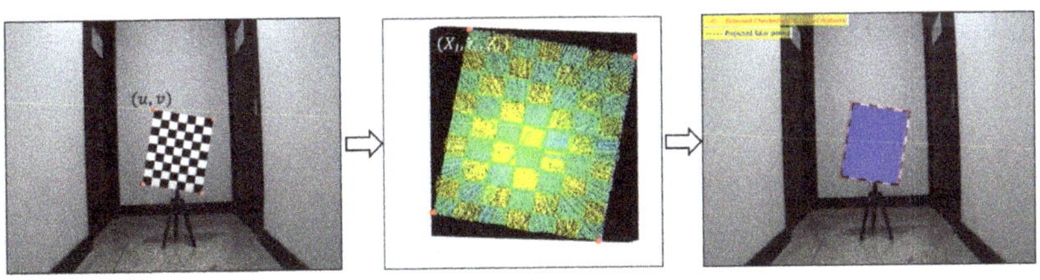

(a)

Figure 5. *Cont.*

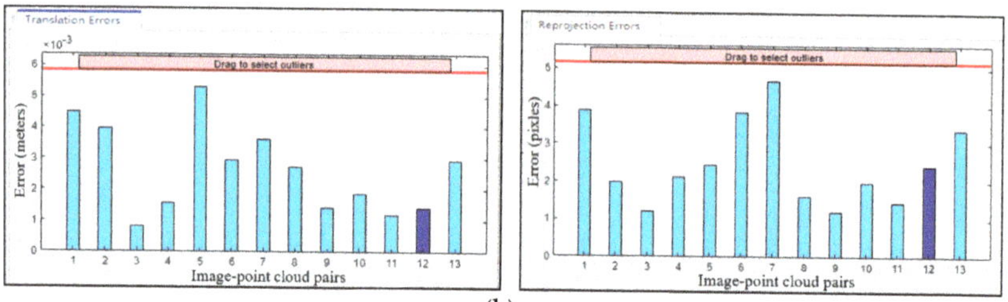

(**b**)

Figure 5. Calibration of LiDAR and camera. (**a**) The collected image, point cloud, and results of point cloud to image projection and (**b**) translation error and reprojection error.

4.2. 3D Reconstruction Using Point Cloud and Image Fusion

Following the methodological framework in Figure 1, this section presents the adopted LiDAR–camera–IMU fused 3D mapping method. Multi-sensor fusion for 3D reconstruction and localization has been a key research focus in robotics in recent years. LiDAR, cameras, and IMUs are the most commonly used sensors, and SLAM methods employing these sensors individually or in combination have proliferated. Representative examples include visual odometries such as LSD-SLAM and DSO, visual-inertial odometries like ORB-SLAM3, and LiDAR-inertial odometries such as LOAM-SLAM and FAST-LIO [31–33]. While these methods have been extensively validated in laboratory tests and outdoor urban environments, real-world scenarios, particularly bridge surroundings, often suffer from sparse structural features and texture-deficient fields of view, posing new challenges to maintaining algorithmic robustness. Fully leveraging the complementary strengths of LiDAR, cameras, and IMUs is critical to ensuring mapping stability when single-sensor feature detection capabilities degrade. Recent advances in multi-sensor fused 3D mapping include R3LIVE++, R3LIVE, LVI-SAM, and FAST-LIVO [34–36].

Among these, R3LIVE++ is an enhanced algorithm derived from R3LIVE and R2LIVE. It consists of two independent subsystems: LIO and VIO modules. The LIO subsystem, directly adopting the FAST-LIO2 algorithm, leverages LiDAR and IMU measurements to establish the geometric structure of the global map. The VIO subsystem, a novel contribution of R3LIVE++, reconstructs the map's texture using visual-inertial data. This VIO tracks camera poses by minimizing photometric discrepancies between sparse pixel sets in current images and precomputed radiation maps. Frame-to-map alignment mitigates odometric drift, while sparse pixel sampling constrains computational overhead. Additionally, the VIO estimates camera exposure time online to recover true environmental radiance.

Similarly, LVI-SAM integrates LIO and VIO subsystems. Its VIO initializes using LIO estimates and employs LiDAR measurements to extract depth information for visual features, thereby improving accuracy. Conversely, the LIO utilizes VIO-derived pose estimates as initial values for point cloud registration. This framework uses VIO to construct the global 3D map and perform loop closure detection, while LIO optimizes the results. The system maintains mapping continuity even if one subsystem fails, ensuring robustness in textureless or feature-scarce environments.

In contrast to the above vision–LiDAR fusion approaches, FAST-LIVO prioritizes computational efficiency, enabling real-time operation on low-performance Advanced RISC Machines (ARM) boards. Its LIO subsystem incrementally appends newly scanned raw points to a dynamically updated point cloud map. The VIO subsystem aligns new images by minimizing direct photometric errors using image patches associated with the point cloud map, bypassing explicit visual feature extraction. The workflow of this method is illustrated in Figure 6.

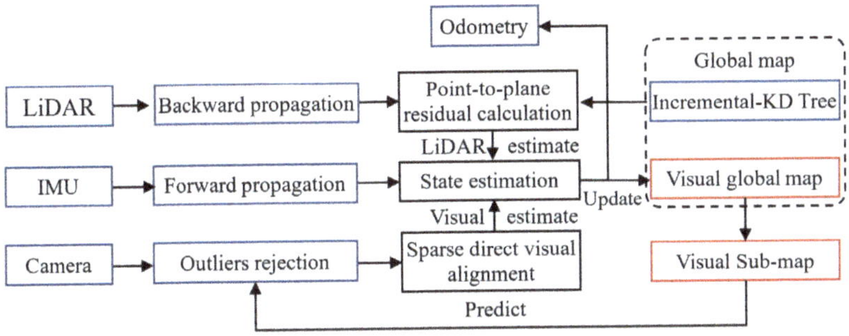

Figure 6. Flow of the reconstruction methodology employed.

The state estimation method in FAST-LIVO employs a tightly coupled error-state iterative Kalman filter (ESIKF) to fuse LiDAR, camera, and IMU data. This method defines the IMU state at the first timestamp as the global coordinate system and estimates subsequent frame poses through a state transition model. The state and covariance matrix at the $(i + 1)$-*th* timestamp are derived via forward propagation. For the LiDAR measurement model in the frame-to-map update phase, the system first removes motion distortion from the input point cloud. During frame-to-map registration, newly observed points are assumed to lie on planes formed by their nearest neighbors in the map. To identify these planes, prior pose estimates project the points into the map, where the five closest points to each projected point are retrieved to fit a local plane.

For the visual measurement model, the method performs sparse direct frame-to-map image alignment in a coarse-to-fine manner by minimizing photometric errors. In map management, the LiDAR global map follows the incremental KD-tree approach from FAST-LIO2, providing interfaces for point querying, insertion, and deletion. The visual global map adopts an axis-aligned voxel-based data structure, where voxel hash tables enable the rapid identification of map points relevant to the current frame. After aligning a new image frame, image patches from the current frame are appended to map points within the field of view, ensuring uniform distribution of effective patches across viewing angles.

This workflow allows the VIO subsystem in FAST-LIVO to reuse LiDAR point clouds while bypassing visual feature extraction, triangulation, and optimization, significantly reducing computational demands. Consequently, the algorithm maintains robust performance in sensor-degraded environments with limited processing resources. Given FAST-LIVO's superior runtime efficiency and mapping stability compared to other algorithms like R3LIVE++ and LVI-SAM, this study adopts FAST-LIVO as the LiDAR–camera fused 3D mapping framework.

4.3. Bridge Component Segmentation Using Image Segmentation and Point Cloud–Image Matching

After obtaining the 3D point cloud of the bridge through the aforementioned steps, the automated extraction of bridge geometry requires identifying key structural components from the large-scale point cloud, whose edge geometries collectively define the bridge's linear form. The specific geometry of interest varies across bridge types and operational scenarios. For example, during the construction phase of suspension bridges, the cable geometry is a critical measurement target, necessitating the segmentation of main cable point clouds. In load-testing scenarios, the deck geometry becomes the focus, requiring the segmentation of girder point clouds. As this study focuses on a concrete arch bridge, the segmentation target in this section is the bridge arch. Once the arch is segmented from the images captured during UAV flights, the 2D mask of the arch in the images must be projected back into the 3D point cloud to extract the 3D coordinates of its geometry. This reconstruction leverages the pose information of each image, camera calibration parameters, and LiDAR–camera calibration results from prior steps. By reprojecting the arch mask into the 3D point cloud, the edge coordinates of the corresponding 3D points directly represent the bridge's critical geometry. Unlike the existing algorithms based on point cloud processing, the proposed method is essentially a 2D image segmentation method that first segments the bridge's components from the 2D image. Then it iterates through all the point clouds projected to the image plane according to the transform matrix calibrated from the image and the point cloud. Then the points projected in the masks of the components of the bridge are the point clouds of the segmented components. This method not only makes full use of the image and point cloud data contained in the system but also avoids the direct processing of large-scale point clouds, which reduces the algorithm's requirements for computer arithmetic power. The algorithmic workflow for this section is illustrated in Figure 7.

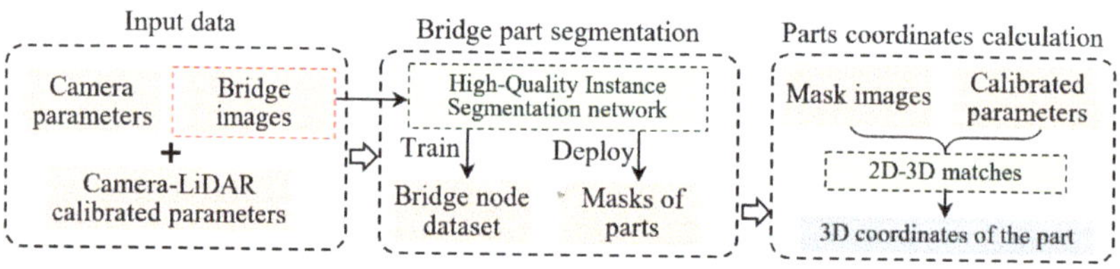

Figure 7. Procedure for the calculation method of bridge geometric shape.

As shown in Figure 7, the proposed method can be divided into two main steps: (1) Fine-grained segmentation of bridge components in images. (2) Three-dimensional coordinate resolution of these components. The core of the first step lies in addressing the identification and segmentation of bridge components using a high-precision segmentation network. Instance segmentation, a critical research direction in computer vision, aims to detect targets in input images and assign category labels to each pixel of the targets. Unlike conventional image segmentation tasks such as human body recognition or structural defect segmentation, where coarse mask edges are acceptable, most existing instance segmentation networks produce masks with rough boundaries, particularly in low-contrast scenarios where target and background differentiation are minimal. This limitation leads to degraded edge recognition accuracy.

In this study, the edge segmentation precision of bridge components directly affects the segmentation results of corresponding point clouds and consequently the accuracy of 3D coordinate resolution. Therefore, when applying instance segmentation networks, the focus is on improving the segmentation accuracy of component edges.

To address this challenge, we propose a transformer-based instance segmentation network designed to refine edge recognition and segmentation. The architecture of this network is illustrated in the accompanying figure. Building on the initial segmentation results from existing networks, our method employs a transformer structure for secondary prediction, specifically targeting edge regions [37]. For the critical edge areas of segmentation targets, a quadtree structure is constructed to represent discrete points across multiple hierarchical levels. A transformer-based segmentation network is utilized to predict instance labels for each tree node, given their distribution in discontinuous spatial domains. This network comprises three modules: a node encoder, a sequence encoder, and a pixel decoder. The node encoder enriches the feature representation of each point, the sequence encoder processes node sequences through a transformer-based architecture, and the pixel decoder predicts instance labels for each point. By serializing sparse feature points across scales and leveraging multi-head attention mechanisms, the model dynamically models multi-scale sparse features while enabling long-range comparisons and cross-hierarchy associations among feature points.

As illustrated in Figure 8, the network first constructs a quadtree comprising information loss nodes at different hierarchical levels. These nodes originate from a region of interest (RoI) pyramid but exhibit varying feature granularities due to their hierarchical differences. Information loss nodes from lower-level RoIs (28 × 28 resolution) correspond to four child nodes in their adjacent higher-level RoIs (56 × 56 resolution). To reduce computational complexity, pixels predicted as loss nodes are further decomposed into upper layers, with the maximum tree depth set to 3. The detection of information loss regions follows a bottom-up cascaded design. As shown on the right side of Figure 8, the lowest-level RoI features (28 × 28) and initial object mask predictions are first fed into a lightweight fully convolutional network (3 × 3 convolution) to predict quadtree root nodes. Each root node is decomposed into four child nodes in the adjacent higher-level RoI layer, extending from 28 × 28 to 56 × 56 resolutions. For higher-level RoI features, the mask from the preceding loss region detection is upsampled and concatenated with the current RoI features. A single 1 × 1 convolutional layer then predicts finer-grained information loss nodes, ensuring the detection module remains computationally efficient.

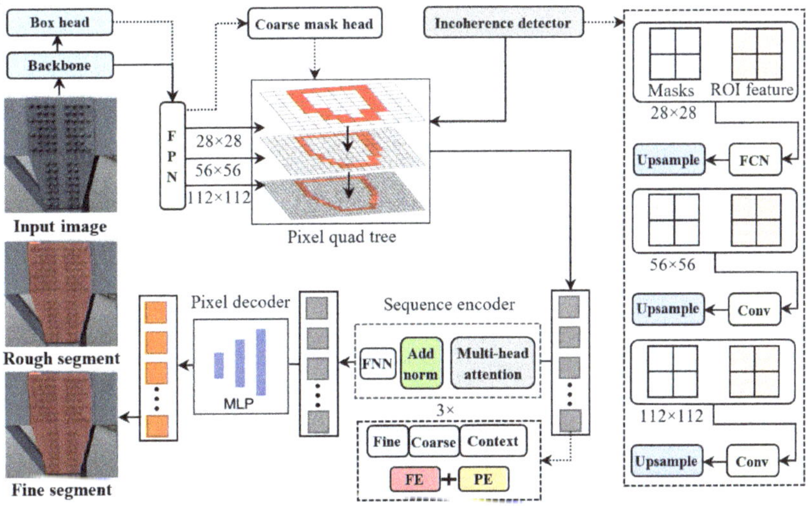

Figure 8. Structure of the hierarchical bridge component extraction network.

Since the proposed method is applied to the geometry measurement of a concrete arch bridge in the testing phase, this section focuses on arch geometry measurement as a case study. A dataset of 300 UAV-captured images of bridge arches under varying illumination conditions was compiled. These images were manually annotated to create the training and evaluation dataset, as exemplified in Figure 9.

The network was trained on hardware comprising an i7-11700K CPU from Intel Company, USA, 32 GB RAM, and an RTX 3090 GPU from NVIDIA Company, USA. The PyTorch (version 1.13.0) framework is used. Training parameters included 1000 iterations with a learning rate of 0.001. During training, the test accuracy of the predicted masks evolved as shown in Figure 10a, ultimately reaching 94.23% upon completion. The trained network was then applied to automatically segment arch regions in UAV-captured images of the bridge, with segmentation results demonstrated in Figure 10b. For a 4000 × 3000-pixel image, the automatic segmentation process required approximately 0.5 s.

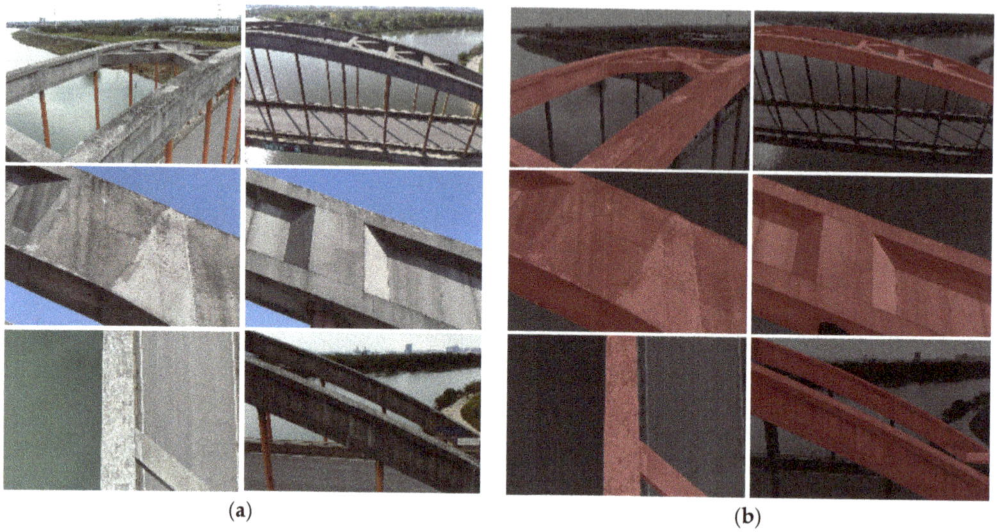

Figure 9. The established bridge arch segmentation dataset. (**a**) Raw images captured by UAV and (**b**) manual labeling results.

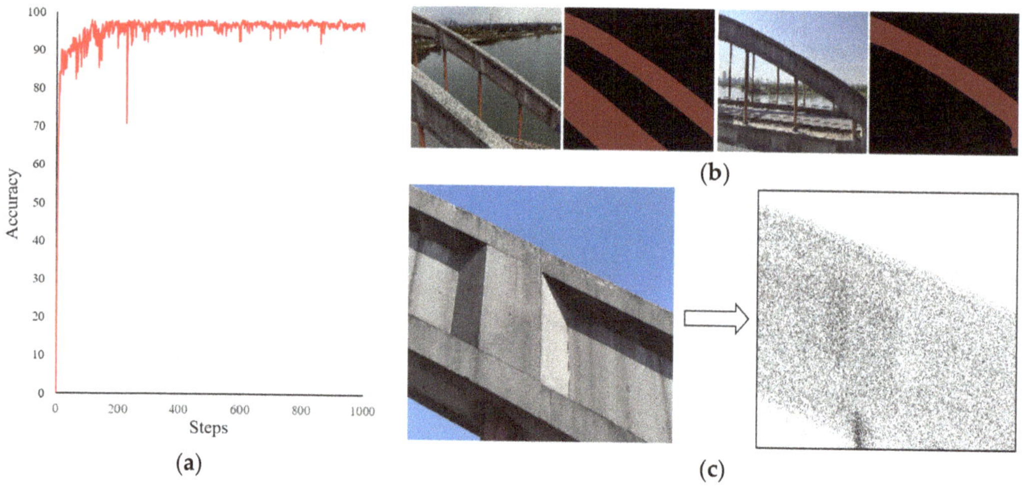

Figure 10. Training results of the hierarchical bridge component segmentation network. (**a**) Accuracy during training, (**b**) segmentation results of the trained network, and (**c**) segmented point cloud after applying the image to point cloud matching.

Subsequently, following the workflow proposed in Figure 7, the fused LiDAR–camera system's images and point clouds were processed. Segmentation results are illustrated in Figure 10c. For an input image and a LiDAR point cloud frame (with a 100 ms integration time), the total processing time from automatic arch segmentation in the image to point cloud segmentation averaged 2 s.

5. Field Test

The tested bridge is a concrete arch structure, as shown in Figure 11. With a main span of 140 m and a bridge width of 8 m, this structure was constructed in the 1970s. After decades of service, the bridge exhibits issues such as arch deformation and lateral tilting, necessitating precise 3D geometry inspection to assess its structural condition.

Figure 11. Overview of the test bridge.

Three methods were employed to reconstruct the 3D model of the bridge during testing: (1) A high-precision TLS was used for multi-station scanning across the entire bridge. The scanning setup comprised 14 stations, with 40 retroreflective targets pre-installed on the bridge surfaces to facilitate accuracy validation. The TLS device (VZ-400i from RIEGL Company, Austria) offers a specified accuracy of 3 mm at 50 m, making its acquired geometry the reference ground truth. (2) A UAV-mounted mapping camera captured multi-view images of the bridge, which were processed using commercial 3D reconstruction software to generate a point cloud. (3) The proposed LiDAR–camera fused system performed rapid scanning while the UAV flew along predefined trajectories outside the bridge. The TLS field workflow and the resulting point cloud are shown in Figure 12. The weather for data collection was a clear day with sufficient light and a wind speed of category 3, and collection began at 12:00 noon to minimize shadows on the bridge surfaces and to avoid degradation of the image quality of the camera due to backlighting. The UAV photogrammetry method collected 2068 images, with the flight paths, operational procedures, and reconstructed point cloud detailed in Figure 13. The proposed system's testing process and acquired point cloud are illustrated in Figure 14.

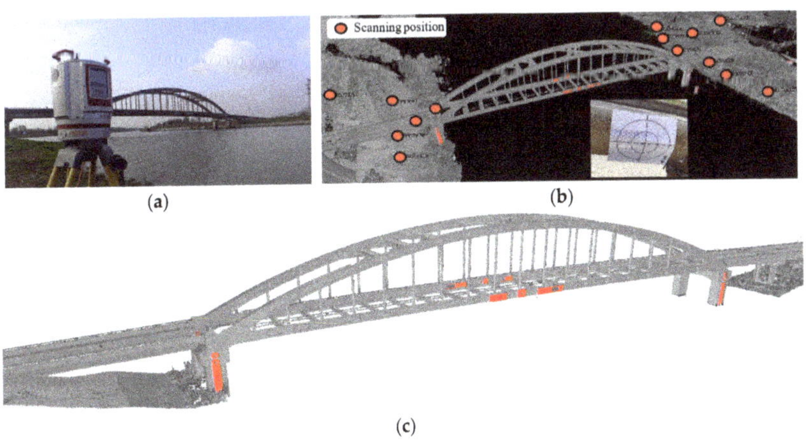

Figure 12. Point clouds of the bridge reconstructed by high-precision TLS. (**a**) VZ-400i TLS, (**b**) scanning station layout and reflector settings, and (**c**) acquired 3D point cloud of the bridge.

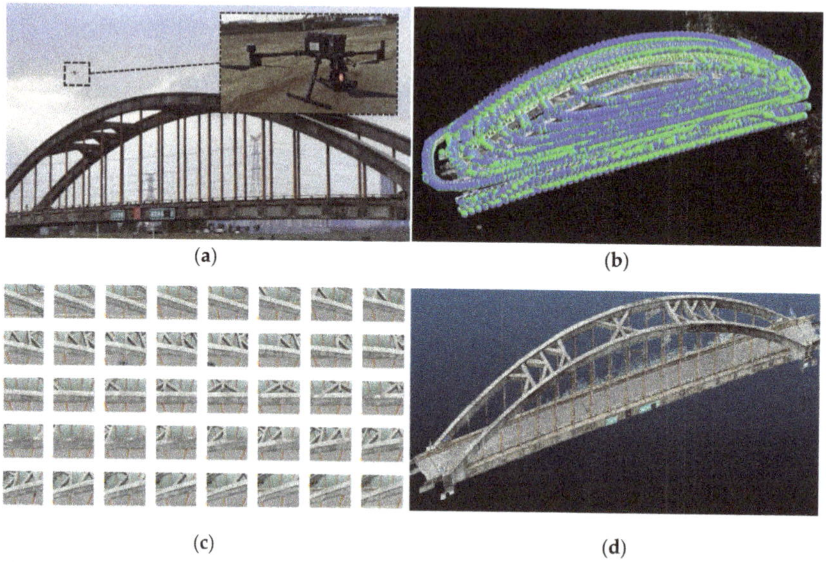

Figure 13. Three-dimensional reconstruction results of the bridge using UAV oblique photogrammetry. (**a**) UAV flight photography process, (**b**) location of images acquired by the UAV, (**c**) images of the bridge arch, and (**d**) 3D reconstruction results of the bridge.

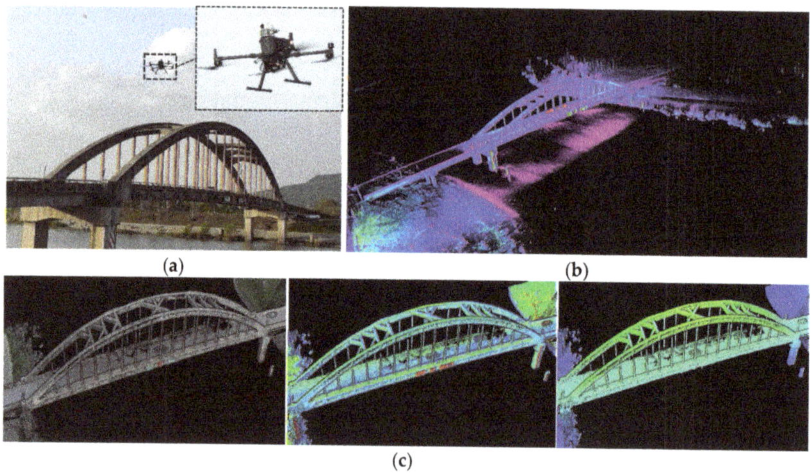

Figure 14. Three-dimensional point cloud of the bridge reconstructed using the proposed method. (**a**) UAV flight scanning process, (**b**) reconstruction result displayed in real time during the flight, and (**c**) reconstructed 3D point cloud. Left: true color, middle: reflectivity assignment, right: height assignment.

For TLS scanning, the angular resolution was set to 0.02°, with each station requiring 3 min. During scanning of the bridge's four sides, one station per side was selected for high-resolution scanning at 0.01° angular resolution, extending the scan time to 7 min per station. The scanner operated at a laser pulse repetition rate of 1200 kHz, achieving a maximum range of 250 m. Field scanning totaled approximately 2 h, followed by 2 h of data registration comprising manual coarse alignment and automated fine registration. The entire TLS-based workflow required 4 h, yielding an initial point cloud of 882 million points. Manual segmentation isolated the bridge structure, resulting in a refined point cloud of 193 million points.

For the UAV oblique photogrammetry-based 3D reconstruction method, data acquisition involved a combination of flight path planning software and manual localized photography to capture images of the bridge. The collected images included both long-range and close-range shots, with the latter achieving a pixel resolution of approximately 1.22 mm/pixel. The UAV platform used was a DJI M300 RTK equipped with a DJI P1 mapping camera. The UAV can provide 40 min of endurance per session, while the camera has a sensor measuring 35.9 mm × 24 mm with an effective pixel count of 45 megapixels and can capture images of 8192 × 5460 pixels. The camera automatically takes pictures at a rate of one per second while in flight. A close-range image of one arch is shown in Figure 13c, while Figure 13b illustrates the spatial distribution of images. The total image acquisition time was approximately 5 h, including an additional 3 h for preliminary data collection and the rough 3D model reconstruction required for flight path planning. After acquiring 2068 images, existing mature multi-view 3D reconstruction software was employed to generate the bridge point cloud. Considering computational efficiency, the open-source COLMAP software (version 3.12.0) was selected, which logged time costs for each processing stage: 0.2 h for feature detection, 11.67 h for feature matching, 21.45 h for sparse point cloud reconstruction, and 16.35 h for dense point cloud generation. The processing hardware comprised an Intel i7-11700K CPU, RTX 3090 GPU, and 64 GB RAM, resulting in a total processing time of 57.67 h. The final point cloud contained 255 million points, which was manually segmented to remove background noise, yielding a refined bridge point cloud of 167 million points.

The 3D scanning process using the proposed method is depicted in Figure 14a. During the test, the acquisition frequency of the LiDAR and the camera was 10 Hz, the frequency of the IMU was 200 Hz, the point cloud acquisition rate of the LiDAR was 240,000 points/second, and the size of the image acquired by the camera was 4000 × 3036 pixels. Since the LiDAR–camera fusion system operates continuously from UAV takeoff, unaffected by translational or rotational motions, the drone only needed to be on the bridge at an appropriate distance (15 m in this test) for at least one complete cycle. In this test, two flights were conducted, totaling 10 min from takeoff to landing. Real-time point cloud updates were displayed via wireless video transmission from the onboard computer, gradually revealing the bridge's whole geometry as coverage expanded (Figure 14b). The reconstructed point cloud (Figure 14c) contained 18.7448 million raw points, reduced to 12.0692 million after manual background removal.

The three point cloud acquisition methods were compared in terms of accuracy and processing speed, and 80 retroreflective targets (40 per bridge side) were analyzed for accuracy validation. Coordinates extracted from TLS data served as ground truth, as shown in Figure 15a, while the proposed method directly identified targets via reflectivity, as shown in Figure 15b. UAV photogrammetry required manual target localization by averaging the corner coordinates of each target, as shown in Figure 15c.

Since the point clouds from the three methods resided in different coordinate systems, the relative coordinates of each target were calculated to analyze errors across methods. Using the first target as the reference point, the relative coordinates of other points were computed to align all measurements into a unified coordinate system. Figures 16 and 17 illustrate the positional errors of UAV oblique photogrammetry and the proposed method relative to the high-precision laser scanning reference. The results reveal significantly larger errors in photogrammetry-derived target coordinates compared to the proposed method. Specifically, for X-axis measurements, the mean error of photogrammetry was 0.452 m, whereas the proposed method achieved 0.083 m, corresponding to 18.36% of photogrammetry's error. For Y-axis measurements, photogrammetry exhibited a 0.602 m error against 0.103 m for the proposed method, representing 17.11% of the former. For Z-axis measurements, errors were 0.184 m for photogrammetry and 0.061 m for the proposed method, equivalent to 33.15%. Overall, the proposed method reduced the average error by 77.13% compared to oblique photogrammetry.

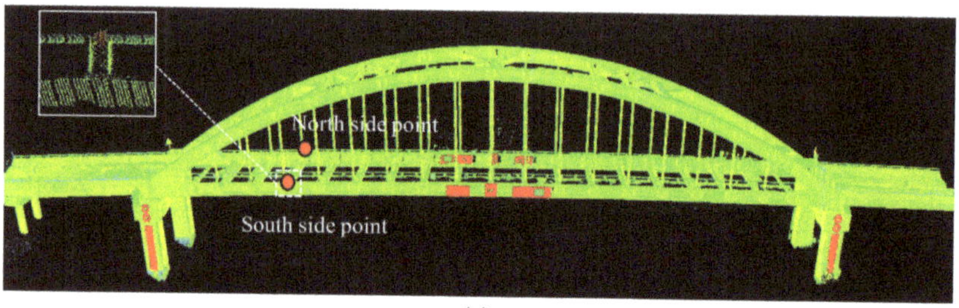

(**a**)

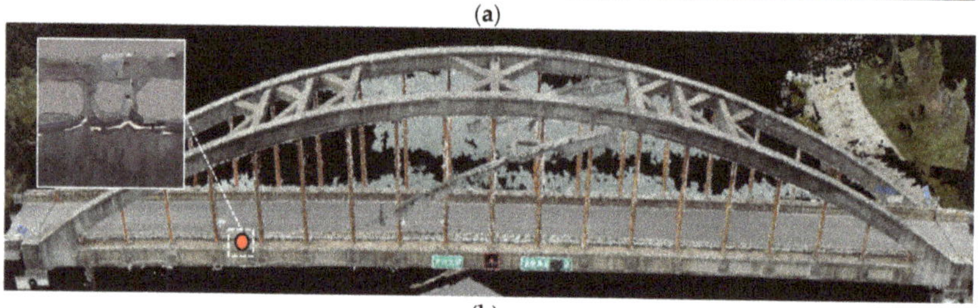

(**b**)

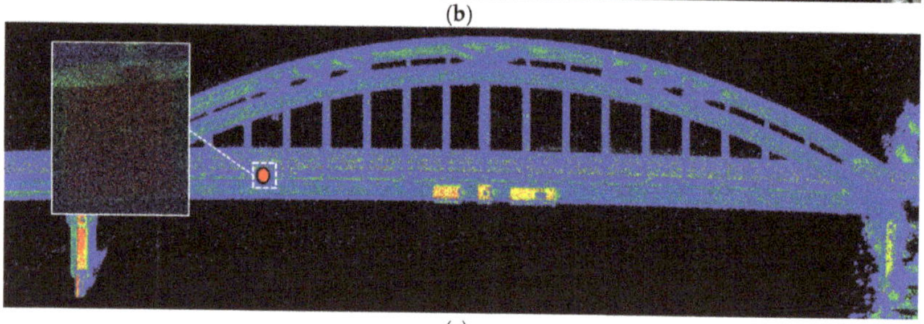

(**c**)

Figure 15. Reflective targets observed in point clouds acquired by three methods. (**a**) Reflective targets observed by high-precision TLS, (**b**) reflective targets observed by UAV oblique photogrammetry, and (**c**) reflective targets observed by the proposed method.

The reconstruction efficiency of the three methods was further compared based on field and computational processing times, as summarized in Table 1. The total time for the proposed method amounted to merely 4.18% of the laser scanning workflow, while photogrammetry required 14.42 times longer than laser scanning. However, the proposed method generated fewer points due to the LiDAR module's lower acquisition rate of 240,000 points per second compared to the laser scanner's 1.2 million points per second, combined with significantly shorter scanning durations. Despite this, the point cloud density from the proposed method proved sufficient for geometry measurement. Notably, the laser scanner produced sparse point clouds at mid-span regions due to excessive scanning distances, whereas the proposed method maintained uniform density across all bridge sections through consistent proximity during scanning. Additionally, both laser scanning and photogrammetry demanded substantial computational memory and slower processing speeds, often requiring point cloud down-sampling for analysis on low-performance systems.

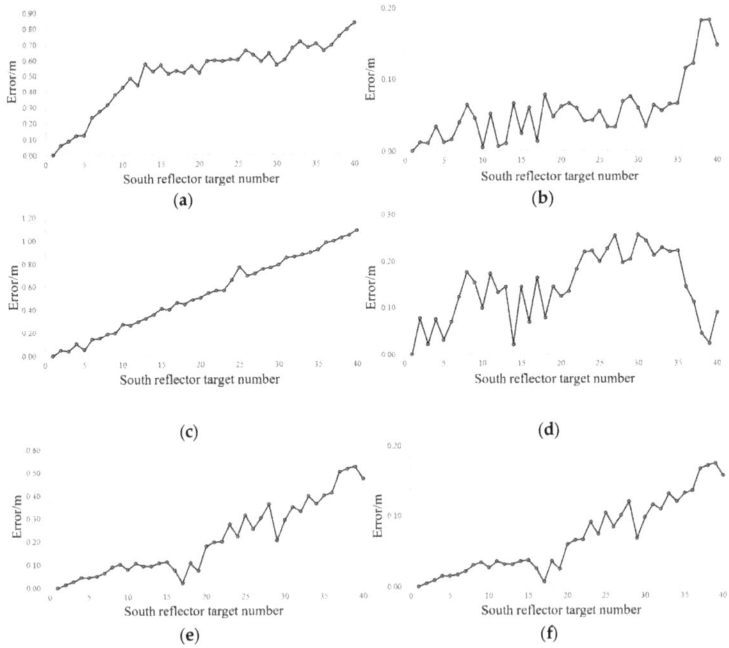

Figure 16. Measurement errors of the three methods compared on the south side of the bridge. (**a**) x-coord errors in oblique photography, (**b**) x-coord errors in the proposed method, (**c**) y-coord errors in oblique photography, (**d**) y-coord errors in the proposed method, (**e**) z-coord errors in oblique photography, and (**f**) z-coord errors in the proposed method.

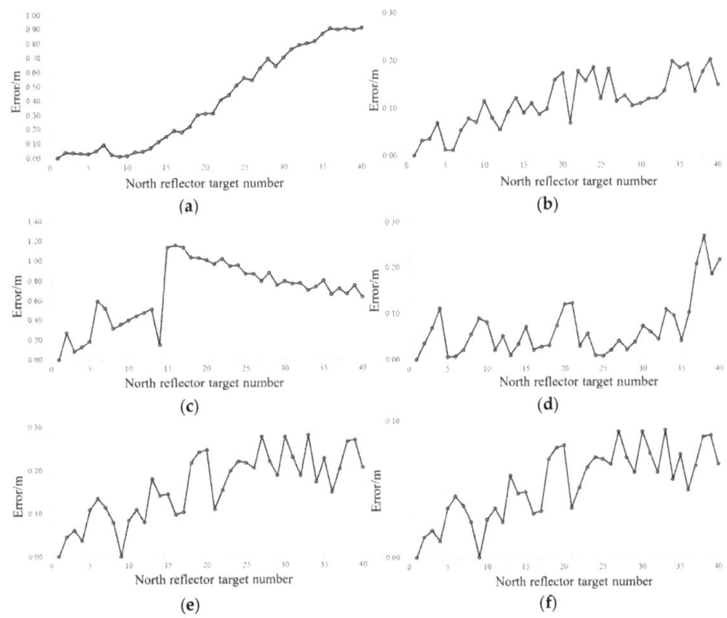

Figure 17. Measurement errors of the three methods compared on the north side of the bridge. (**a**) x-coord errors in oblique photography, (**b**) x-coord errors in proposed method, (**c**) y-coord errors in oblique photography, (**d**) y-coord errors in proposed method, (**e**) z-coord errors in oblique photography, and (**f**) z-coord errors in proposed method.

Table 1. Comparison of the time consumption of the three methods in the reconstruction.

Methods	Data Collection/h	Data Processing/h	Total Time Consumption/h	Number of Point Clouds	Number of Effective Point Clouds
TLS	2	2	4	882 million	193 million
Oblique photography	8	49.67	57.67	255 million	167 million
Proposed method	0.167	0	0.167	19 million	12 million

6. Conclusions

This study proposes a LiDAR–camera fused 3D reconstruction and automated component segmentation method for bridge geometry measurement. By synergizing the strengths of LiDAR point clouds and images, the proposed method achieves higher accuracy and efficiency in 3D reconstruction than image-based approaches. Unlike stationary high-precision laser scanning, it leverages image–point cloud correspondences for rapid segmentation, bypassing the direct processing of massive point clouds. Field tests on in-service bridge geometry inspection yield the following conclusions:

(1) For LiDAR–camera data fusion, the method employs inertial navigation-based point cloud de-distortion, calibration target-based sensor alignment, and tightly coupled image–point cloud fusion to reconstruct large-scale 3D environments dynamically. The integrated system demonstrates rapid and stable 3D point cloud reconstruction for inspecting an in-service bridge.

(2) An image-guided segmentation framework is proposed to address challenges in automated component segmentation from large-scale bridge point clouds. It maps 2D segmentation results to 3D point clouds, incorporating a coarse-to-fine segmentation network that achieves 94.23% accuracy in bridge component identification.

(3) In the validation of an in-service bridge, it is demonstrated that the measurement accuracy of the proposed method is 83 mm in the X-axis direction, 103 mm in the Y-axis direction, and 61 mm in the Z-axis direction. The average error is reduced by 77.13% compared to the widely used method of UAV-based oblique photography. The total processing time is equivalent to 4.18% of the method using stationary laser scanning and 0.29% of the technique of UAV-based oblique photography.

When applying the proposed method to inspect the geometric shape of in-service bridges, there are limitations that need to be overcome and improved in future research. Firstly, the LiDAR in the constructed system is a low-cost, non-repeating scanning LiDAR, which is less accurate at the hardware level than the expensive, high-precision LiDAR. This resulted in an error of 8.3 cm in the reconstructed bridge compared to the static TLS scanner. The selection of higher accuracy LiDAR in the future is expected to improve the accuracy of the proposed method further. Secondly, the automated segmentation network of bridge components currently trained is only for the segmentation of arches of concrete arch bridges, so the proposed method can only realize the detection of the arch's line shape at present. And because the proposed method for segmenting the point cloud of the components of a bridges is not to process the point cloud of the bridge directly but to segment the image first, and then to segment the point cloud projected in the mask of the components according to the calibration results of the image and the point cloud, the method can only be applied to the data collected by the proposed LiDAR–camera fusion system rather than the data collected by the other devices. In subsequent research, the datasets including bridge cables, box girder girders, and bridge towers will be added, so that the proposed method can be applied to more types of bridges and bridge components for geometric shape measurement.

Author Contributions: S.J.: Conceptualization, Methodology, Validation, Writing—original draft, Writing—review and editing, Funding Acquisition. Y.Y.: Methodology, Validation, Writing—original draft, Writing—review and editing. S.G.: Conceptualization, Validation. J.L.: Methodology, Validation. Y.H.: Conceptualization, Funding Acquisition. All authors have read and agreed to the published version of the manuscript.

Funding: The research presented was financially supported by the Key Laboratory of Target Cognition and Application Technology (No.: 2023-CXPT-LC-005), the Key Laboratory of Road and Railway Engineering Safety and Security (No.: STDTKF202303), and the Innovative training program for university students (No.: 202410304059Z).

Data Availability Statement: The data supporting this study's findings are available from the corresponding author upon reasonable request.

Conflicts of Interest: The authors declare no conflicts of interest.

References

1. Jiang, S.; Cheng, Y.; Zhang, J. Vision-guided unmanned aerial system for rapid multiple-type damage detection and localization. *Struct. Health Monit.* **2023**, *22*, 319–337. [CrossRef]
2. Jiang, S.; Zhang, J.; Gao, C. Bridge Deformation Measurement Using Unmanned Aerial Dual Camera and Learning-Based Tracking Method. *Struct. Control Health Monit.* **2023**, *2023*, 4752072. [CrossRef]
3. Tian, Y.; Zhang, C.; Jiang, S.; Zhang, J.; Duan, W. Noncontact cable force estimation with unmanned aerial vehicle and computer vision. *Comput.-Aided Civ. Infrastruct. Eng.* **2021**, *36*, 73–88. [CrossRef]
4. Hu, F.; Zhao, J.; Huang, Y.; Li, H. Structure-aware 3D reconstruction for cable-stayed bridges: A learning-based method. *Comput.-Aided Civ. Infrastruct. Eng.* **2021**, *36*, 89–108. [CrossRef]
5. Tian, X.; Liu, R.; Wang, Z.; Ma, J. High-quality 3D reconstruction based on fusion of polarization imaging and binocular stereo vision. *Inf. Fusion* **2022**, *77*, 19–28. [CrossRef]
6. Zhang, S. High-speed 3D shape measurement with structured light methods: A review. *Opt. Lasers Eng.* **2018**, *106*, 119–131. [CrossRef]
7. Hosamo, H.H.; Hosamo, M.H. Digital twin technology for bridge maintenance using 3d laser scanning: A review. *Adv. Civ. Eng.* **2022**, *2022*, 2194949. [CrossRef]
8. Ma, Z.; Choi, J.; Sohn, H. Continuous bridge displacement estimation using millimeter-wave radar, strain gauge and accelerometer. *Mech. Syst. Signal Process.* **2023**, *197*, 110408. [CrossRef]
9. Wang, F.; Zou, Y.; del Rey Castillo, E.; Ding, Y.; Xu, Z.; Zhao, H.; Lim, J.B. Automated UAV path-planning for high-quality photogrammetric 3D bridge reconstruction. *Struct. Infrastruct. Eng.* **2024**, *20*, 1595–1614. [CrossRef]
10. Chen, S.; Laefer, D.F.; Mangina, E.; Zolanvari, S.M.I.; Byrne, J. UAV bridge inspection through evaluated 3D reconstructions. *J. Bridge Eng.* **2019**, *24*, 05019001. [CrossRef]
11. Verykokou, S.; Ioannidis, C. An overview on image-based and scanner-based 3D modeling technologies. *Sensors* **2023**, *23*, 596. [CrossRef] [PubMed]
12. Lin, Y.C.; Liu, J.; Cheng, Y.T.; Hasheminasab, S.M.; Wells, T.; Bullock, D.; Habib, A. Processing strategy and comparative performance of different mobile lidar system grades for bridge monitoring: A case study. *Sensors* **2021**, *21*, 7550. [CrossRef] [PubMed]
13. Lu, R.; Brilakis, I. Recursive segmentation for as-is bridge information modelling. In Proceedings of the LC3 2017—Joint Conference on Computing in Construction (JC3), Heraklion, Greece, 4–7 July 2017; Volume I, pp. 209–217. [CrossRef]
14. Mohammadi, M.; Rashidi, M.; Yu, Y.; Samali, B. Integration of TLS-derived Bridge Information Modeling (BrIM) with a Decision Support System (DSS) for digital twinning and asset management of bridge infrastructures. *Comput. Ind.* **2023**, *147*, 103881. [CrossRef]
15. Shao, J.; Zhang, W.; Mellado, N.; Grussenmeyer, P.; Li, R.; Chen, Y.; Wan, P.; Zhang, X.; Cai, S. Automated markerless registration of point clouds from TLS and structured light scanner for heritage documentation. *J. Cult. Herit.* **2019**, *35*, 16–24. [CrossRef]
16. Sun, L. Practical, fast and robust point cloud registration for scene stitching and object localization. *IEEE Access* **2022**, *10*, 3962–3978. [CrossRef]
17. Tarabini, M.; Giberti, H.; Giancola, S.; Sgrenzaroli, M.; Sala, R.; Cheli, F. A moving 3D laser scanner for automated underbridge inspection. *Machines* **2017**, *5*, 32. [CrossRef]
18. Feroz, S.; Abu Dabous, S. UAV-based remote sensing applications for bridge condition assessment. *Remote Sens.* **2021**, *13*, 1809. [CrossRef]

19. Wang, D.; Gao, L.; Zheng, J.; Xi, J.; Zhong, J. Automated recognition and rebar dimensional assessment of prefabricated bridge components from low-cost 3D laser scanner. *Measurement* **2025**, *242*, 115765. [CrossRef]
20. Zhu, Y.; Brigham, J.C.; Fascetti, A. LiDAR-RGB Data Fusion for Four-Dimensional UAV-Based Monitoring of Reinforced Concrete Bridge Construction: Case Study of the Fern Hollow Bridge Reconstruction. *J. Constr. Eng. Manag.* **2025**, *151*, 05024016. [CrossRef]
21. Li, J.; Peng, Y.; Tang, Z.; Li, Z. Three-dimensional reconstruction of railway bridges based on unmanned aerial vehicle–terrestrial laser scanner point cloud fusion. *Buildings* **2023**, *13*, 2841. [CrossRef]
22. Huang, Y.; Du, Y.; Shi, W. Fast and accurate power line corridor survey using spatial line clustering of point cloud. *Remote Sens.* **2021**, *13*, 1571. [CrossRef]
23. Xu, Y.; Zhang, J. UAV-based bridge geometric shape measurement using automatic bridge component detection and distributed multi-view reconstruction. *Autom. Constr.* **2022**, *140*, 104376. [CrossRef]
24. Kim, H.; Narazaki, Y.; Spencer, B.F., Jr. Automated bridge component recognition using close-range images from unmanned aerial vehicles. *Eng. Struct.* **2023**, *274*, 115184. [CrossRef]
25. Xia, T.; Yang, J.; Chen, L. Automated semantic segmentation of bridge point cloud based on local descriptor and machine learning. *Autom. Constr.* **2022**, *133*, 103992. [CrossRef]
26. Yang, X.; del Rey Castillo, E.; Zou, Y.; Wotherspoon, L.; Tan, Y. Automated semantic segmentation of bridge components from large-scale point clouds using a weighted superpoint graph. *Autom. Constr.* **2022**, *142*, 104519. [CrossRef]
27. Jing, Y.; Sheil, B.; Acikgoz, S. Segmentation of large-scale masonry arch bridge point clouds with a synthetic simulator and the BridgeNet neural network. *Autom. Constr.* **2022**, *142*, 104459. [CrossRef]
28. Yang, W.; Gong, Z.; Huang, B.; Hong, X. Lidar with velocity: Correcting moving objects point cloud distortion from oscillating scanning lidars by fusion with camera. *IEEE Robot. Autom. Lett.* **2022**, *7*, 8241–8248. [CrossRef]
29. Cui, J.; Niu, J.; Ouyang, Z.; He, Y.; Liu, D. ACSC: Automatic calibration for non-repetitive scanning solid-state LiDAR and camera systems. *arXiv* **2020**, arXiv:2011.08516. [CrossRef]
30. Mannel, F.; Aggrawal, H.O.; Modersitzki, J. A structured L-BFGS method and its application to inverse problems. *Inverse Probl.* **2024**, *40*, 045022. [CrossRef]
31. Campos, C.; Elvira, R.; Rodriguez, J.J.G.; Montiel, J.M.M.; Tardos, J.D. ORB-SLAM3: An accurate open-source library for visual, visual–inertial, and multimap slam. *IEEE Trans. Robot.* **2021**, *37*, 1874–1890. [CrossRef]
32. Yi, S.; Lyu, Y.; Hua, L.; Pan, Q.; Zhao, C. Light-LOAM: A lightweight lidar odometry and mapping based on graph-matching. *IEEE Robot. Autom. Lett.* **2024**, *9*, 3219–3226. [CrossRef]
33. Wu, F.; Beltrame, G. Direct sparse odometry with planes. *IEEE Robot. Autom. Lett.* **2021**, *7*, 557–564. [CrossRef]
34. Lin, J.; Zhang, F. R3LIVE++: A Robust, Real-time, Radiance Reconstruction Package with a Tightly-coupled LiDAR-Inertial-Visual State Estimator. *IEEE Trans. Pattern Anal. Mach. Intell.* **2024**, *46*, 11168–11185. [CrossRef] [PubMed]
35. Han, L.; Shi, Z.; Wang, H. A localization and mapping algorithm based on improved LVI-SAM for vehicles in field environments. *Sensors* **2023**, *23*, 3744. [CrossRef]
36. Zheng, C.; Xu, W.; Zou, Z.; Hua, T.; Yuan, C.; He, D.; Zhou, B.; Liu, Z.; Lin, J.; Zhu, F.; et al. FAST-LIVO2: Fast, direct lidar-inertial-visual odometry. *IEEE Trans. Robot.* **2024**, *41*, 326–346. [CrossRef]
37. Wang, J.; Yu, X.; Gao, Y. Feature fusion vision transformer for fine-grained visual categorization. *arXiv* **2021**, arXiv:2107.02341. [CrossRef]

buildings

MDPI

Review

Urban Viaduct Structural Health Monitoring: A Review of Wireless Sensor Approaches

Tianli Wang [1], Tong Cui [1], Xin Qi [1] and Xintao Huan [2,*]

1 School of Sciences for the Human Habitat, University of Chinese Academy of Sciences, Beijing 101408, China; wangtianli211@mails.ucas.ac.cn (T.W.); cuitong@ucas.ac.cn (T.C.); 13901129392@163.com (X.Q.)
2 School of Cyberspace Science and Technology, Beijing Institute of Technology, Beijing 100081, China
* Correspondence: xintao.huan@bit.edu.cn

Abstract: Urban viaducts have been extensively built in large cities due to traffic pressure on main roads. Although urban viaducts provide relief and reduce traffic, they also pose potential threats to citizens' safety due to the possibility of collapse. However, most collapses do not happen instantly. Structural health monitoring (SHM) systems can effectively detect structural damage to prevent collapses; however, they are not widely deployed due to the high cost of sensor installation. With wireless sensor technology thriving due to its accuracy and cost-efficiency, the wireless monitoring of massive urban viaducts has become possible. In this paper, we review the full-chain methods of employing wireless sensor technology for urban viaduct structural health monitoring. By proposing a general wireless structural health monitoring approach as a framework, we first review the sensor types and wireless transmission technologies, then the data processing and modal identification methods, and finally the damage detection methods. This paper aims to provide a full-chain review of the mainstream methods in wireless sensor approaches for the SHM of urban viaducts.

Keywords: structural health monitoring; wireless sensor network

1. Introduction

Population expansion in large cities has significantly increased traffic pressure, necessitating the construction of urban viaducts to relieve congestion, as shown in Figure 1. Urban viaducts offer important transportation infrastructure, but they also pose threats to public safety, mainly due to their possible collapse caused by inefficient structural health monitoring (SHM) [1]. Ensuring the serviceability and safety of urban viaducts is critical for people's livelihoods, as well as the stability and progress of the smart city vision [2].

(a) (b) (c)

Figure 1. Urban viaducts are prevalent in large cities with (**a**) large amounts of traffic, and their (**b**) collapse can cause serious casualties. (**c**) SHM based on wireless sensor networks (WSNs) can help monitor the health condition of these urban viaducts.

SHM involves the continuous or regular assessment of a structure's condition throughout its lifetime [3] to assess its health and performance, as well as identify possible damage

https://doi.org/10.3390/buildings15101619

or deterioration [1,4]. SHM requires the installation of a network of sensors strategically positioned across the entire viaduct to collect real-time data on structural responses to environmental pressures, traffic, or other external effects and establish a comprehensive view of its health and performance.

It also requires continuous data collection and analysis, which frequently involve the use of modern algorithms and machine learning techniques [5,6] to identify small changes in structural responses. However, the high cost of sensor installation, along with data transmission difficulties, has become a major obstacle to effective and efficient SHM for urban viaducts in recent decades. Advancements in downsizing and integration have resulted in the development of compact wireless sensors with enhanced capabilities. This has made wireless sensors more affordable for large-scale urban viaduct deployments.

On the other hand, the energy efficiency of existing wireless sensors, along with optimized power management strategies, has significantly extended the lifespan of wireless sensors. Efficient wireless communication protocols such as LoRa [7,8] provide reliable transmission of data while also minimizing energy consumption and mitigating the requirement for extensive signal repeaters [2]. The integration of data analytics with cloud computing has resulted in enhanced efficiency in data processing and analysis, leading to a decrease in computational requirements for individual sensors and a reduction in expenses. The combination of these new developments has made wireless sensors easier to install and has resulted in fewer disturbances during operations.

Continuous wireless sensor monitoring can detect various changes or anomalies in urban viaducts through different structural parameters [9]; for example, vibration or excessive movement of the viaduct, structural integrity affected by temperature and humidity, and even cracks that require repair or maintenance. Wireless sensor-enhanced SHM [10] is expected to significantly improve the structural health condition of urban viaducts. This step forward may reduce or completely prevent the possibility of urban viaduct collapses, improving safety.

Wireless sensor technology is thriving, offering a promising prospect for improving the monitoring and maintenance of urban viaducts. However, the wide variety of sensor types, deployment methodologies, analysis approaches, and practical obstacles pose difficulties in establishing practical SHM in actual deployments [11].

This study provides a full-chain review of the state-of-the-art methods in wireless sensor SHM approaches for urban viaducts. Our contributions are summarized as follows:

- We present a general wireless sensor approach (see Figure 2) for SHM in urban viaducts as a framework and further provide a full-chain review of the fundamental sensor types and wireless protocols for sensor data transmission, as well as the data processing, modal identification, and damage detection methods.
- We provide a comparative review, illustrating the differences in the functions and selection of sensors and wireless transmission technologies and highlighting the mainstream SHM structural health assessment methods in the frequency and time domains, with suggestions for deployments.
- We review the state-of-the-art SHM structural damage detection methods, where we compare the representative methods in terms of modal parameters and signal processing. We also review the possibility of integration with machine learning. We further examine the pros and cons of vibration-based SHM methods, with a focus on forced vibration testing and ambient vibration testing. Finally, we outline the future direction of wireless sensor SHM approaches.

This study will provide insights for professionals in both architecture and engineering who are involved in the development of efficient wireless sensor-based structural health monitoring. The rest of this paper is organized as follows. Section 2 reviews SHM data collection, including wireless sensors and data transmission. Section 3 reviews the SHM data processing and modal identification methods. Section 4 focuses on reviewing SHM structural damage detection. Section 5 provides a comparative review of vibration-based SHM methods. Section 6 provides our concluding remarks and outlines our future work.

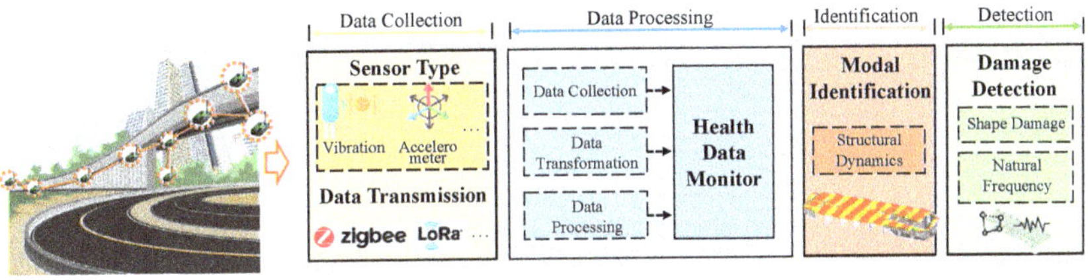

Figure 2. A general structural health monitoring approach, which consists of four processes.

2. SHM Data Collection

In this section, we describe the data collection for SHM in urban viaducts and review the functions and selection of sensor types and wireless data transmission technologies.

2.1. Sensor Types

Sensors are summarized in this subsection, in which we focus on those for maintaining the SHM of viaducts. We also discuss the factors involved in sensor selection, including compatibility with wireless networks, power consumption, and suitability for detection algorithms.

2.1.1. Accelerometers

An accelerometer is a sensor that monitors the rate of change in the vibration of an object or structure [12]. A vibration of a viaduct may occur due to dynamic parameters, including traffic movement, wind forces, or seismic activity [13–15]. Structural damage can be detected by monitoring the vibrations, since changes in the bridge's stiffness or mass can be identified by detection methods [16,17]. The terms viaduct and bridge are used interchangeably in this review.

The detection of vibrations with low amplitude and low frequency is necessary for their usage in applications related to SHM. Piezoelectric accelerometers and Micro-Electro-Mechanical System (MEMS) accelerometers [18] are two types of accelerometers employed in concrete structures for identifying vibration and excitation through dynamic parameters [19].

Current state-of-the-art MEMS accelerometer sensors have demonstrated high measurement accuracy compared to classic wired or wireless piezoelectric accelerometers [18]. M.J. Whelan et al. [13] proposed a MEMS accelerometer (see Figure 3a) and a Bridge Diagnostic Index strain transducer. Structural behavior under ambient and traffic loads can be assessed using accelerometers to investigate the natural frequencies, damping ratios, and mode shapes of the viaducts. Structural excitation is primarily caused by low ambient environmental stresses and traffic. The transmission of data is completed with minimal loss. The validation of the proprietary wireless sensor approach in an operational service context is confirmed through the successful acquisition of high-rate, lossless data on highway bridges.

Huynh et al. [20] deployed a MEMS accelerometer on the Hwamyung Bridge, a cable-stayed bridge located in Busan, South Korea. The accelerometers were strategically positioned on several components of the bridge, including the deck, pylon, and stay cables. The sensors obtained data pertaining to the acceleration, velocity, and displacement of the bridge. The results of the investigation indicated that the method effectively determined the structural parameters of the bridge, including its natural frequencies, mode shapes, and damping ratios. The method demonstrated its ability to identify alterations in structural characteristics resulting from typhoon events.

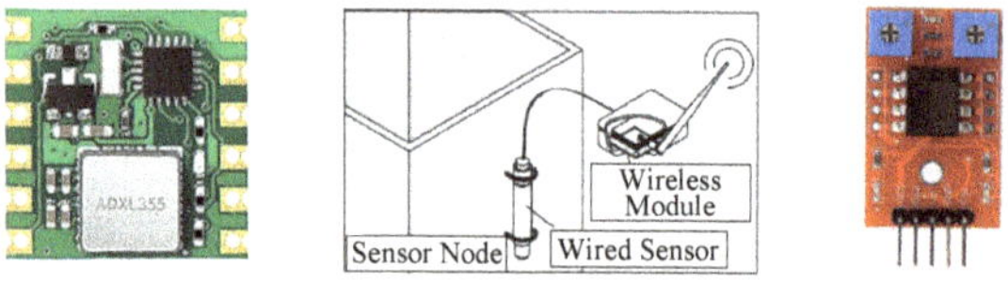

Figure 3. Examples of sensors used in the literature for SHM: (**a**) MEMS accelerometer [13], (**b**) vibrating wire sensor [21], and (**c**) MEMS inclinometer [22].

2.1.2. Strain Sensors

Strain sensors can detect structural damage in viaducts and are generally positioned at critical areas on bridge structures to measure stress and strain levels.

When a load is applied to a viaduct, strain sensors detect the ensuing deformation and transmit the information to a monitoring system. This information can be utilized to arrange maintenance activities in advance of the damage becoming serious [23]. Embedded stress sensors and piezoelectric sensors are two types of strain sensors commonly used for monitoring concrete-based structure health.

Piezoelectric sensors, also known as cement-based strain sensors, have proven useful in the health monitoring of concrete structures and can be embedded in structures [24]. Vibrating wire (VW) sensors (see Figure 3b) were employed by Yanbin Shen et al. [21] for deployment at the Olympic Stadium of Beijing, China. They can collect long-term strain measurements due to their stability, durability, and resistance to electromagnetic components, making them suitable for steel and concrete components.

After attaching a VW sensor to the surface of a measured component, the wires in a VW sensor vibrate and experience fluctuations in the frequency due to tension and compression. Note that a plucking and pickup coil is responsible for exciting the wires and capturing the resonance frequency. A temperature sensor is also integrated into VW sensors, since temperature is one of the critical factors that affect strain measurements [25]. To this end, Sahar Hassani et al. [26] proposed a temperature compensation method that mitigates the temperature effect on strain measurements.

2.1.3. Inclinometers

An inclinometer is a sensor used to measure the angle of inclination of a structure. This sensor can be applied to measure the deflection of a viaduct by monitoring the change in the angle of inclination of the viaduct deck.

Yan Yu et al. [22] employed a MEMS inclinometer, as shown in Figure 3c, to measure the deflection of a viaduct by measuring the change in the angle of inclination, which was achieved by installing an inclinometer on the viaduct deck and positioning it at the location expected to reveal the highest degree of deflection. Note that the inclinometer was calibrated to ensure that the reading displayed a value of zero.

When a bridge undergoes deflection, a pendulum or tilt sensor undergoes rotation, and the readout device displays the magnitude of deflection. The measurement of viaduct deflection can also be established by utilizing a combination of inclinometers. A pair of inclinometers can be mounted to the bridge deck, positioned at opposite ends of the viaduct. The difference in the readings obtained from the two inclinometers is equivalent to the amount of deflection of the viaduct [22,27–29].

A tri-axis accelerometer can also be employed to enable the determination of the inclination of an object or civil structure based on the measurement of its acceleration along three spatial axes. The gravitational force, which indicates a continuous direction toward the Earth's center, influences the acceleration of the object or structure. Therefore, by measuring the structure's acceleration along the three spatial axes, the object's or structure's inclination angle in relation to the horizontal plane can be determined.

To determine the inclination of an object or civil structure using a tri-axis accelerometer, the accelerometer must be attached to the object or structure in a position that ensures its alignment with the horizontal plane. The acceleration measurement in the three spatial axes of an object or structure can be conducted once the accelerometer is properly positioned.

Recently, a diverse array of sensors has become available, utilizing various technologies, each with a wide range of capabilities to support SHM. Among the various options, integrating an accelerometer with vibration-based techniques has emerged as a promising approach for SHM [30].

Overall, sensors offer the capability to detect minor changes in the dynamic response of viaducts, potentially indicating signs of damage or deterioration. Sensor data can provide deep insights into the status of a viaduct, enabling early warnings, maintenance decisions, and safety procedures. However, note that environmental factors may affect sensor readings and further affect modal identification: temperature and humidity can cause hardware variations such as frequency drifting [31], which can lead to data fluctuations, while heavy operational loads can also cause hardware thermal variations.

Moreover, the novel sensor fusion strategy, which combines multiple sensor data, can further enhance the robustness and accuracy of SHM. Ref. [32] integrated sensor readings from distributed devices and converted these readings into actionable information for SHM. Ref. [33] proposed multi-sensor data fusion to enhance the consistency and robustness of monitoring systems. Ref. [34] explored the adoption of an advanced encoder transformer for handling data fusion in SHM. Multi-sensor data fusion is a promising approach for achieving robust and highly accurate SHM.

2.2. *Wireless Sensor Data Transmission*

Wireless networks employed for SHM sensor data transmission can be categorized into four main types: Wireless Personal Area Networks (WPANs), Wireless Local Area Networks (WLANs), Wireless Neighborhood Area Networks (WNANs), and Wireless Wide-Area Networks (WWANs) [35].

2.2.1. WPANs

WPANs are designed for short-range communication, where transmissions are often restricted to a few meters. WPANs are typically used for short-range data transmission within a group of devices and require minimal infrastructure.

Wireless communication protocols in WPANs, including Radio-Frequency Identification (RFID), Bluetooth, and Zigbee [36], are widely employed. RFID is a cutting-edge technology that utilizes radio waves to precisely identify and monitor various sensor data; it can be employed for diverse purposes, such as monitoring inventories, managing access control systems, and identifying locations for viaduct SHM.

On the other hand, Bluetooth is a widely used technology that has multiple purposes for SHM in viaducts, such as sensor data streaming. ZigBee is a wireless technology specifically developed for wireless sensor and control networks that demand low power consumption, low cost, and low data rates. Note that ZigBee is extensively deployed in multi-hop viaduct SHM.

2.2.2. WLANs

WLANs are a group of wireless protocols designed to efficiently transmit data between devices within a range of tens to hundreds of meters. Mesh WLANs, a more advanced generation of ordinary WLANs, employ several access points to expand network coverage and improve performance in larger regions, ensuring uninterrupted connectivity across expansive locations.

The coverage of WLANs is greater than that of WPANs but significantly smaller than that of wide-area wireless communications; thus, the employment of WLANs in viaduct SHM is primarily suitable for medium-scale viaducts.

2.2.3. WWANs

WWANs have a greater coverage area than WLANs. For example, cellular networks such as 4G and 5G are built to facilitate the rapid transmission of data over long distances, ranging from a few to tens of kilometers. However, data transmission through cellular networks is quite expensive in terms of hardware and energy costs.

A special category of WWANs is the Low-Power Wide-Area Network (LPWAN) [37], which is tailored for long-distance communication and requires low power consumption at the expense of limited bandwidth. LPWANs have diverse implementations, and in Table 1, we summarize some of them, including LoRaWAN, Sigfox, and NB-IoT [37–39]. LPWANs are an ideal solution for SHM of large-scale viaducts, and their low-power, wide-area transmission is essential for achieving real-time, long-term, and continuous SHM of urban viaducts.

Table 1. Comparison of LoRaWAN, Sigfox, and NB-IoT [37].

Feature	LoRaWAN	Sigfox	NB-IoT
Modulation	Chirp	GFSK/UNB	OFDM
Data Rate	290 bps–50 kbps	100 bps	100 kbps
Battery Lifetime	Years	Years	Years
Power Efficiency	Low	Low	High
Range	2–5 km urban	3–10 km urban	3 km urban

2.2.4. LPWANs

We chose LoRaWAN from Table 1 to illustrate its advantages for SHM of large-scale urban viaducts due to its long-distance communication ability and low power consumption [40]. LoRaWAN allows long-range communication, often ranging from several kilometers to tens of kilometers.

An example deployment of a LoRaWAN for viaduct SHM is illustrated in Figure 4a. The LoRa nodes are made up of sensors, actuators, batteries, and other essential components. They perform diverse tasks according to different SHM requirements, providing the capability to generate, send, and receive SHM sensor data to and from other nodes and gateways.

In LoRaWANs, gateways are essential components that function as a connection between end nodes and network servers. They transmit packets in both the uplink and downlink directions, guaranteeing sensor data and controlling data transmission. The gateways also perform tasks, such as inspecting bit error rates to guarantee the integrity of the data, managing admissions to control network access, and implementing security measures that protect the data.

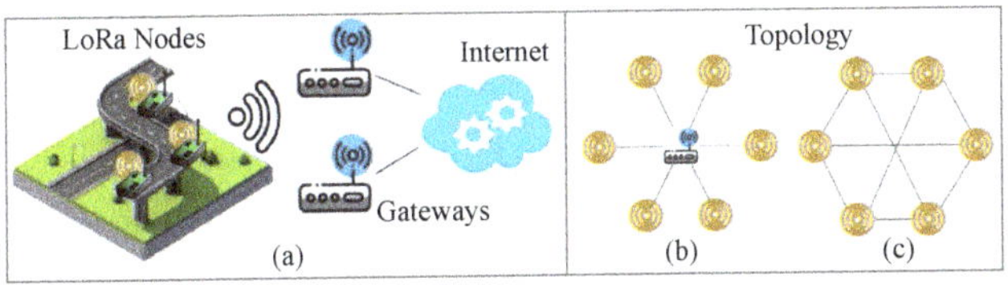

Figure 4. The data transmission structure of (**a**) LoRa-based SHM, with (**b**) star topology and (**c**) mesh topology.

LoRaWANs commonly utilize a star topology (see Figure 4b) in their network architecture to achieve low power consumption. In a star topology, each node establishes its connection directly with the gateway, avoiding the need for additional devices to relay data; all its communications with other nodes have to pass through the gateway. Therefore, data transmission in such a topology usually requires only a single-hop communication from the node to the gateway. This centralized structure simplifies the administrative process in SHM.

Moreover, the star topology in LoRaWANs offers the benefit of enabling easy fault detection. Each node in the network is directly connected to the gateway, ensuring that the failure of one node does not cause interruption to the entire network. This further enhances the efficiency of troubleshooting and helps to ensure the reliability and accuracy of data collection in SHM.

In actual deployments, LoRa nodes are distributed across the monitoring scene; thus, their time must be synchronized for accurate data ordering in modal identification in SHM. A number of highly accurate time synchronization methods [8,40] have been proposed for LoRa. The more accurate the time synchronization among the LoRa nodes, the higher the accuracy and validity of the sensor data ordering, and the higher the accuracy of modal identification in the SHM system [41,42].

Of particular note is that, besides the above wireless transmission protocols, there are also several energy-harvesting communications to further enhance the deployment lifetime in the SHM of viaducts. Ref. [43] revealed the feasibility of integrating energy-harvesting WSNs for structural health monitoring. Ref. [44] summarized the energy sources and harvesting techniques for energy harvesting in WSNs. Ref. [45] further presented how energy harvesting can be achieved for wireless sensor nodes in viaduct SHM.

2.2.5. Topology and Scalability

Scalability is a vital aspect in designing SHM for urban viaducts. The star topology is particularly effective in such cases: when there is a need to add new SHM nodes to the network, they can be easily integrated by establishing a connection with the gateway. Achieving scalability in this way is beneficial in situations where the monitored viaducts may experience expansions in the future.

Although the star topology has many benefits, it is important to emphasize its drawbacks, particularly its reliance on the gateway. A major constraint is the limitation on the number of devices that can be connected to the network, which is defined by the gateway's capacity. The gateway acts as a central hub for all communications in the star topology, and its capacity determines the maximum size the network can achieve. When connected devices reach or exceed the hosting capacity of a gateway, congestion and possible interruptions may be encountered in data transmission.

When conducting SHM of urban viaducts, the failure or damage of a single node has only a minor effect, affecting only that particular node; however, in the case of a breakdown in the gateway, the entire network suffers a total disruption. This single point of vulnerability presents a fatal risk to the reliability of SHM, especially in scenarios where continuous monitoring is required.

In contrast, the mesh topology, as shown in Figure 4c, has been proposed to counter the single-point failure. Each node is interconnected with multiple neighboring nodes, creating a mesh of communication paths. Unlike the star topology, where nodes connect to a gateway or central hub, the mesh topology allows nodes to transmit directly with each other. Each node can directly communicate with other nodes, thus avoiding the need for a central gateway. The mesh topology is considered a more reliable network topology, but it is complex, and its establishment requires a large amount of power and resources, which, unfortunately, contradicts the primary need for energy efficiency in the SHM of urban viaducts.

Moreover, in large SHM coverage involving many long-distance transmissions, deployment faces interoperability issues, while scalability faces major challenges in terms of data loss, noise, and errors. To handle interoperability, cross-technology is often employed, where different wireless protocols can translate their packet contents for communication [46,47]. Regarding scalability, we use LoRa wireless sensor transmission as an example to illustrate how the data loss and error correction are handled in the literature. Ref. [48] studied the path loss model, which guides path loss estimation to reduce data loss in long-distance transmission. Ref. [49] provided an in-depth study on the performance of channel access to alleviate data loss. Ref. [50] investigated the error correction technique for maximizing data availability. Ref. [51] studied the error correction under interferences. Experimental evaluations in [52] further revealed the impact of transmission parameter configuration on packet reception performance.

3. SHM Data Processing and Modal Identification

Modal identification (MI) is a widely employed technique used in mechanical engineering to understand a system's dynamic behavior independently of the applied loads and the system's response. Recently, MI has been introduced into viaduct SHM, where its methods of experimental estimation and condition identification through modal parameters are employed to investigate structural health and assess performance. We categorize existing methods that analyze structural data into frequency-domain and time-domain methods and review them in this section.

3.1. Frequency-Domain Method

The frequency-domain method is employed in SHM to examine a structure's dynamic behavior. This method is especially useful in structural damage detection. The frequency-domain method involves converting the structural responses from the time domain to the frequency domain. Such a transformation enables the examination of the dynamic properties of the structure, including key properties such as natural frequencies, damping ratios, and mode shapes.

Various Experimental Modal Analysis (EMA), or input–output modal analysis, and Operational Modal Analysis (OMA), or output-only modal analysis, techniques are utilized, which estimate the Frequency Response Functions (FRFs) [53] or Impulse Response Functions (IRFs) obtained using Fast Fourier Transform (FFT) techniques.

An FRF quantifies the degree to which the output response of a structure corresponds to the applied force. Furthermore, the FFT algorithm is utilized to convert the output response or observed time-domain data—e.g., displacement, velocity, and acceleration—of a structure from the time domain to the frequency domain [54].

Singh et al. [55] utilized the FRF method to capture the dynamic response of a two-story building model to assess its resilience to external disturbances such as earthquakes.

An accelerometer detects vibrations and updates the Finite Element Model (FEM) to forecast behavior and detect damage during seismic occurrences. FRFs are employed to forecast the structural dynamic response under different damage scenarios because of the intricate connection between the input and output signals. The study concluded that natural frequencies are affected by the location and level of damage, resulting in greater frequency reductions in higher modes.

On the other hand, the FFT and other methods were utilized by Amezquita-Sanchez et al. [56] in a civil structure to determine the dynamic response during artificial seismic activity using a shaker. Their results showed that the FFT is an effective tool for understanding the impact of seismic activity on civil infrastructure and assisting in the construction of more resilient structures.

3.1.1. Peak Picking Method (PP)

PP is a straightforward and efficient method employed to determine MI in the frequency-time domain. This method has been adopted for various reasons over the years. The principle of proximity placement states that a structure has notable responses at its natural frequencies when subjected to external stimuli. The peaks in the Power Spectral Densities (PSDs), calculated from the time histories gathered at the measurement locations, can help identify these frequencies.

This method involves determining the natural frequencies by identifying the peaks in the averaged normalized PSDs. The natural frequencies are directly obtained from the PSD plot at the peak. Kotsakos et al. [57] employed PP by deploying an accelerometer and using the FRF to achieve SHM. Then, frequency-domain decomposition was introduced, which is particularly useful when only output data are measured, making it valuable for structural health monitoring and modal analysis.

3.1.2. Frequency-Domain Decomposition (FDD)

FDD is a method typically employed in output-only modal analysis and involves estimating the modal characteristics of a structure only from its response data without requiring knowledge of the input forces. To determine the mode shapes and frequencies, the process involves breaking down the PSD matrix of the structure's response into singular values and vectors using Singular-Value Decomposition (SVD).

FDD can determine damping by converting the auto-spectral density functions of the identified Single-Degree-of-Freedom (SDOF) systems back to the time domain using the inverse FFT, analyzing free decay, and computing the damping ratio and natural frequencies. FDD is known for its dependability and efficiency in estimating modes, particularly in scenarios involving intricate structures and scarce data on external forces. FDD offers a methodical and verified way to estimate damping by modifying the modal assurance criterion limit and assessing the quality using basic graphs [58–60].

3.2. *Time-Domain Method*

The time-domain method utilizes an SDOF for calculations but processes time-series data. The SVD of the correlation matrix output is employed to isolate the undamped mode shapes related to the sensor locations. PP has also been employed in the time domain to determine the natural frequencies and damping ratios from the SVD signal, following the acquisition of the mode shapes. Multiple unique algorithms are utilized in this method.

3.2.1. Natural Excitation Technique (NExT)

As a theoretical basis for NExT, James et al. [61] showed that random inputs can cause a multiple-input multiple-output, multiple-mode system to produce autocorrelation and cross-correlation functions that look like damped sinusoids [61]. These damped sinusoids must share identical frequencies and decay with the structural modes. The correlation functions have a combination of sinusoids as their modal forms within a smooth modal basis algorithm.

A universal solution for a structure made up of discrete spatial elements should be proposed for using this method. Then, the cross-correlation function between results should be computed, and the situations where the inputs are random should be further handled. The theoretical justification for NExT can be established for a broad range of random inputs, complex modes, and white-noise inputs with known harmonic inputs. This advancement will only apply to exact inputs, actual modes, and no harmonics.

The estimation of modal parameters solely from multiple-channel responses under ambient excitation presents challenges, particularly when handling asynchronous responses due to sensor errors. Yang et al. [62] proposed a novel approach to address this issue, introducing a method based on NExT, where the power spectra of monitoring responses are initially calculated and transformed into correlation functions through the inverse Fourier transform.

By comparing the phase characteristics of power spectra and the linear dependency of modal components in correlation functions for synchronous and asynchronous responses, Yang enhanced NExT. Through a meticulous process involving the minimization of phase slope and maximization of linear dependency of modal components, the enhanced NExT synchronizes multiple-channel responses, aligning them to yield high-precision mode shapes. The efficacy of Yang's approach was validated through numerical examples and real-world applications, demonstrating its potential to advance operational modal identification in viaduct health monitoring.

3.2.2. Autoregression Moving Average (ARMA)

According to [63], the ARMA model is a time-domain method used in Operational Modal Analysis (OMA) to determine modal parameters from structural responses. This method predicts the present values of a time series by utilizing past values and a prediction error, which is beneficial for analyzing linear systems. The ARMA model expands on the idea of a linear time-invariant system driven by white noise, under the assumption that the observed output is stationary.

Vector ARMA models are used when there are many input excitations. The prediction error is commonly used with ARMA models to determine modal parameters by reducing the discrepancy between the estimated response from the model and the actual measured response. Despite its effectiveness, the ARMA technique has limitations: it can be computationally intensive and may not always reach convergence in every scenario.

ARMA methods have been employed in civil engineering applications in the past, but their popularity has decreased because they require a large amount of computation time. Several ARMA methods, including the instrumental variable, linear multi-stage, and two-stage least squares techniques, have been created, all of which require significant computational resources.

3.2.3. Stochastic Subspace Identification (SSI)

According to [64], SSI methods can be categorized into two primary groups: covariance-driven (Cov-SSI) and data-driven (DD-SSI). Cov-SSI utilizes the covariance matrices of the measured data to detect system modes, whereas DD-SSI directly analyzes the observed data to create a data matrix. SVD is a crucial component of both Cov-SSI and DD-SSI; it decomposes the data matrix into its singular vectors and singular values, which correspond to the fundamental spatial modes and modal frequencies of the structure.

SVD results allow for the extraction of modal data for SHM, including natural frequencies, mode shapes, and damping ratios. Chang et al. [65] employed and validated the application of SSI. SSI must allocate computational resources to process a substantial amount of data. The proposed order selection criteria were used to establish the number of structural modes, which, in turn, led to the determination of the structure's MI.

SSI has also been employed to analyze the practicality of a real-time application, where it was applied to an actual viaduct [66]. The results demonstrated that the enhanced SSI effectively reduced computational time while maintaining qualitative modal parameters. We summarize the pros and cons of all of these methods in Table 2.

In summary, various factors should be considered when selecting a processing method in a practical SHM application. First, the selection should be based on the computational resources of the underlying processing platform. The computational complexity of PP is the lowest among all the methods, that of SSI is the highest, and that of FDD, NExT, and ARMA is moderate. Second, the selection should consider the application scenario. PP is suitable for rapid screening and preliminary modal identification, FDD is suitable for multi-modal separation and offline analysis, NExT is suitable for non-stationary data preprocessing, ARMA is suitable for online processing of short data segments, and SSI is suitable for detailed modeling of complex structures. Note that, in practice, the actual selection is more complicated and should jointly consider many factors such as sensor data, deployment requirements, platform, and environment. The processing methods are compared in Table 2.

Table 2. Summary of modal identification methods.

Method	Type	Advantages	Disadvantages
PP [57]	Frequency Domain	The most basic, direct, and computationally efficient method	Not applicable to systems with closely spaced modes, which is the case in most real structures
FDD [58]	Frequency Domain	Precise identification of natural frequencies and mode shapes that are tightly spaced is possible	Calculating damping ratios is not possible
NExT [61]	Time Domain	Robust basic structural dynamic properties	Relies on structured, consistent processes.
ARMA [63]	Time Domain	Easy setup	Extensive computations needed
SSI [65]	Time Domain	High level of parameter estimation accuracy	Mathematically complex method

4. SHM Structural Damage Detection

As illustrated in Figure 5, structural damage detection methods in SHM are commonly classified into two categories: global methods and local methods. Note that local SHM methods aim to identify and assess structural damage on a smaller scale, without relying on or analyzing the structural vibration response.

The classification of most Non-Destructive Testing and Evaluation (NDTE) methods as local is primarily due to their limited detection range. Ultrasonic testing, acoustic emissions, infrared thermography, visual testing, and various other techniques are employed to examine, assess, and appraise structural components and assemblies within local areas of infrastructure [67–70]. In addition, the expertise of a trained technician is required to conduct the examination, establish threshold values for signal processing, and interpret the results.

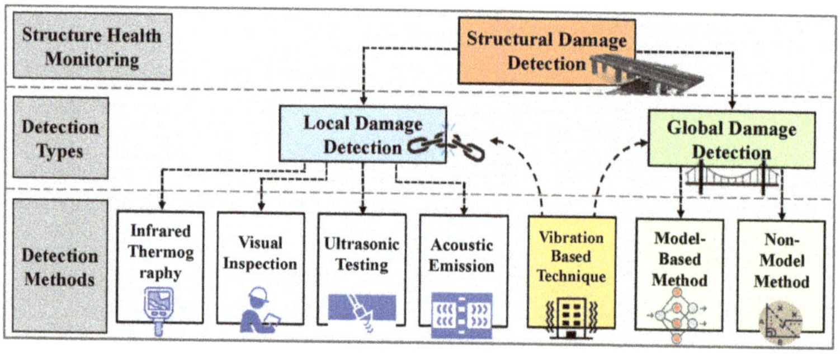

Figure 5. The classification of structural damage detection methods.

Monitoring large structures using traditional NDTE methods is often costly and time-consuming. A review of NDTE methods and other local methods for damage detection, localization, and quantification is not included in this paper. This section exclusively focuses on vibration-based global SHM techniques. Nevertheless, it is crucial to acknowledge that local techniques are hardly suitable for the long-term SHM of large structures like viaducts and high-rise buildings. Consequently, global SHM methods are discussed due to the size and complexity of these structures.

Vibration-based techniques are popular and reliable in SHM and include four main steps. Due to their non-destructive nature, which allows for continuous long-term monitoring without process interruptions, these methods have gained popularity and demonstrated reliability in SHM in most industries. Vibration-based techniques can acquire structural dynamic parameters such as natural frequencies, mode shapes, and damping ratios from the dynamic response of a structure. These structural dynamic parameters, also known as modal parameters, can change due to external excitation, which may indicate structural damage or degradation. The dynamic response of a structure can be monitored using sensors such as accelerometers, strain sensors, or fiber optic sensors, but this review focuses on accelerometers due to their low power and reliability, as mentioned in Section 2.1.

On the other hand, tri-axis acceleration monitoring is a dedicated methodology that involves the simultaneous analysis of vibrations along the X-, Y-, and Z-axes. Vibration-based techniques offer a comprehensive understanding of how a structure reacts to external forces or dynamic forces such as wind, traffic loads, and earthquakes. This dynamic excitation will induce a dynamic response in the structure, which is essentially for identifying modal parameters [71–74].

The extraction of features from collected vibration signals is a challenging task, as it is influenced by the presence of noise [75]. The construction of a complex data processing and analysis system is crucial for obtaining precise data related to the health of the structure. By comparing the healthy dynamic response of a structure to the damaged dynamic response, vibration-based methods can detect damage to the structure. Changes in the dynamic response of a structural parameter, captured with respect to an established reference state, can easily be used as indicators of structural damage, damage location, and damage severity [76].

Comprehensive research has been conducted on structural damage detection in viaduct SHM, which relies on vibration measurements from viaducts. These studies aim to identify variations in the dynamic response parameters, including the natural frequency, modal strain energy, mode form curvature, and dynamic flexibility [77]. For instance, a decrease in natural frequencies indicates structural degradation or damage caused by an intense event, which leads to a decrease in stiffness [78].

According to [77], natural frequencies are considered highly significant indicators in vibration-based SHM methods for detecting damage in a structure. A large number of research studies have reported on the use of natural frequency characteristics in the deployment of vibration-based damage detection systems for viaduct structures. Hou et al. [78] defined structural damage as alterations in the parameters of structural characteristics, such as the dynamic response parameters, that harm the structural integrity of the system. This includes changes in the mass, damping, and stiffness of the structure. Therefore, an in-depth review is necessary.

4.1. Natural Frequencies

Magalhaes et al. [79] employed a vibration-based SHM method to analyze the changes in dynamic parameters, particularly the natural frequencies, of a concrete arch bridge. Their research focused on monitoring the first 12 natural frequencies of the bridge over a duration exceeding two years. They aimed to reduce the impact of environmental and operational factors on the inherent frequencies of the bridge, thereby facilitating the precise detection of damage. In their research, the ability of a monitoring system to detect anomalous behavior was influenced by the amount of data collected before the occurrence of the unusual event. Note that a larger dataset can result in a more accurate determination of reference values.

Further, Fan et al. [80] studied prestressed concrete viaducts. Extensive experiments were conducted on highway viaducts to identify and measure damage by analyzing alterations in natural frequencies and mode shapes. Empirical observations demonstrated that alterations in dynamic properties can be used to identify damage in its initial stages. Nevertheless, natural frequencies alone are insufficient to accurately measure the extent and specific area of damage to the structure.

4.2. Finite Element Model

The FEM is a computational tool employed in SHM to model and examine the response of complex structures to different excitation sources. This method involves breaking down a complex system into smaller and more manageable components, known as finite elements, and material characteristics, geometric properties, and boundary conditions are assigned to each element, enabling the accurate estimation of the structural response to external excitation sources. FEM is employed to build a model of a structure—e.g., a viaduct or building—that accurately captures its physical characteristics, including its mass, damping, and stiffness. Afterward, the model is used to forecast the behavior of the structure in different scenarios.

Over time, a structure may experience alterations due to aging, climatic conditions, or damage. These alterations can lead to changes in the structure's physical characteristics from those indicated in the original FEM analysis. To address this, FEM is revised using recorded vibration responses, which is referred to as model updating. By updating the FEM, it can be compared to the original model of the structure that has not been damaged. Any significant differences between the updated model and the reference model may suggest the possibility of damage. This approach serves as a means of monitoring and detecting deterioration [81].

Cheng et al. [82] conducted experiments on large transmission towers in China. These towers exhibit high sensitivity to environmental pressures, which showcases the importance of ensuring their safety. The researchers created a FEM-based SHM method for a transmission tower that is 131 m tall. The process involved detailed modeling, manual tuning, model updating using the response surface method, and validation. Before the installation of sensors, in situ measurements were taken to determine the dynamic characteristics, yielding crucial data for the FEM. By manually adjusting and updating the model, the researchers further enhanced the accuracy of the FEM. A seismic analysis demonstrated that model updating accurately represented the tower's seismic performance, resulting in alterations in displacement and stress distribution.

Table 3 presents a summary of vibration-based SHM methods. Note that the modal parameters of a structure, especially its natural frequencies, possess high sensitivity to even minor damage. Besides the conventional methods discussed above, the advent of machine learning has enabled significant improvements in modal identification and damage detection. Ref. [1] provided a comprehensive review of the effectiveness of machine learning in damage clustering, regression, and classification for diverse structures. State-of-the-art research has further expanded the application of machine learning in damage detection. Ref. [5] employed machine learning to detect structural energy time-history responses for damage detection, while [6] designed a hybrid transfer learning-based neural network to detect damage in bridges. With the development of new artificial intelligence methods, structural damage detection solutions will continue to flourish.

Table 3. State-of-the-art vibration-based SHM methods.

References	Modal Parameter	Signal Processing	Description
Alamdari et al. (2016) [83]	Natural Frequency	FRF	Sydney Harbor Bridge
Khatir et al. (2018) [15]	Natural Frequency	FRF	Beam-like structure constructed from steel
Yang D et al. (2010) [84]	Natural Frequency, Displacement, Temperature	ARMA	Runyang Suspension Bridge in China
Alves et al. (2021) [85]	Natural Frequency	SSI	Tianjin Yonghe Bridge
Magalhães et al. (2010) [79]	Natural Frequency, Temperature	FDD, SSI	Concrete bridge
Saidin et al. (2023) [86]	Natural Frequency, Mode Shape	FDD, SSI	Ultra-high-performance concrete bridge
Cabboi et al. (2014) [87]	Natural Frequency, Mode Shape	FDD, SSI-COV	San Michele iron arch bridge
Cheng et al. (2018) [82]	Natural Frequency, Mass, Damping, and Stiffness	FEM	Transmission tower in China
Fan et al. (2011) [80]	Natural Frequency, Mode Shape	FDD	Prestressed concrete bridges
Wang et al. [77]	Natural Frequency, Mode Shape	FFT	Victoria Bridge
Chang et al. [65]	Natural Frequency	SSI	Cable-stayed bridge in Taiwan
Kotsakos et al. [57]	Natural Frequency	PP, FRF, FFT	Multi-layer buildings
Mezquita-Sanchez et al. [56]	Natural Frequency	FFT	Cube-shaped structure

5. Vibration-Based SHM Methods: Pros and Cons

Having reviewed the state-of-the-art vibration-based SHM methods, we provide more insights into two of these methods in this section. It is worth mentioning that vibration testing for dynamic analysis in SHM involves the process of regularly evaluating a structure's dynamic response to external forces or operational conditions. Excitations can be applied as forced or ambient vibrations to observe the dynamic response of a structure and obtain modal parameters, such as natural frequencies and mode shapes.

Although the vibration-based methods described below utilize different input excitation methods, they are two common approaches used in dynamic analysis to evaluate the health of structures such as viaducts, buildings, dams, and pipelines.

5.1. Forced Vibration Testing (FVT)

FVT employs a controlled external excitation or dynamic load to a structure at particular frequencies or amplitude levels. This excitation can be produced by mechanical shakers, electromagnetic actuators, impulse hammers, or other controlled artificial vibration sources. Conversely, ambient vibration testing monitors a structure's dynamic response caused by external uncontrolled excitation like wind, traffic, or machinery. Ambient vibration testing utilizes the natural vibrations present in a structure's environment during regular operation, unlike forced vibration testing, which requires controlled excitation sources. Both methods aim to efficiently detect anomalies or damage to prevent major breakdowns and ensure structural safety and reliability.

Existing research has investigated the effectiveness of FVT in detecting structural damage by analyzing how structures respond dynamically to controlled input excitation. These methods include applying a controlled external artificial excitation to the structure and then analyzing the dynamic response to obtain important information about the structural health and integrity of the structure. During FVT, researchers can adjust the input excitation parameters, including the frequency, amplitude, and duration, to meet the experiment's specific requirements.

Structural health can be assessed by comparing the modal parameters of undamaged structures with those of damaged structures. Chen et al. [88] applied a static load using a 30-ton hydraulic jack equipped with a donut-shaped load cell. Following each static level of stress, the load apparatus was taken away to allow room for the dynamic test setup. A full-scale beam with fiber-reinforced polymer sheets was tested in the study. It was shown that the natural frequencies of the beam were lower at the start of cracking and then stabilized. The unstrengthened beam's frequency decreased continuously as it underwent concrete cracking and reinforcement yielding, but the proposed technique may not be suitable for assessing significant cracks or extensive damage that has persisted over an extended period.

Orhan et al. [89] employed the FEM of a beam to simulate a crack on the beam. The study revealed that the natural frequencies of a beam decreased with constant crack location but increased with crack depth. The dominant response occurred at 96 Hz for a single crack on the top and bottom surfaces, while additional frequencies appeared for two cracks on the top and bottom surfaces. The study revealed that degradation or damage can be detected using the FVT method by examining the natural frequencies of the structure.

However, this method cannot pinpoint the exact location of the damage. A full-scale synthetic two-story high-performance building was tested under ambient and forced vibrations in a controlled laboratory environment by Lamarche et al. [90]. The building was subjected to simulated artificial earthquake damage for forced vibration testing, and ambient vibration testing was also performed with human walking as an input excitation.

In the study, FVT was regarded as a reliable method for experimentally evaluating the modal parameters of a structure or building. Table 4 illustrates the pros and cons of FVT.

FVT has been applied to in situ structures in several studies [91–93], showing that the dynamic responses of the structures were accurate.

Table 4. Pros and cons of forced vibration testing.

Advantages	Disadvantages
1. Customizes EMA by adjusting input excitation parameters to target various aspects of structural health. 2. Precise damage detection, such as cracks and material degradation. 3. Fast and straightforward approach to excite the structure.	1. Not suitable for assessing significant cracks or extensive damage that has persisted over an extended period. 2. Does not provide precise details about the exact location of the damage within the structure. 3. Specialized equipment is often complex and costly. 4. Improperly conducted forced vibration testing may lead to increased stress or harm to the structure. 5. Testing may interfere with regular operations, particularly for structures that are already in use.

5.2. Ambient Vibration Testing (AVT)

As the equipment required for forced excitation has become cumbersome and costly, employing natural excitation sources such as wind and waves has led to the development of alternative techniques for MI. AVT has emerged as a practical solution, which avoids the need for mechanical excitation devices and is particularly beneficial for large structural bridge applications.

Alamdari et al. [83] proposed a spectral-based clustering technique that aims to find the characteristics in the FRFs of the Sydney Harbor Bridge structure that were damaged. The authors emulated several damage scenarios by selectively removing cables and hangers from the bridge and then monitoring and extracting the dynamic responses of the bridge under uncontrolled traffic loading. Accelerometers placed on the bridge deck and arch employed the wavelet transform to extract spectral features, which were used to determine the natural frequencies of the structure. K-means clustering was then applied to group these spectral features. The results showed that the location of damage can be effectively identified and localized.

Saidin et al. [86] also proposed a method using AVT for the SHM of an ultra-high-performance concrete (UHPC) bridge to acquire the natural frequencies and mode shapes. They compared the modal parameters with the FEM of the bridge and used simple linear regression to evaluate and predict the bridge's condition. Their results showed that the vibration-based method was effective and reliable for detecting and evaluating the structural condition of the UHPC bridge.

Alves et al. [85] proposed an SHM system for the Tianjin Yonghe Bridge, where the acceleration data were normalized to eliminate the effects of scaling and offset. In addition, they applied a wavelet transform to break down the data into different frequency ranges and identified those that were relevant for damage identification. They also extracted the natural frequencies of the structure, which were utilized to indicate the precise locations of structural damage with various levels of severity.

Yang et al. [84] proposed an approach that employed extended monitoring data and statistical modeling methods to identify abnormal changes based on natural frequencies and displacement, thus mitigating the impact of temperature-related issues on the Runyang Suspension Bridge in China. The approach included a number of sensors, such as accelerometers, displacement sensors, and temperature sensors. The authors employed a statistical modeling approach to establish the relationship between the natural frequencies, temperature, and displacement of the bridge. However, the steps and methods used to extract the modal frequency of the bridge were not provided. The findings showed that the proposed method can detect anomalous changes in the natural frequencies and displacement of the bridge.

Magalhaes et al. discussed the use of vibration-based SHM approaches for bridges, specifically focusing on modal parameter tracking. Output-only modal analysis was highlighted as a way to identify a structure's modal parameters under operational conditions, with natural frequencies being commonly used as damage indicators. AVT was employed to identify natural frequencies. Then, regression analysis and principal component analysis were employed to reduce the influence of environmental and operational factors. The proposed method showed that the capability of SHM to detect abnormal behavior depends on the amount of data collected before the occurrence of an event. The time needed to detect damage is inversely proportional to the extent of the damage.

Cabboi et al. [87] conducted research to compare damage detection by obtaining the natural frequencies and mode shapes of the San Michele iron arch bridge. Although no damage was detected in the research, it was determined that using AVT to acquire the modal parameters was successful.

In summary, the advantages and limitations of AVT, as outlined in Table 5, underscore its practicality compared to FVT, particularly due to its feasibility during traffic operations, minimal requirement for costly excitation equipment, provision of reliable data, and lower total cost of experimental work. While FVT may offer more precise MI findings, the simplicity, cost-effectiveness, and practicality of AVT make it preferable for various scenarios, particularly for large and flexible viaducts where regulated excitation is challenging.

Table 5. Pros and cons of ambient vibration testing.

Advantages	Disadvantages
1. A convenient method for measuring structures under servicing loading. 2. No special excitation devices are required. 3. Non-destructive testing. 4. The cost of the overall experiment is lower. 5. Can be performed during the normal operation of structures. 6. Practical for use in large and heavy structures.	1. Tedious setup, using hundreds of meters of electrical cables. 2. Does not provide precise details about the exact location of the damage within a structure. 3. Tough and shielded. Need to ensure minimal signal loss and interference over long distances. 4. Estimated damping values are prone to errors. 5. High-sensitivity sensors are required.

Note that the applicability of AVT is highly dependent on the characteristics of the structure itself and the environmental excitation conditions; thus, there are scenarios that may lead to its inapplicability or reduced reliability. For example, a lightweight structure where the vibration caused by environmental excitation is too small.

6. Conclusions

In this paper, we conduct a full-chain review of SHM, from data collection to data processing, modal identification, and finally structural damage detection. In our review, we first examine the wireless sensors and wireless networks suitable for SHM. We then investigate the existing modal identification methods. We further provide a review of structural damage detection methods, with a focus on state-of-the-art vibration-based methods. Finally, we review the pros and cons of the vibration-based methods for the SHM of viaducts.

Based on our review, we identify that the combination of long-range WSNs and vibration-based SHM is a promising method for attaining effective long-term viaduct structural monitoring. This combination addresses the inherent limitations of traditional wired SHM systems for urban viaducts, improving monitoring effectiveness and precision. Moreover, long-range wireless sensor-based SHM is highly beneficial for monitoring large viaducts or those located in remote areas.

Note that combining the wireless sensor approach with the vibration method provides non-destructive and cost-effective monitoring for the SHM of urban viaducts. Through the wireless examination of a structure's dynamic response to excitations, these methods allow for the early detection of damage in the initial phase, enabling quicker responses and avoiding severe collapse. Overall, the integration of long-range WSNs with vibration-based methods is a cost- and energy-efficient solution for the long-term SHM of urban viaducts.

Structural health monitoring technology continues to progress. We envisage four key developments in the future: energy-harvesting sensor networks, enhanced SHM deployments, advanced machine learning-driven SHM data processing, and fusion of vision and wireless sensing-enabled SHM damage detection.

Author Contributions: Conceptualization, T.W. and X.H.; methodology, T.W. and X.H.; writing—original draft preparation, T.W. and X.H.; writing—review and editing, T.W., X.H., T.C., and X.Q.; supervision, T.C. and X.Q. All authors have read and agreed to the published version of the manuscript.

Funding: This research was funded by the Beijing Young Elite Scientists Sponsorship Program under grant number BYESS2023161.

Data Availability Statement: No new data were created or analyzed in this study. Data sharing is not applicable to this article.

Conflicts of Interest: The authors declare no conflicts of interest.

References

1. Flah, M.; Nunez, I.; Ben Chaabene, W.; Nehdi, M.L. Machine learning algorithms in civil structural health monitoring: A systematic review. *Arch. Comput. Methods Eng.* **2021**, *28*, 2621–2643. [CrossRef]
2. Ali, A.I.; Partal, S.Z.; Kepke, S.; Partal, H.P. ZigBee and LoRa based wireless sensors for smart environment and IoT applications. In Proceedings of the 2019 1st Global Power, Energy and Communication Conference (GPECOM), Nevsehir, Turkey, 12–15 June 2019; IEEE: Piscataway, NJ, USA, 2019; pp. 19–23.
3. Xu, Z.D.; Liu, M.; Wu, Z.; Zeng, X. Energy Damage Detection Strategy Based on Strain Responses for Long-Span Bridge Structures. *J. Bridge Eng.* **2011**, *16*, 644–652. [CrossRef]
4. Farrar, C.R.; Worden, K. *Structural Health Monitoring: A Machine Learning Perspective*; John Wiley & Sons: Hoboken, NJ, USA, 2012.
5. Tao, Y.; Xu, Z.D.; Wei, Y.; Liu, X.Y.; Dong, Y.R.; Dai, J. Integrating Deep Learning Into an Energy Framework for Rapid Regional Damage Assessment and Fragility Analysis Under Mainshock-Aftershock Sequences. *Earthq. Eng. Struct. Dyn.* **2025**, *54*, 1678–1697. [CrossRef]
6. Yu, Y.; Rashidi, M.; Dorafshan, S.; Samali, B.; Farsangi, E.N.; Yi, S.; Ding, Z. Ground penetrating radar-based automated defect identification of bridge decks: A hybrid approach. *J. Civ. Struct. Health Monit.* **2025**, *15*, 521–543. [CrossRef]
7. Lavric, A.; Popa, V. Internet of things and LoRa™ low-power wide-area networks: A survey. In Proceedings of the 2017 International Symposium on Signals, Circuits and Systems (ISSCS), Iasi, Romania, 13–14 July 2017; IEEE: Piscataway, NJ, USA, 2017; pp. 1–5.
8. Huan, X.; Chen, W.; Wang, T.; Hu, H.; Zheng, Y. A One-Way Time Synchronization Scheme for Practical Energy-Efficient LoRa Network Based on Reverse Asymmetric Framework. *IEEE Trans. Commun.* **2023**, *71*, 6468–6481. [CrossRef]
9. Farrar, C.R.; Worden, K. An introduction to structural health monitoring. *Philos. Trans. R. Soc. A Math. Phys. Eng. Sci.* **2007**, *365*, 303–315. [CrossRef]

10. Wan, J.; Yu, Y.; Wu, Y.; Feng, R.; Yu, N. Hierarchical Leak Detection and Localization Method in Natural Gas Pipeline Monitoring Sensor Networks. *Sensors* **2012**, *12*, 189–214. [CrossRef]
11. Khan, S.M.; Atamturktur, S.; Chowdhury, M.; Rahman, M. Integration of Structural Health Monitoring and Intelligent Transportation Systems for Bridge Condition Assessment: Current Status and Future Direction. *IEEE Trans. Intell. Transp. Syst.* **2016**, *17*, 2107–2122. [CrossRef]
12. Xu, Z.D.; Wu, Z. Energy damage detection strategy based on acceleration responses for long-span bridge structures. *Eng. Struct.* **2007**, *29*, 609–617. [CrossRef]
13. Whelan, M.J.; Gangone, M.V.; Janoyan, K.D.; Jha, R. Real-time wireless vibration monitoring for operational modal analysis of an integral abutment highway bridge. *Eng. Struct.* **2009**, *31*, 2224–2235. [CrossRef]
14. Avci, O.; Abdeljaber, O.; Kiranyaz, S.; Hussein, M.; Gabbouj, M.; Inman, D.J. A review of vibration-based damage detection in civil structures: From traditional methods to Machine Learning and Deep Learning applications. *Mech. Syst. Signal Process.* **2021**, *147*, 107077. [CrossRef]
15. Khatir, S.; Dekemele, K.; Loccufier, M.; Khatir, T.; Wahab, M.A. Crack identification method in beam-like structures using changes in experimentally measured frequencies and Particle Swarm Optimization. *C. R. Mécanique* **2018**, *346*, 110–120. [CrossRef]
16. Sohn, H.; Farrar, C.R. Damage diagnosis using time series analysis of vibration signals. *Smart Mater. Struct.* **2001**, *10*, 446. [CrossRef]
17. Torbol, M.; Gomez, H.; Feng, M. Fragility analysis of highway bridges based on long-term monitoring data. *Comput.-Aided Civ. Infrastruct. Eng.* **2013**, *28*, 178–192. [CrossRef]
18. Sabato, A.; Niezrecki, C.; Fortino, G. Wireless MEMS-based accelerometer sensor boards for structural vibration monitoring: A review. *IEEE Sens. J.* **2016**, *17*, 226–235. [CrossRef]
19. Zacchei, E.; Lyra, P.H.; Lage, G.E.; Antonine, E.; Soares Jr, A.B.; Caruso, N.C.; de Assis, C.S. Structural health monitoring of a Brazilian concrete bridge for estimating specific dynamic responses. *Buildings* **2022**, *12*, 785. [CrossRef]
20. Huynh, T.C.; Park, J.H.; Kim, J.T. Structural identification of cable-stayed bridge under back-to-back typhoons by wireless vibration monitoring. *Measurement* **2016**, *88*, 385–401. [CrossRef]
21. Shen, Y.; Yang, P.; Zhang, P.; Luo, Y.; Mei, Y.; Cheng, H.; Jin, L.; Liang, C.; Wang, Q.; Zhong, Z. Development of a multitype wireless sensor network for the large-scale structure of the national stadium in China. *Int. J. Distrib. Sens. Netw.* **2013**, *9*, 709724. [CrossRef]
22. Yu, Y.; Liu, H.; Li, D.; Mao, X.; Ou, J. Bridge deflection measurement using wireless mems inclination sensor systems. *Int. J. Smart Sens. Intell. Syst.* **2013**, *6*, 38. [CrossRef]
23. Xu, Z.D.; Wu, K.Y. Damage Detection for Space Truss Structures Based on Strain Mode under Ambient Excitation. *J. Eng. Mech.* **2012**, *138*, 1215–1223. [CrossRef]
24. Sun, M.q.; Liew, R.J.; Zhang, M.H.; Li, W. Development of cement-based strain sensor for health monitoring of ultra high strength concrete. *Constr. Build. Mater.* **2014**, *65*, 630–637. [CrossRef]
25. Liu, C.; DeWolf, J.T. Effect of temperature on modal variability of a curved concrete bridge under ambient loads. *J. Struct. Eng.* **2007**, *133*, 1742–1751. [CrossRef]
26. Hassani, S.; Mousavi, M.; Gandomi, A.H. Structural health monitoring in composite structures: A comprehensive review. *Sensors* **2021**, *22*, 153. [CrossRef]
27. Roberts, G.W.; Meng, X.; Dodson, A.H. Integrating a global positioning system and accelerometers to monitor the deflection of bridges. *J. Surv. Eng.* **2004**, *130*, 65–72. [CrossRef]
28. Zhang, W.; Sun, L.; Sun, S. Bridge-deflection estimation through inclinometer data considering structural damages. *J. Bridge Eng.* **2017**, *22*, 04016117. [CrossRef]
29. Sousa, H.; Cavadas, F.; Henriques, A.; Bento, J.; Figueiras, J. Bridge deflection evaluation using strain and rotation measurements. *Smart Struct. Syst* **2013**, *11*, 365–386. [CrossRef]
30. Sun, L.; Shang, Z.; Xia, Y.; Bhowmick, S.; Nagarajaiah, S. Review of bridge structural health monitoring aided by big data and artificial intelligence: From condition assessment to damage detection. *J. Struct. Eng.* **2020**, *146*, 04020073. [CrossRef]
31. Huan, X.; Hao, Y.; Miao, K.; He, H.; Hu, H. Carrier Frequency Offset in Internet of Things Radio Frequency Fingerprint Identification: An Experimental Review. *IEEE Internet Things J.* **2024**, *11*, 7359–7373. [CrossRef]
32. Alzughaibi, A.A.; Ibrahim, A.M.; Na, Y.; El-Tawil, S.; Eltawil, A.M. Community-Based Multi-Sensory Structural Health Monitoring System: A Smartphone Accelerometer and Camera Fusion Approach. *IEEE Sens. J.* **2021**, *21*, 20539–20551. [CrossRef]
33. Stoian, I.; Kovacs, I.; Ignat, S.; Ghiran, O.; Oltean, G.; Gordan, M. Multisensorial data fusion and correlation for dams structural health monitoring: (Multi Scale, Uncertainty-Sensitive Heterogeneous Data Integration). In Proceedings of the 2018 IEEE International Conference on Automation, Quality and Testing, Robotics (AQTR), Cluj-Napoca, Romania, 24–26 May 2018; pp. 1–6. [CrossRef]

34. Li, H.; Wang, J.; Wu, S.; Gao, Y.; Wang, Y.; Nie, G. Transforming Structural Health Monitoring: Leveraging Multisource Data Fusion With Two-Stage Encoder Transformer for Bridge Deformation Prediction. *IEEE Trans. Instrum. Meas.* **2025**, *74*, 1–13. [CrossRef]
35. Chaudhari, B.S.; Zennaro, M. *LPWAN Technologies for IoT and M2M Applications*; Academic Press: Cambridge, MA, USA, 2020.
36. Ramya, C.M.; Shanmugaraj, M.; Prabakaran, R. Study on ZigBee technology. In Proceedings of the 2011 3rd International Conference on Electronics Computer Technology, Kanyakumari, India, 8–10 April 2011; IEEE: Piscataway, NJ, USA, 2011; Volume 6, pp. 297–301.
37. Stanco, G.; Botta, A.; Frattini, F.; Giordano, U.; Ventre, G. On the performance of IoT LPWAN technologies: The case of Sigfox, LoRaWAN and NB-IoT. In Proceedings of the ICC 2022-IEEE International Conference on Communications, Seoul, Republic of Korea, 16–20 May 2022; pp. 2096–2101.
38. Zhou, Q.; Zheng, K.; Hou, L.; Xing, J.; Xu, R. Design and implementation of open LoRa for IoT. *IEEE Access* **2019**, *7*, 100649–100657. [CrossRef]
39. Suji Prasad, S.; Thangatamilan, M.; Suresh, M.; Panchal, H.; Rajan, C.A.; Sagana, C.; Gunapriya, B.; Sharma, A.; Panchal, T.; Sadasivuni, K.K. An efficient LoRa-based smart agriculture management and monitoring system using wireless sensor networks. *Int. J. Ambient Energy* **2022**, *43*, 5447–5450. [CrossRef]
40. Huan, X.; Chen, W.; Wang, T.; Hu, H. A Microsecond Energy-Efficient LoRa Time Synchronization Based on Low-Layer Timestamping and Asymmetric Time Translation. *IEEE Trans. Veh. Technol.* **2024**, *73*, 7328–7332. [CrossRef]
41. Wang, X. High-Precision Time Synchronization Method for Smart Sensor Network for Structural Health Monitoring. In Proceedings of the 2010 International Conference on Measuring Technology and Mechatronics Automation, Changsha, China, 13–14 March 2010; Volume 2, pp. 394–397. [CrossRef]
42. Abdaoui, A.; Ahmed, M.H.; El Fouly, T.M. Impact of time synchronization error on the mode-shape calculation in wireless sensor networks for structural health monitoring. In Proceedings of the 2016 IEEE Wireless Communications and Networking Conference, Doha, Qatar, 3–6 April 2016; pp. 1–6. [CrossRef]
43. Cammarano, A.; Spenza, D.; Petrioli, C. Energy-harvesting WSNs for structural health monitoring of underground train tunnels. In Proceedings of the 2013 IEEE Conference on Computer Communications Workshops (INFOCOM WKSHPS), Turin, Italy, 14–19 April 2013; pp. 75–76. [CrossRef]
44. Philip, M.S.; Singh, P. Review of Energy Harvesting in LoRa based Wireless Sensor Network. In Proceedings of the 2020 2nd International Conference on Advances in Computing, Communication Control and Networking (ICACCCN), Greater Noida, India, 18–19 December 2020; pp. 338–341. [CrossRef]
45. Riaz, A.; Mohamed, R.; Sarker, M.R. Development and Analysis of Vibrational based Piezoelectric Micro Energy Harvester for WSN Sensor Node Capturing Abundant Vibration of Suspension Bridge. In Proceedings of the 2023 Second International Conference on Smart Technologies for Smart Nation (SmartTechCon), Singapore, 18–19 August 2023; pp. 825–830. [CrossRef]
46. Chi, Z.; Li, Y.; Sun, H.; Yao, Y.; Zhu, T. Concurrent Cross-Technology Communication Among Heterogeneous IoT Devices. *IEEE/ACM Trans. Netw.* **2019**, *27*, 932–947. [CrossRef]
47. Jiang, W.; Yin, Z.; Liu, R.; Li, Z.; Kim, S.M.; He, T. Boosting the Bitrate of Cross-Technology Communication on Commodity IoT Devices. *IEEE/ACM Trans. Netw.* **2019**, *27*, 1069–1083. [CrossRef]
48. Liu, L.; Yao, Y.; Cao, Z.; Zhang, M. DeepLoRa: Learning Accurate Path Loss Model for Long Distance Links in LPWAN. In Proceedings of the IEEE INFOCOM 2021-IEEE Conference on Computer Communications, Virtual, 10–13 May 2021; pp. 1–10. [CrossRef]
49. Loh, F.; Mehling, N.; Hoßfeld, T. Towards LoRaWAN without Data Loss: Studying the Performance of Different Channel Access Approaches. *Sensors* **2022**, *22*, 691. . [CrossRef]
50. Chache, F.M.; Maxon, S.; Narayanan, R.M.; Bharadwaj, R. Effects on Age of Information From Forward Error Correction in LoRa Transmissions. *IEEE Sens. J.* **2024**, *24*, 30853–30862. [CrossRef]
51. Jiannqi, L.; Chunyan, A.; Linchang, M.; Yuandong, L.; Qingjiang, H. Splitting data and forward error correction methods for solving the interference in LoRa networks. In Proceedings of the 2020 12th IEEE PES Asia-Pacific Power and Energy Engineering Conference (APPEEC), Nanjing, China, 20–23 September 2020; pp. 1–5. [CrossRef]
52. Guo, Q.; Yang, F.; Wei, J. Experimental Evaluation of the Packet Reception Performance of LoRa. *Sensors* **2021**, *21*, 1071. . [CrossRef]
53. Sulaiman, M.; Yunus, M.; Bahari, A.; Rani, M.A. Identification of damage based on frequency response function (FRF) data. In Proceedings of the MATEC Web of Conferences, the 2nd International Conference on Automotive Innovation and Green Vehicle (AiGEV 2016), Cyberjaya, Malaysia, 2–3 August 2016; EDP Sciences: Paris, France, 2017; Volume 90, p. 01025.
54. Xu, Y.; Zhu, W. Efficient and accurate calculation of discrete frequency response functions and impulse response functions. *J. Vib. Acoust.* **2016**, *138*, 031003. [CrossRef]
55. Singh, T.; Sehgal, S.; Prakash, C.; Dixit, S. Real-time structural health monitoring and damage identification using frequency response functions along with finite element model updating technique. *Sensors* **2022**, *22*, 4546. [CrossRef]

56. Amezquita-Sanchez, J.; Osornio-Rios, R.; Romero-Troncoso, R.; Dominguez-Gonzalez, A. Hardware-software system for simulating and analyzing earthquakes applied to civil structures. *Nat. Hazards Earth Syst. Sci.* **2012**, *12*, 61–73. [CrossRef]
57. Kotsakos, D.; Sakkos, P.; Kalogeraki, V.; Gunopulos, D. SmartMonitor: Using smart devices to perform structural health monitoring. *Proc. VLDB Endow.* **2013**, *6*, 1282–1285. [CrossRef]
58. Brincker, R.; Ventura, C.E.; Andersen, P. Damping estimation by frequency domain decomposition. In Proceedings of the IMAC 19: A Conference on Structural Dynamics, Kissimmee, FL, USA, 5–8 Februar 2001; Society for Experimental Mechanics: Bethel, CT, USA, 2001; pp. 698–703.
59. Altunişik, A.C.; Bayraktar, A.; Sevim, B. Operational modal analysis of a scaled bridge model using EFDD and SSI methods. *Indian J. Eng. Mater. Sci.* **2012**, *19*, 320–330.
60. Saidin, S.S.; Jamadin, A.; Abdul Kudus, S.; Mohd Amin, N.; Anuar, M.A. An overview: The application of vibration-based techniques in bridge structural health monitoring. *Int. J. Concr. Struct. Mater.* **2022**, *16*, 1–17. [CrossRef]
61. James, G.H. The natural excitation technique (NExT) for modal parameter extraction from operating structures. *J. Anal. Exp. Modal. Anal.* **1995**, *10*, 260.
62. Yang, X.M.; Yi, T.H.; Qu, C.X.; Li, H.N. Modal identification of bridges using asynchronous responses through an enhanced natural excitation technique. *J. Eng. Mech.* **2021**, *147*, 04021106. [CrossRef]
63. Carden, E.P.; Brownjohn, J.M. ARMA modelled time-series classification for structural health monitoring of civil infrastructure. *Mech. Syst. Signal Process.* **2008**, *22*, 295–314. [CrossRef]
64. Peeters, B.; De Roeck, G. Reference-based stochastic subspace identification for output-only modal analysis. *Mech. Syst. Signal Process.* **1999**, *13*, 855–878. [CrossRef]
65. Chang, C.M.; Loh, C.H. Improved stochastic subspace system identification for structural health monitoring. *Proc. J. Phys. Conf. Ser.* **2015**, *628*, 012010. [CrossRef]
66. Hassani, H. Singular spectrum analysis: Methodology and comparison. *MPRA Pap.* **2007**, *5*, 239–257. [CrossRef]
67. Dwivedi, S.K.; Vishwakarma, M.; Soni, A. Advances and researches on non destructive testing: A review. *Mater. Today Proc.* **2018**, *5*, 3690–3698. [CrossRef]
68. Wang, B.; Zhong, S.; Lee, T.L.; Fancey, K.S.; Mi, J. Non-destructive testing and evaluation of composite materials/structures: A state-of-the-art review. *Adv. Mech. Eng.* **2020**, *12*, 1687814020913761. [CrossRef]
69. Helal, J.; Sofi, M.; Mendis, P. Non-destructive testing of concrete: A review of methods. *Electron. J. Struct. Eng.* **2015**, *14*, 97–105. [CrossRef]
70. Gholizadeh, S. A review of non-destructive testing methods of composite materials. *Procedia Struct. Integr.* **2016**, *1*, 50–57. [CrossRef]
71. Radzieński, M.; Krawczuk, M.; Palacz, M. Improvement of damage detection methods based on experimental modal parameters. *Mech. Syst. Signal Process.* **2011**, *25*, 2169–2190. [CrossRef]
72. Kim, J.T.; Stubbs, N. Improved damage identification method based on modal information. *J. Sound Vib.* **2002**, *252*, 223–238. [CrossRef]
73. Maia, N.; Silva, J.; Almas, E.; Sampaio, R. Damage detection in structures: From mode shape to frequency response function methods. *Mech. Syst. Signal Process.* **2003**, *17*, 489–498. [CrossRef]
74. Villacorta, J.J.; Del-Val, L.; Martínez, R.D.; Balmori, J.A.; Magdaleno, Á.; López, G.; Izquierdo, A.; Lorenzana, A.; Basterra, L.A. Design and validation of a scalable, reconfigurable and low-cost structural health monitoring system. *Sensors* **2021**, *21*, 648. [CrossRef]
75. Zhang, C.; Mousavi, A.A.; Masri, S.F.; Gholipour, G.; Yan, K.; Li, X. Vibration feature extraction using signal processing techniques for structural health monitoring: A review. *Mech. Syst. Signal Process.* **2022**, *177*, 109175. [CrossRef]
76. Yan, W.J.; Zhao, M.Y.; Sun, Q.; Ren, W.X. Transmissibility-based system identification for structural health Monitoring: Fundamentals, approaches, and applications. *Mech. Syst. Signal Process.* **2019**, *117*, 453–482. [CrossRef]
77. Wang, F.; Chan, T. Review of vibration-based damage detection and condition assessment of bridge structures using structural health monitoring. In Proceedings of the Second Infrastructure Theme Postgraduate Conference: Rethinking Sustainable Development-Planning, Infrastructure Engineering, Design and Managing Urban Infrastructure, Brisbane, Australia, 26 March 2009; Queensland University of Technology: Brisbane, Australia, 2009; pp. 35–47.
78. Hou, R.; Xia, Y. Review on the new development of vibration-based damage identification for civil engineering structures: 2010–2019. *J. Sound Vib.* **2021**, *491*, 115741. [CrossRef]
79. Magalhães, F.; Cunha, Á.; Caetano, E. Vibration based structural health monitoring of an arch bridge: From automated OMA to damage detection. *Mech. Syst. Signal Process.* **2012**, *28*, 212–228. [CrossRef]
80. Fan, W.; Qiao, P. Vibration-based damage identification methods: A review and comparative study. *Struct. Health Monit.* **2011**, *10*, 83–111. [CrossRef]
81. Dhatt, G.; Lefrançois, E.; Touzot, G. *Finite Element Method*; John Wiley & Sons: Hoboken, NJ, USA, 2012.

82. Cheng, X.; Dong, J.; Han, X.; Fei, Q. Structural health monitoring-oriented finite-element model for a large transmission tower. *Int. J. Civ. Eng.* **2018**, *16*, 79–92. [CrossRef]
83. Alamdari, M.M.; Rakotoarivelo, T.; Khoa, N.L.D. A spectral-based clustering for structural health monitoring of the Sydney Harbour Bridge. *Mech. Syst. Signal Process.* **2017**, *87*, 384–400. [CrossRef]
84. Yang, D.; Youliang, D.; Aiqun, L. Structural condition assessment of long-span suspension bridges using long-term monitoring data. *Earthq. Eng. Eng. Vib.* **2010**, *9*, 123–131. [CrossRef]
85. Alves, V.; Cury, A. A fast and efficient feature extraction methodology for structural damage localization based on raw acceleration measurements. *Struct. Control Health Monit.* **2021**, *28*, e2748. [CrossRef]
86. Saidin, S.S.; Kudus, S.A.; Jamadin, A.; Anuar, M.A.; Amin, N.M.; Ya, A.B.Z.; Sugiura, K. Vibration-based approach for structural health monitoring of ultra-high-performance concrete bridge. *Case Stud. Constr. Mater.* **2023**, *18*, e01752. [CrossRef]
87. Cabboi, A.; Gentile, C.; Saisi, A. Vibration-based SHM of a centenary bridge: A comparative study between two different automated OMA techniques. In Proceedings of the 9th International Conference on Structural Dynamics (Eurodyn 2014), Porto, Portugal, 30 June–2 July 2014; pp. 1461–1468.
88. Chen, G.; Yang, X.; Ying, X.; Nanni, A. Damage detection of concrete beams using nonlinear features of forced vibration. *Struct. Health Monit.* **2006**, *5*, 125–141. [CrossRef]
89. Orhan, S. Analysis of free and forced vibration of a cracked cantilever beam. *Ndt E Int.* **2007**, *40*, 443–450. [CrossRef]
90. Lamarche, C.P.; Mousseau, S.; Paultre, P.; Proulx, J. A comparison of Ambient and Forced-Vibration Testing of a Full-Scale Concrete Structure. In Proceedings of the 22nd International Modal Analysis Conference IMA-XXII, Dearborn, MI, USA, 26–29 January 2004.
91. Barrett, A.R.; Avci, O.; Setareh, M.; Murray, T.M. Observations from vibration testing of in-situ structures. In Proceedings of the Structures Congress 2006: Structural Engineering and Public Safety, St. Louis, MO, USA, 18–21 May 2006; pp. 1–10.
92. Omenzetter, P.; Beskhyroun, S.; Shabbir, F.; Chen, G.W.; Chen, X.; Wang, S.; Zha, A. Forced and Ambient Vibration Testing of Full Scale Bridges. A Report Submitted to Earthquake Commission Research Foundation (Project No. UNI/578). 2013. Available online: https://www.naturalhazards.govt.nz/assets/Publications-Resources/383-Forced-and-ambient-vibration-testing-of-full-scale-bridges-compressed.pdf (accessed on 3 May 2025).
93. De Sortis, A.; Antonacci, E.; Vestroni, F. Dynamic identification of a masonry building using forced vibration tests. *Eng. Struct.* **2005**, *27*, 155–165. [CrossRef]

buildings

MDPI

Article

Analyzing Influencing Factors of Low-Carbon Technology Adoption in Hospital Construction Projects Based on TAM-TOE Framework

Lei Jin [1], Dezhi Li [2,3,*], Yubin Zhang [1,4] and Yi Zhao [5]

1 The First Affiliated Hospital with Nanjing Medical University (Jiangsu Province Hospital), Nanjing 210029, China; jinlei5222@jsph.org.cn (L.J.); 230239371@seu.edu.cn (Y.Z.)
2 School of Civil Engineering, Southeast University, Nanjing 211189, China
3 Engineering Research Center of Building Equipment, Energy, and Environment, Ministry of Education, Nanjing 211189, China
4 School of Energy and Environment, Southeast University, Nanjing 211189, China
5 School of Civil Engineering, Southeast University Chengxian College, Nanjing 210088, China; zhaoyi880704@outlook.com
* Correspondence: njldz@seu.edu.cn

Abstract: Hospitals rank among the most energy-intensive public building typologies and offer substantial potential for carbon mitigation. However, their construction phase has received limited scholarly attention within China's 'dual carbon' agenda. To address this research gap, this study develops and empirically validates an integrated Technology Acceptance Model and Technology-Organization-Environment framework tailored for hospital construction projects. The study not only identifies 12 critical adoption factors but also offers recommendations and discusses the relevance to multiple Sustainable Development Goals. This research provides both theoretical and practical insights for promoting sustainable hospital construction practices.

Keywords: low-carbon technology; hospital construction projects; technology acceptance model; technology-organization-environment framework; structural equation modeling

1. Introduction

Nowadays, addressing climate change and energy shortages has emerged as an urgent global issue. As the primary driver of global climate change, carbon emissions have prompted an international consensus on implementing effective energy-saving and carbon reduction measures to mitigate rising global temperatures. Against this backdrop, world leaders adopted the Paris Agreement in 2015, establishing a global framework for achieving carbon neutrality and emphasizing the importance of international cooperation in addressing climate change. This initiative aligns closely with the United Nations Sustainable Development Goals (SDGs). In response to the commitments outlined in the Paris Agreement, China has proposed its 'dual carbon' strategic objective: to achieve 'carbon peaking' by 2030 and 'carbon neutrality' by 2060. The construction industry is a significant contributor to carbon emissions, accounting for 37% of global carbon emissions [1] and 47.1% of national carbon emissions in China. Furthermore, emissions from the construction industry continue to escalate [2]. Hospital buildings, as large-scale public facilities with complex mechanical and electrical systems, high occupancy rates, and extended operating hours, consume substantial amounts of energy resources. According to statistics, hospitals exhibit a carbon emission intensity that is 1.6–2 times higher than that of general public

https://doi.org/10.3390/buildings15152703

buildings [3], while contributing to approximately 4% of total emissions [4]. This places them among the highest emitters in this category, with significant potential for reducing their carbon emissions [5]. According to the China Health Statistics Yearbook, the total number of hospitals in China has steadily increased. The early stage construction phase, which encompasses hospital project design and implementation activities, not only impacts its own phase but also has a significant impact on the carbon emissions during the later operational phase after the hospital is completed. Therefore, it becomes imperative to embrace low-carbon technology in hospital construction projects.

Currently, an increasing number of countries are focusing on the adoption of low-carbon technology in hospitals. Most scholars concentrate on developing low-carbon technology during the operational phase of hospitals, including waste emission management, water resource management, and the utilization of higher-quality chemicals. Their aim is to reduce carbon emissions during hospital operations in countries such as the United States [6], Germany [7–9], Greece [10], Poland [11], and China [12]. However, some scholars also pay attention to low-carbon technology development during the construction phase of hospitals. For instance, countries like the UK [13] and Turkey [14] strive to decrease carbon emissions during hospital construction by enhancing ventilation conditions, researching appropriate material choices, selecting suitable building orientations and window-to-wall ratios, and adopting photovoltaic power generation schemes. Furthermore, certain scholars dedicate their efforts to studying carbon emissions throughout all lifecycle stages of hospitals. Countries like Canada [15], Australia [16–19] and China [20,21] calculate and evaluate carbon emissions from hospital projects' lifecycles, with the aim of formulating plans to reduce them.

Existing research on low-carbon technology in hospitals has predominantly focused on qualitative analyses of specific technologies. However, limited studies have explored the managerial and behavioral factors influencing low-carbon technology adoption. Staff in the Infrastructure Construction Department of Hospitals (ICDH) are directly responsible for managing design and implementation processes, selecting technologies, and ensuring compliance with relevant low-carbon policies and standards. Given their central role in construction-related decision-making, they are an appropriate and relevant group for evaluating factors influencing low-carbon technology adoption in hospital construction projects. To address this gap, a staff perspective within ICDH is adopted, and an integrated model combining the Technology Acceptance Model (TAM) and the Technology-Organization-Environment (TOE) framework is proposed to examine factors influencing low-carbon technology adoption in hospital construction projects. Furthermore, integrating the TAM-TOE framework with legal, regulatory, and environmental sustainability frameworks is crucial for ensuring its practical applicability in hospital construction projects [22].

In China, policies such as the 'Assessment standard for green building' (GB/T 50378-2019) [23] and the 'dual-carbon' strategy have provided institutional and regulatory foundations that shape technology adoption behavior. Internationally, the Paris Agreement and the Sustainable Development Goals (SDGs) also offer strategic direction and standardized benchmarks for low-carbon practices in the construction sector. Therefore, embedding the TAM-TOE model within such regulatory and normative contexts enhances its explanatory power and policy relevance in driving sustainable transitions in the hospital construction industry. Although this study was conducted exclusively in economically developed provinces in China, its findings may serve as a reference for regions at comparable stages of development. However, broader applicability across diverse regions requires further empirical verification.

Following this analysis, the study identifies key factors and provides concrete recommendations. The paper is structured as follows: Section 2 proposes the TAM-TOE theoretical framework and develops the research hypotheses. Section 3 details the survey design and data collection methodology. Section 4 validates the hypothesized paths using structural equation modeling. Finally, Section 5 proposes recommendations and implementation strategies for enhancing low-carbon technology adoption in hospital construction projects, based on the validated model.

2. Theoretical Framework and Research Hypothesis

2.1. TAM-TOE Framework

2.1.1. TAM

The Technology Acceptance Model (TAM), proposed by Davis [24], is widely used to investigate individual-level technology adoption. TAM posits that adoption behavior is primarily determined by behavioral intention, encompassing key variables such as perceived ease of use (PEOU), perceived usefulness (PU), and attitude towards usage. TAM has been extensively applied across various domains, including higher education to examine artificial intelligence adoption [25,26], the business sector to explore factors influencing accounting information systems [27], agriculture to study the impact of digital extension services on productivity and resource efficiency [28], and architecture to investigate adoption factors for building information modeling [29] and analyze the impact mechanisms of virtual reality in design [30]. Furthermore, within healthcare, TAM has been employed to study nurses' adoption of internet-based care services [31]. Based on TAM's established applicability and the supporting literature, this study adopts the model to investigate technology adoption by staff in the ICDH within hospital construction projects. Consequently, four core TAM indicators—PEOU, PU, attitude toward usage, and behavioral intention—are utilized for this investigation.

2.1.2. TOE

The Technology-Organization-Environment (TOE) framework, proposed by Tornatzky and Fleischer [32], is widely adopted in academic research as an effective theory for analyzing organizational-level adoption of new technologies. The TOE framework posits that technology diffusion is influenced not only by the characteristics of the technology itself but also by organizational and environmental factors. This framework has been applied across diverse sectors: within business, to examine impacts of successful digital transformation initiatives [33]; in government, to explore factors influencing local governments' e-participation adoption decisions [34]; in education, to investigate factors affecting student adoption of electronic platforms [35]; and in the construction industry, to identify factors influencing employee adoption of smart building technologies [36]. Consequently, this study adopts the three core dimensions of the TOE framework—technological, organizational, and environmental factors—as key analytical factors for understanding low-carbon technology adoption in hospital construction projects.

2.1.3. TAM-TOE

The Technology Acceptance Model (TAM) has undergone ongoing expansion, with research on technology adoption increasingly investigating its integration with complementary models or frameworks. In TAM, external variables that influence user perception are not explicitly constrained. However, TOE framework addresses this limitation by incorporating higher-level external factors, thereby aligning more closely with behavioral adoption patterns in the complex context of low-carbon technology adoption [37]. Compared with models such as UTAUT and DOI, TAM features a simpler structure and clearer variable definitions. It is particularly suitable for studying the adoption behavior of non-standardized and non-platform-based technologies, such as low-carbon technology, and can be more effectively integrated with the TOE framework. The integration of TAM and TOE frameworks enables simultaneous consideration of individual attitudes, organizational factors, and external environmental influences, thereby forming a more comprehensive and explanatory theoretical framework [38].

Building upon the TOE framework, Songer [39] proposed incorporating economic factors as an additional dimension. This recommendation arises from the recognition that economic considerations, particularly high initial investment requirements, exert a direct influence on decision-making behavior. Consequently, this study extends the TOE framework by introducing economic factors as a fourth category. This modification aims to enable a more comprehensive analysis of the factors influencing low-carbon technology adoption specifically within hospital construction projects.

The integrated theoretical framework (TAM-TOE) is illustrated in Figure 1.

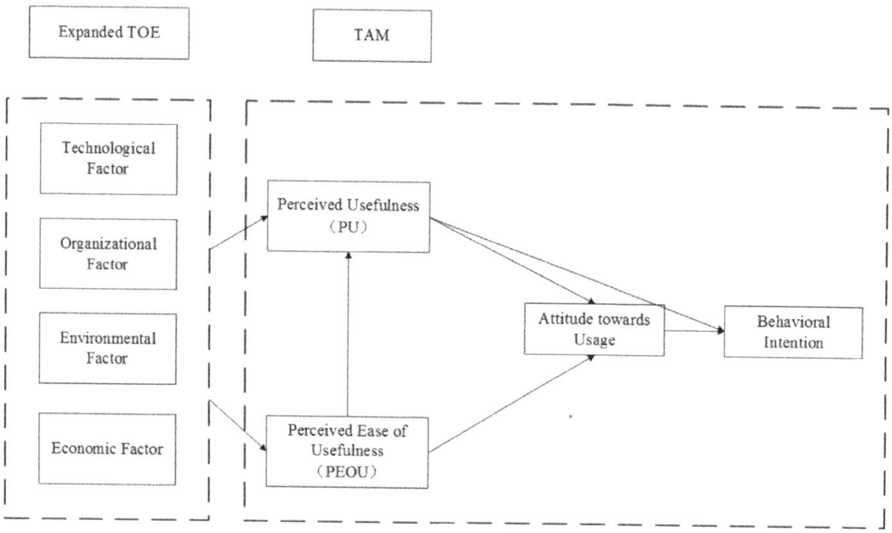

Figure 1. TAM-TOE theoretical framework.

2.2. Research Hypotheses

2.2.1. Identification of Influencing Factors

A literature review method was employed in this study to identify and synthesize relevant research related to 'hospital, low-carbon technology, TAM and the TOE framework'. Based on the TOE framework presented in the previous chapter, the factors influencing the adoption of low-carbon technology by staff in the ICDH in construction projects were categorized into technological, organizational, environmental, and economic dimensions. The classification results will be employed in the subsequent construction of hypotheses and the conduct of survey design.

2.2.2. Hypotheses Based on the TAM-TOE

Low-carbon technology exhibits significant growth potential in hospital construction projects. PU reflects users' belief that a new technology will enhance task performance, increase efficiency, and facilitate organizational goal attainment. In this study, PU represents the extent to which low-carbon technology aids ICDH staff in the following: (1) understanding carbon emissions generated during project construction, (2) reducing those emissions, and (3) improving project design and construction processes. PEOU denotes the user's subjective assessment of the effortlessness involved in adopting a new technology. Here, PEOU encompasses ICDH staff's following: (1) comprehension of low-carbon technology concepts and content, (2) perceived ease of learning the technology itself, and (3) understanding of carbon emission calculation principles and processes. TAM proposes that ICDH staff's evaluations of low-carbon technology—based on PU and PEOU—significantly influence their attitude toward usage, thereby positively affecting their behavioral intention [40–42]. Furthermore, TAM posits that PEOU positively influences PU [43–45].

Based on these premises, the following hypotheses were proposed:

H1. *Attitude towards usage exerts a significant positive impact on behavioral intention;*

H2. *PU exerts a significant positive impact on behavioral intention;*

H3. *PU exerts a significant positive impact on attitude towards usage;*

H4. *PEOU exerts a significant positive impact on PU;*

H5. *PEOU exerts a significant positive impact on attitude towards usage.*

The TOE framework proposes that attitudes toward usage and behavioral intention are influenced by technological, organizational, environmental, and economic factors [46]. These external variables impact the PEOU and PU. Prior to constructing a low-carbon technology adoption model for hospital construction projects, it is imperative to analyze how these external variables affect the PEOU and PU.

Technological factors encompass the maturity level of carbon emission calculation software, compatibility with other architectural software, and operational stability. These factors contribute to a sense of ease in utilizing low-carbon technology among staff in the ICDH and serve as critical factors influencing its convenient implementation in hospital construction projects. Consequently, they can positively impact PEOU [41,47].

Organizational factors include policies and financial support from hospital senior managers, as well as the professional capabilities and project experience of low-carbon technology consulting teams in hospital construction. These factors play a critical role in facilitating project implementation, instilling confidence among staff in the ICDH to adopt low-carbon technology, and consequently positively influencing perceived usefulness PU and PEOU [48].

Environmental factors encompass policy requirements and incentives provided by government departments, as well as the adoption of low-carbon technology by other hospitals. Staff in ICDH are often influenced by external environmental factors and the choices made by their counterparts in neighboring institutions. The adoption of low-carbon technology by nearby hospitals enhances the perceived importance of such technology among staff in ICDH, thereby influencing PU [49,50].

Economic factors encompass costs associated with low-carbon technology consulting teams, costs related to the implementation of these technologies, and reduced operational maintenance costs following the completion of hospital construction projects. Low-cost and high-return technologies can mitigate adoption risks, thereby influencing staff's assessment of the practical applicability of low-carbon technology. Furthermore, economic factors can also impact PU [48,51].

Therefore, the following hypotheses were proposed:

H6. *Technological factors exert a significant positive impact on PEOU;*

H7. *Organizational factors exert a significant positive impact on PU;*

H8. *Organizational factors exert a significant positive impact on PEOU;*

H9. *Environmental factors exert a significant positive impact on PU;*

H10. *Economic factors exert a significant positive impact on PU.*

In summary, this study incorporates the attributes of low-carbon technology to introduce technological, organizational, environmental, and economic factors from the TOE model. These factors enhance the explanatory power of the hypothetical model, which is grounded in TAM variables such as PU, PEOU, attitude toward usage, and behavioral intention. Furthermore, based on presumed relationships among these influencing factors, we construct a hypothetical model for the adoption of low-carbon technology in hospital construction projects, as illustrated in Figure 2.

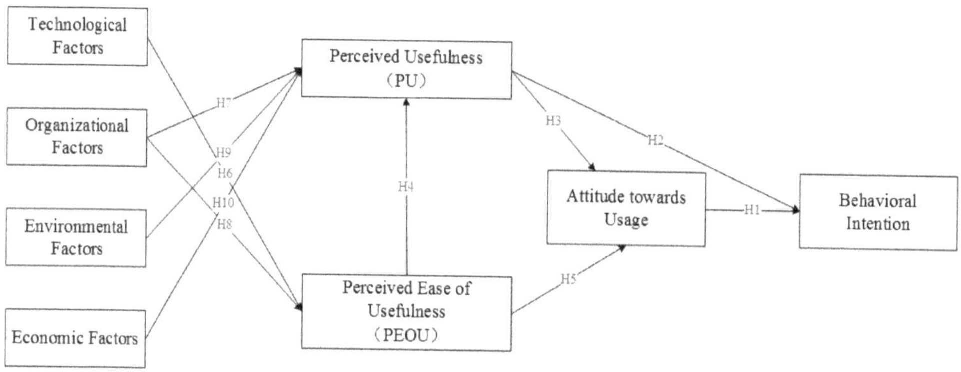

Figure 2. Hypothetical model based on TAM-TOE framework.

3. Survey Design and Data Collection

3.1. Survey Design

In order to validate the hypothesis path, it is imperative to conduct a questionnaire survey among relevant personnel. The questionnaire comprises two sections. The first section encompasses fundamental information about the respondents, primarily encompassing gender, age, educational attainment, professional designation, years of work experience, understanding of low-carbon policies, etc. The second section employs a 5-level Likert scale to gauge the PU, PEOU, technological factors, organizational factors, environmental factors, economic factors, attitude towards usage, and behavioral intention of the respondents. The scale ranges from 'strongly disagree' to 'strongly agree'. To measure the eight factors in our model accurately, we designed 23 specific items as observation variables based on an extensive review of pertinent literature (refer to Table 1). These items aim at assessing the relationship between influencing factors in low-carbon technology adoption within hospital construction projects. To ensure comprehensiveness and effectiveness of our survey questionnaire, it underwent rigorous scrutiny and revision by experts in hospital construction, low-carbon technology consulting teams, and university researchers. After multiple rounds of discussions and modifications, the clarity and precision of questions have been significantly enhanced.

Table 1. Latent variables and observed variables in the hypothetical model.

Latent Variable	Observed Variable	Source
PU	PU1: Through the application of low-carbon technology, staff in the ICDH can gain a comprehensive understanding of carbon emissions across various stages of hospital construction projects.	[24,52]
PU	PU2: The application of low-carbon technology can effectively mitigate carbon emissions in hospital construction projects.	
PU	PU3: The application of low-carbon technology contributes to optimizing the design and construction processes of hospital construction projects, thereby enhancing their overall implementation quality.	
PEOU	PEOU1: Low-carbon technology and the primary technologies it encompasses are characterized by clarity and precision.	[24,53]
PEOU	PEOU2: Low-carbon technology demonstrates high acceptability and comprehensibility in terms of conceptual understanding and content mastery.	
PEOU	PEOU3: The principles and processes of carbon emission calculation offer a relatively high level of accessibility in terms of learning and understanding.	
Technological Factors	TF1: Currently, the development of carbon emission calculation software has reached a relatively mature stage.	[40,54–57]
Technological Factors	TF2: Carbon emission calculation software possesses excellent compatibility and can be effectively integrated with building-related software such as CAD, BIM, and engineering quantity list preparation.	
Technological Factors	TF3: Carbon emission calculation software exhibits excellent stability and high reliability during practical applications.	
Organizational Factors	OF1: Hospital senior managers demonstrate strong support for the application of low-carbon technologies in terms of economic resource allocation and policy guidance.	[51,55]
Organizational Factors	OF2: The members of the low-carbon technology consulting team possess strong professional expertise and comprehensive overall competence.	
Organizational Factors	OF3: Members of the low-carbon technology consulting team possess extensive engineering practice experience similar to hospital construction projects.	
Environmental Factors	ENF1: Government departments issue relevant policies mandatorily requiring the implementation of low-carbon technology during the construction project implementation process.	[40,44,45,56]
Environmental Factors	ENF2: Government departments grant corresponding rewards to construction entities that actively apply low-carbon technology in construction projects.	
Environmental Factors	ENF3: Other hospitals have actively incorporated low-carbon technology into their construction projects.	
Economic Factors	EOF1: The cost associated with introducing a low-carbon technology consulting team into a hospital construction project is acceptable.	[48,58]
Economic Factors	EOF2: The cost of incorporating low-carbon technology into hospital construction projects is deemed acceptable.	
Economic Factors	EOF3: The utilization of low-carbon technology can lead to reduced operational maintenance costs following the completion of hospital construction projects.	

Table 1. *Cont.*

Latent Variable	Observed Variable	Source
Attitude Towards Usage	ATU1: The low-carbon technology holds great promise and exhibits significant potential for development.	[24,59]
Attitude Towards Usage	ATU2: I advocate the prudent adoption of low-carbon technology in hospital construction projects.	
Attitude Towards Usage	ATU3: I am committed to dedicating time towards comprehending and acquiring knowledge in the field of low-carbon technology.	
Behavioral Intention	BI1: In the event of new hospital construction projects, I will employ low-carbon technology to ensure adherence to sustainable practices.	[24]
Behavioral Intention	BI2: I will advocate for the adoption of low-carbon technology in hospital construction projects to other hospitals.	

3.2. Data Collection

To ensure the representativeness and relevance of the sample, purposive sampling (also known as judgmental sampling) was adopted. This method was chosen because the study aims to target a specific group of professionals—staff in the ICDH.

Jiangsu Province was selected as the sampling region because it is located in the Yangtze River Delta, one of the most economically developed and urbanized regions in China. Tertiary A-grade hospitals in this province typically have well-established infrastructure departments and are more likely to understand advanced low-carbon technology due to stronger financial and policy support.

Before administering the formal survey, a preliminary list of target hospitals was created based on public records of Tertiary A-grade hospitals. Staff in the ICDH were then contacted through hospital administrative offices and professional engineering associations. The questionnaires were distributed via an online platform from August to September 2024, and participants were invited through internal communication channels, receiving informed consent prior to participation. The flowchart to illustrate the full workflow of the research study is shown in Figure 3.

This purposive approach ensures that the sample consists of informed, relevant, and decision-involved professionals, which enhances the validity of the results. A total of 320 responses were obtained; following rigorous screening, 309 valid datasets were retained, resulting in a valid response rate of 96.6%. The demographic characteristics of the respondents (as shown in Table 2) demonstrate broad coverage in terms of gender, age, education level, professional titles, and years of experience, suggesting that the sample is reasonably diverse and representative of the hospital construction projects.

Table 2. Respondent demographic information.

Characteristics	Category	Frequency	Proportion (%)
Gender	Male	212	68.6
	Female	97	31.4
Age	Under 30	54	17.5
	31–40	100	32.4
	41–50	98	31.7
	Over 50	57	18.4

Table 2. *Cont.*

Characteristics	Category	Frequency	Proportion (%)
Education level	Associate degree or less	17	5.5
	Undergraduate degree	191	61.8
	Postgraduate degree or more	101	32.7
Professional title	None	63	20.4
	Junior professional designation	21	6.8
	Intermediate professional designation	102	33
	Senior professional designation	123	39.8
Years of work experience	0–5	56	18.1
	6–10	33	10.7
	11–20	99	32
	Over 20	121	39.2
Do you concur with the comprehension of the 'dual carbon' policy?	Disagree	17	5.5
	Neutral	166	53.7
	Agree	93	30.1
	Strongly agree	33	10.7
Do you concur with the comprehension of low-carbon technology?	Disagree	25	8.1
	Neutral	175	56.6
	Agree	89	28.8
	Strongly agree	20	6.5

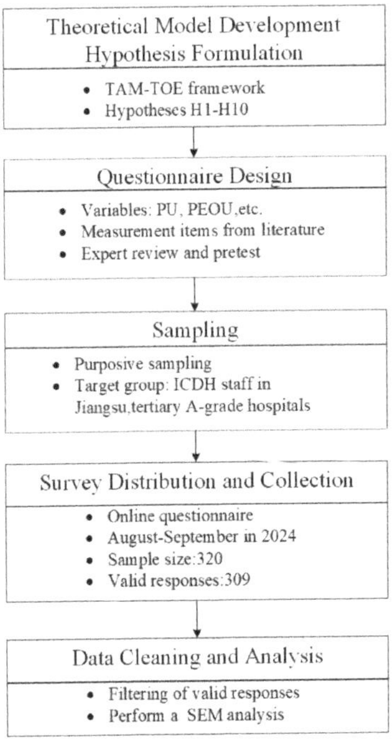

Figure 3. The flowchart for the full workflow.

4. Results

The study utilizes the structural equation modeling (SEM) approach and employs AMOS 21.0 software (IBM Corp, Armonk, NY, USA) to estimate the hypothetical model developed in Section 2 and analyze the data collected in Section 3. The two-step approach recommended by Anderson [60] was adopted in this study. The first step is to test the validity and reliability of the measurement model using confirmatory factor analysis (CFA). The second step is to test whether the hypotheses fit the theoretical model using the structural model.

4.1. Measured Model

Reliability is a statistical method employed to assess the consistency and stability of questionnaire results, typically measured by Cronbach's alpha. This coefficient ranges from 0 to 1, with values above 0.7 indicating favorable outcomes [61]. In this study, SPSS 25.0 software (IBM Corp, Armonk, NY, USA) was used to compute the Cronbach's alpha coefficients for each variable and dimension, as presented in Table 3. The obtained coefficients for each latent variable ranged from 0.84 to 0.923, signifying satisfactory reliability for this measurement model.

Table 3. Reliability and validity test of measure model.

Latent Variable	Items	FL	Mean	SD	Alpha	CR	AVE
PU	PU1	0.877	3.41	1.061	0.891	0.888	0.726
	PU2	0.868	3.47	1.037			
	PU3	0.810	3.53	1.049			
PEOU	PEOU1	0.870	3.21	0.992	0.878	0.879	0.707
	PEOU2	0.847	3.27	0.978			
	PEOU3	0.805	3.61	1.009			
Technological Factors	TF1	0.902	3.59	1.004	0.923	0.923	0.8
	TF2	0.912	3.59	1.008			
	TF3	0.869	3.68	1.101			
Organizational Factors	OF1	0.876	3.73	1.020	0.904	0.904	0.759
	OF2	0.861	3.62	1.114			
	OF3	0.876	3.75	1.087			
Environmental Factors	ENF1	0.855	3.66	1.062	0.874	0.875	0.7
	ENF2	0.853	3.78	1.095			
	ENF3	0.801	3.54	1.024			
Economic Factors	EOF1	0.841	3.46	0.988	0.896	0.895	0.74
	EOF2	0.849	3.54	1.040			
	EOF3	0.890	3.58	1.025			
Attitude towards Usage	ATU1	0.783	3.49	0.976	0.872	0.876	0.702
	ATU2	0.873	3.48	1.031			
	ATU3	0.855	3.25	0.964			
Behavioral Intention	BI1	0.846	3.55	1.152	0.84	0.813	0.684
	BI2	0.808	3.35	1.004			

Validity is conducted to evaluate the stability and structure of the measurement variables for each latent variable, encompassing both convergent validity and discriminant validity [62]. Convergent validity refers to the extent of correlation between the measurement variables and their corresponding latent variables. As presented in Table 3, all composite reliabilities (CR) for the latent variables in the measurement model surpass 0.7, while the average variance extracted (AVE) meets the recommended threshold of 0.5. Moreover, standardized factor loadings (FL) for each measurement variable range from 0.783 to 0.912, satisfying the requirement of being greater than 0.5 and achieving statistical significance at a level of $p < 0.001$. It indicates that the convergent validity of the measurement model meets the requirements [63].

Discriminant validity pertains to non-correlation among different latent variables' measurement variables. As demonstrated in Table 4, all correlation coefficients between latent variables vary from 0.272 to 0.571, with each square root AVE value exceeding its respective correlation coefficient with other latent variables. It indicates that the discrimination validity of the measurement model meets the requirements. These results validate that the measurement model has sufficient discriminant validity, ensuring that each latent variable captures distinct constructs. This provides a robust foundation for subsequent structural analysis, fulfilling a key methodological requirement of the study.

Table 4. Correlations among latent variables.

Latent Variable	PU	PEOU	Technological Factors	Organizational Factors	Environmental Factors	Economic Factors	Attitude Towards Usage	Behavioral Intention
PU	0.852							
PEOU	0.382	0.841						
Technological Factors	0.433	0.508	0.895					
Organizational Factors	0.349	0.396	0.387	0.871				
Environmental Factors	0.483	0.466	0.443	0.451	0.837			
Economic Factors	0.372	0.398	0.364	0.571	0.403	0.860		
Attitude towards Usage	0.383	0.293	0.272	0.417	0.331	0.396	0.838	
Behavioral Intention	0.420	0.404	0.420	0.401	0.381	0.385	0.369	0.827

4.2. Structural Model

4.2.1. Model Fitness Test

The structural model is used to estimate the relationships between latent variables in the research framework. In order to ensure accurate estimation of these relationships, a goodness-of-fit test needs to be conducted. The results of the test are shown in Table 5, and all fit indices meet the recommended values, indicating a good fit for the hypothetical model. Minor improvements could be made by addressing any residual covariance or ensuring even larger and more diverse samples in future studies.

Table 5. The recommended and actual value of fit indices.

Index	Recommended Value	Actual Value
X^2/df	<3	1.515
GFI	>0.9	0.921
AGFI	>0.8	0.898
NFI	>0.9	0.936
IFI	>0.9	0.977
CFI	>0.9	0.977
TLI	>0.9	0.973
RMSEA	>0.08	0.041

4.2.2. Hypotheses Test

Utilizing SPSS 25.0 (IBM Corp, Armonk, NY, USA) for constructing a structural equation model, Table 6 presents the standardized coefficient table, while Figure 4 illustrates the standardized model.

The results reveal that H1 ($\beta = 0.261$, $p < 0.001$), H2 ($\beta = 0.389$, $p < 0.001$), H3 ($\beta = 0.359$, $p < 0.001$), H6 ($\beta = 0.472$, $p < 0.001$), H8 ($\beta = 0.257$, $p < 0.001$), and H9 ($\beta = 0.397$, $p < 0.001$) are validated at the significant level of $p < 0.001$. Also, H4 ($\beta = 0.159$, $p < 0.05$), H5 ($\beta = 0.192$, $p < 0.05$), and H10 ($\beta = 0.174$, $p < 0.05$) are accepted at the significant level of $p < 0.05$. However, H7 ($\beta = 0.035$, $p = 0.669$) is found to be negligible and does not reach statistical significance ($p > 0.05$). Therefore, the effect size is not only small ($\beta = 0.035$) but also statistically unsupported, indicating that organizational factors do not exert a meaningful influence on PU in this context. Overall, 9 out of the 10 hypotheses (H1–H10) received empirical support, except for H7. Notably, the strongest effects are observed between technological factors and PEOU ($\beta = 0.472$, $p < 0.001$), and between PU and behavioral intention ($\beta = 0.389$, $p < 0.001$), aligning with the theoretical expectations of the TAM-TOE framework. Although economic factors show a smaller standardized path coefficient (β = 0.174), the effect remains statistically significant ($p < 0.05$), suggesting its practical relevance should not be disregarded in policy considerations.

Table 6. Hypotheses test of the theoretical model.

Hypothesis	Path	Standardized Path Coefficient	SE	t	*p*	Result
H1	Attitude towards Usage → Behavioral Intention	0.261	0.073	3.822	***	Supported
H2	PU → Behavioral Intention	0.389	0.062	5.534	***	Supported
H3	PU → Attitude towards Usage	0.359	0.055	5.324	***	Supported
H4	PEOU → PU	0.159	0.071	2.499	0.012 *	Supported
H5	PEOU → Attitude towards Usage	0.192	0.061	2.919	0.004 **	Supported
H6	Technological Factors → PEOU	0.472	0.053	7.466	***	Supported
H7	Organizational Factors → PU	0.035	0.077	0.428	0.669	Not supported
H8	Organizational Factors → PEOU	0.257	0.051	4.249	***	Supported
H9	Environmental Factors → PU	0.397	0.077	5.660	***	Supported
H10	Economic Factors → PU	0.174	0.075	2.333	0.02 *	Supported

Note: *** $p < 0.001$, ** $p < 0.01$, * $p < 0.05$.

Figure 4. Result of SEM analysis.

4.2.3. Mediating Effect Test

A total of 5000 Bootstrap tests were conducted on the potential mediating paths in the hypothetical model by employing the bias-corrected bootstrap method. The results (Table 7) indicate that 9 of the 11 tested paths were statistically significant. However, for the two paths involving the influence of organizational factors on PU, the 95% confidence intervals were [−0.041, 0.082] and [−0.01, 0.021]. These findings suggest that organizational factors do not significantly moderate the effects of PU on either attitude toward usage or behavioral intention. Thus, H7 is not supported. The mediating effect test further confirms the central role of PU and PEOU as intermediaries. Organizational factors, while not significantly influencing PU directly (as indicated by the non-significant H7), do exert indirect influence via PEOU.

Table 7. Mediating effect of the theoretical model.

Path	Indirect Effect	SE	Bias-Corrected 95% Confidence Interval		p
			Lower	Upper	
Technological Factors → PEOU → Attitude Towards Usage → Behavioral Intention	0.024	0.014	0.005	0.059	0.007
Technological Factors → PEOU → PU → Behavioral Intention	0.029	0.017	0.003	0.073	0.029
Technological Factors → PEOU → PU → Attitude Towards Usage → Behavioral Intention	0.007	0.004	0.001	0.018	0.016
Organizational Factors → PU → Behavioral Intention	0.013	0.031	−0.041	0.082	0.612
Organizational Factors → PU → Attitude Towards Usage → Behavioral Intention	0.003	0.008	−0.01	0.021	0.569
Organizational Factors → PEOU → Attitude Towards Usage → Behavioral Intention	0.013	0.009	0.002	0.041	0.006
Organizational Factors → PEOU → PU → Behavioral Intention	0.016	0.010	0.002	0.043	0.020
Organizational Factors → PEOU → PU → Attitude Towards Usage → Behavioral Intention	0.004	0.002	0.001	0.011	0.013
Environmental Factors → PU → Behavioral Intention	0.154	0.040	0.085	0.241	0.000
Environmental Factors → PU → Attitude Towards Usage → Behavioral Intention	0.037	0.014	0.016	0.075	0.000
Economic Factors → PU → Behavioral Intention	0.068	0.03	0.015	0.133	0.011
Economic Factors → PU → Attitude Towards Usage → Behavioral Intention	0.016	0.009	0.004	0.041	0.007

5. Discussions and Implications

5.1. Discussions

By conducting a comprehensive analysis and rigorous testing of the TAM-TOE hypothetical model in this study, the following key conclusions can be derived:

(1) Technological factors exhibited both direct and indirect positive effects on PEOU, whereas environmental and economic factors have similar impacts on PU. Organizational factors do not significantly affect PU directly (H7 not supported) but have an indirect impact via PEOU. This may be attributed to ICDH staff perceiving support from senior administration and the exceptional expertise and project experience of low-carbon technology consulting team as key factors that enhance their perception of the ease of use of low-carbon technology, compared to the perceived usefulness alone [64,65]. The factor loadings of TF1, TF2, and TF3 all exceed 0.8, suggesting that, in terms of technological factors, the development maturity, operational stability, and compatibility with other software are all critically important features of carbon emission calculation software. The factor loadings of OF1, OF2, and OF3 all exceed 0.8, suggesting that within organizational factors, the support from hospital senior administration staff, as well as the exceptional expertise and project experience of the low-carbon technology consulting team, are of considerable significance. The factor loadings of ENF1, ENF2, and ENF3 all exceed 0.8, suggesting that laws, regulations, and policy support; incentive measures by government agencies for technology adoption; and the implementation of low-carbon technology by other hospitals effectively capture the underlying dimensions of environmental factors. The factor loadings of EOF1, EOF2, and EOF3 all exceed 0.8, indicating a strong association between the costs of the low-carbon technology consulting teams, the cost of the technology itself, and the operational maintenance costs following the completion of hospital construction projects with economic factors.

(2) PEOU and PU have a direct positive impact on the attitude towards usage. Specifically, PEOU serves as a mediating variable between technological and organizational

factors, and it positively influences the attitude towards usage. Similarly, PU functions as a mediating variable between environmental and economic factors, also exerting a positive influence on the attitude towards usage. Therefore, both variables play significant roles in the model. On the other hand, the relatively low scores for PEOU1 and PEOU2 in the PEOU suggest that staff in ICDH currently lack a sufficient understanding of low-carbon technology. Similarly, the score for ATU3 is relatively low, suggesting that staff in ICDH exhibit limited willingness to invest time in learning low-carbon technology. This implies the need to enhance staff awareness and adoption of low-carbon technology through more accessible and intuitive software tools and effective dissemination strategies.

(3) The total influence coefficients of technological factors, organizational factors, environmental factors, and economic factors on behavioral intention are 0.06, 0.049, 0.191, and 0.084, respectively. This indicates that environmental factors exert the strongest influence on promoting the adoption of low-carbon technology in hospital construction projects. Therefore, enhancing the legal, regulatory, and policy frameworks for low-carbon technology, as well as improving the application environment, are crucial measures for promoting its implementation in hospital construction. This finding aligns with conclusions from previous studies [66].

(4) This study utilized SEM to test complex latent constructs. However, several limitations should be acknowledged. First, SEM assumes linear relationships among variables, which may oversimplify real-world interactions in behavioral adoption processes. Second, SEM requires an adequate sample size to ensure model stability and parameter reliability. While sample size ($N = 309$) meets general SEM criteria, future studies with larger and more diverse samples are encouraged to improve generalizability and robustness.

5.2. Implications

The implications are as follows:(1) Based on the results of the validated TAM-TOE model discussed in Section 5.1, this section proposes practical strategies to enhance the adoption of low-carbon technologies in hospital construction projects. These recommendations are structured around the significant paths identified through SEM analysis and are further categorized by key stakeholder groups to facilitate targeted implementation.

In regards to strengthening environmental incentives (H9), governments should prioritize unified low-carbon regulatory frameworks and provide financial support (e.g., performance-based subsidies) to foster a policy environment that enhances PU [67]. In regards to enhancing technological tool usability (H6), low-carbon consulting teams should improve compatibility, stability, and automation of carbon emission calculation software to improve PEOU and encourage broader use [68]. In regards to ensuring economic feasibility (H10), although the effect size is moderate, cost remains a key barrier. Governments and institutions should develop return on investment (ROI) models and co-funding mechanisms to support adoption [69]. In regards to promoting organizational readiness via PEOU (H8), while organizational factors do not directly affect PU, they shape behavioral intention through PEOU. Staff training, managerial support, and interdepartmental collaboration are essential.

(2) Based on the validated TAM-TOE framework, policy recommendations can be tailored to different stakeholders:

Government departments should establish unified carbon emission policies and implement incentive measures targeting low-carbon technology [70]. Currently, there are significant disparities in carbon emission-related standards and regulations among provinces and cities, which urgently require gradual integration and the establishment of a unified national standard system. Additionally, it is suggested to incorporate carbon emission assessments into the preliminary review process of construction projects, such as integrating them into the environmental impact assessment framework. Furthermore, the government could provide effective incentives to construction entities through a combination of base subsidies and performance-based subsidies.

Academic institutions and industry alliances can bridge the gap between policy and practice through empirical research, case studies, and technical guidance specific to the healthcare construction sector [71,72]. Academic institutions and industry alliances are recommended to regularly organize academic conferences related to low-carbon technology, such as inviting experts to deliver lectures or reports on low-carbon technology at academic conferences like the annual conference on hospital construction. Additionally, the industry is encouraged to actively collaborate with government departments in the formulation of carbon emission policies and standards specific to the hospital sector [73,74].

Hospital senior managers must foster an innovation-oriented culture, allocate resources for technical training, and encourage internal collaboration to support low-carbon adoption [75].

Staff in the ICDH should closely monitor national policy developments and trends in the construction industry while continuously enhancing their understanding of low-carbon technology. They should proactively report the current status and advantages of low-carbon technology to hospital senior managers outside the construction field, specifically highlighting that the application of low-carbon technology during the construction phase can significantly reduce operational and maintenance costs after the completion of hospital facilities.

Low-carbon consulting teams can collaborate with government departments, universities, and project owner to jointly conduct research and establish a low-carbon technology database and inventory database, thereby providing practical low-carbon implementation pathways for hospital construction projects. Additionally, these consulting teams should strengthen the maintenance and updates of carbon emission calculation software to continuously enhance its stability and compatibility with other building-related software [76]. Furthermore, low-carbon technology consulting teams should engage in regular technical exchanges with design institutes and construction companies to assist hospital infrastructure management personnel in selecting cost-effective carbon emission control technologies.

(3) As concluded in Section 5.1, environmental factors have the greatest influence on driving the adoption of low-carbon technology in hospital construction projects. Additionally, the recommendations proposed based on the TAM-TOE framework are highly aligned with certain aspects of the SDGs. Emphasis on software innovation and technical integration supports SDG 9 (Industry, Innovation, and Infrastructure), which promotes sustainable and resilient infrastructure through innovation. The prioritization of cost-effective low-carbon technologies aligns with SDG 12 (Responsible Consumption and Production), encouraging efficient resource use in the construction phase. Strengthening carbon standards, policy frameworks, and government incentives reflects SDG 13 (Climate Action), which advocates urgent actions to combat climate change. Moreover, these strategies contribute to SDG 3 (Good Health and Well-being) by promoting the development of low-carbon hospitals that offer healthier built environments. Therefore, this study not only identifies the key factors affecting the adoption of low-carbon technology but also proposes a series of comprehensive strategies to promote the achievement of multiple sustainable development goals related to hospital development.

6. Conclusions

The implementation of carbon emission reduction has become a global imperative, and hospital buildings, as critical public infrastructure, have emerged as a key focus in achieving the national 'dual carbon' goals. Given China's large population and extensive number of hospitals, promoting the application of low-carbon technologies in hospital construction projects holds significant practical importance.

This study set out to investigate the key factors influencing the adoption of low-carbon technology in hospital construction projects by staff in the ICDH. Through integrating the TAM and the TOE framework, a comprehensive TAM-TOE model was constructed and validated using SEM based on 309 valid responses from Tertiary A-grade hospitals in Jiangsu Province, China. The model confirms that environmental and economic factors significantly influence PU, while technological and organizational factors shape perceived ease of use PEOU. PEOU and PU, in turn, influence users' attitudes and behavioral intention toward adopting low-carbon technology. Among these, environmental factors exerted the strongest total effect on behavioral intention, highlighting the importance of external policy and institutional context. Notably, hypothesis H7 was not supported, suggesting that organizational support enhances adoption primarily through improving PEOU rather than PU directly. These findings successfully fulfill the study's original objective of identifying and empirically validating the behavioral, technological, organizational, environmental, and economic drivers of low-carbon technology adoption in hospital construction.

Several limitations should be acknowledged. First, SEM assumes linear relationships among variables, which may oversimplify real-world interactions in behavioral adoption processes. Future research could explore non-linear or interaction-based models to better capture these dynamics. Second, SEM requires an adequate sample size to ensure model stability and parameter reliability. While sample size (N = 309) meets general SEM criteria, future studies with larger and more diverse samples are encouraged to improve generalizability and robustness. Third, the sample is geographically concentrated in Jiangsu Province and limited to ICDH staff. Broader validation in different regions, and involving other stakeholders such as consultants, designers, or contractors, would enhance the model's generalizability [77].

Nonetheless, this work provides a valuable foundation for future extensions. The TAM-TOE framework can be expanded to explore the full lifecycle carbon performance of hospital buildings, including embodied carbon, greenhouse gas (GHG) accounting, and carbon lifecycle assessment (LCA). This can help align construction-phase technology decisions with operation and maintenance (O&M) strategies, strengthening the hospital's decarburization pathway across its entire lifecycle. Moreover, the framework can contribute to global sustainability agendas, such as SDG 9, SDG 12 and SDG 13, by enabling data-driven and behaviorally informed policies.

In conclusion, this study provides theoretical and empirical insights into hospital construction decarburization, bridges behavioral models with sustainability science, and offers pathways to integrate low-carbon decision-making across construction and operations phases.

Author Contributions: Conceptualization, L.J. and D.L.; methodology, L.J.; software, L.J. and Y.Z. (Yi Zhao); validation, L.J., D.L. and Y.Z. (Yubin Zhang); formal analysis, L.J.; investigation, L.J. and Y.Z. (Yubin Zhang); resources, D.L.; data curation, L.J. and Y.Z. (Yi Zhao); writing—original draft preparation, L.J.; writing—review and editing, D.L.; supervision, D.L.; funding acquisition, D.L. All authors have read and agreed to the published version of the manuscript.

Funding: This research received no external funding.

Institutional Review Board Statement: All methods were carried out in accordance with relevant guidelines and regulations, and all experimental protocols were approved by Jiangsu Province Hospital (jsph20240715) on 15 July 2024.

Informed Consent Statement: Informed consent was obtained from all subjects involved in the study.

Data Availability Statement: The datasets used and analyzed during the current study are available from the corresponding author on reasonable request.

Conflicts of Interest: The authors declare no conflicts of interest.

References

1. 2023 Global Status Report for Buildings and Construction. Available online: https://unfccc.int/ttclear/misc_/StaticFiles/gnwoerk_static/tn_meetings/00cf22a4049c4ece9f414e190def4202/8dff87ea3e1e4e7ba7d349a83ed04cbd.pdf (accessed on 20 June 2024).
2. Huang, G.; Li, D.; Ng, S.T.; Wang, L.; Zhang, Y. Vision-Based Personal Thermal Comfort Modeling under Facial Occlusion Scenarios. *Energy Build.* **2025**, *335*, 115566. [CrossRef]
3. Zhang, J.; Yan, Z.; Bi, W.; Ni, P.; Lei, F.; Yao, S.; Lang, J. Prediction and Scenario Simulation of the Carbon Emissions of Public Buildings in the Operation Stage Based on an Energy Audit in Xi'an, China. *Energy Policy* **2023**, *173*, 113396. [CrossRef]
4. Karliner, J.; Slotterback, S.; Boyd, R.; Ashby, B.; Steele, K.; Wang, J. Health Care's Climate Footprint: The Health Sector Contribution and Opportunities for Action. *Eur. J. Public Health* **2020**, *30*, ckaa165.843. [CrossRef]
5. Li, H.; Wu, Y.; Xin, B.; Xu, M.; Wu, S. Optimal Carbon–Neutral Strategies in the Healthcare System: A Three-Stage Stackelberg Game Model. *Transp. Res. Part E: Logist. Transp. Rev.* **2025**, *198*, 104128. [CrossRef]
6. Alshqaqeeq, F.; Amin Esmaeili, M.; Overcash, M.; Twomey, J. Quantifying Hospital Services by Carbon Footprint: A Systematic Literature Review of Patient Care Alternatives. *Resour. Conserv. Recycl.* **2020**, *154*, 104560. [CrossRef]
7. Quitmann, C.; Sauerborn, R.; Herrmann, A. Gaps in Reporting Greenhouse Gas Emissions by German Hospitals—A Systematic Grey Literature Review. *Sustainability* **2021**, *13*, 1430. [CrossRef]
8. Schmidt, L.; Bohnet-Joschko, S. Planetary Health and Hospitals' Contribution—A Scoping Review. *Int. J. Environ. Res. Public Health* **2022**, *19*, 13536. [CrossRef] [PubMed]
9. Quitmann, C.; Sauerborn, R.; Danquah, I.; Herrmann, A. Reducing the Carbon Footprint of a German University Hospital: Perspectives from Hospital Stakeholders. *J. Clim. Chang. Health* **2023**, *12*, 100247. [CrossRef]
10. Bozoudis, V.; Sebos, I. The Carbon Footprint of Transport Activities of the 401 Military General Hospital of Athens. *Environ. Model. Assess.* **2021**, *26*, 155–162. [CrossRef]
11. Tomiak, W.; Kupis, R.; Kowalik, J.; Skrobisz, K.; Milewska-Augusto, M. Green hospitals: A remedy for the climate change burden on the polish healthcare system. *Prospect. Pharm. Sci.* **2024**, *22*, 45–49. [CrossRef]
12. Guo, S.; Wang, W.; Zhou, Y. Research on Energy Saving and Economy of Old Buildings Based on Parametric Design: A Case Study of a Hospital in Linyi City, Shandong Province. *Sustainability* **2022**, *14*, 16681. [CrossRef]
13. Townsend, K.; Dougherty, W.; Housley, R.; Walker, J.; Cruttenden-Wood, D. Exploring the Carbon Impact of Virtual Wards in a Large Acute Hospital. *Computers* **2025**, *14*, 230. [CrossRef]
14. As, M.; Bilir, T. Enhancing Energy Efficiency and Cost-Effectiveness While Reducing CO_2 Emissions in a Hospital Building. *J. Build. Eng.* **2023**, *78*, 107792. [CrossRef]
15. Cimprich, A.; Young, S.B. Environmental Footprinting of Hospitals Organizational Life Cycle Assessment of a Canadian Hospital. *J. Ind. Ecol.* **2023**, *27*, 1335–1353. [CrossRef]
16. Hensher, M. Health Technology Assessment and Healthcare Environmental Sustainability: Prioritizing Effort and Maximizing Impact. *Int. J. Technol. Assess. Health Care* **2024**, *40*, e25. [CrossRef]
17. Bi, P.; Hansen, A. Carbon Emissions and Public Health: An Inverse Association? *Lancet Planet. Health* **2018**, *2*, e8–e9. [CrossRef] [PubMed]
18. Verlis, K.; Haddock, R.; Barratt, A. The Current State of Sustainable Healthcare in Australia. *Aust. Health Rev.* **2024**, *48*, 489–496. [CrossRef] [PubMed]
19. Malik, A.; Lenzen, M.; McAlister, S.; McGain, F. The Carbon Footprint of Australian Health Care. *Lancet Planet. Health* **2018**, *2*, e27–e35. [CrossRef]
20. Yue, J.; Li, W.; Cheng, J.; Xiong, H.; Xue, Y.; Deng, X.; Zheng, T. A Dynamic Calculation Model of the Carbon Footprint in the Life Cycle of Hospital Building: A Case Study in China. *Eng. Constr. Archit. Manag.* **2023**, *30*, 4735–4761. [CrossRef]
21. Yu, D.; Tan, X.; Liu, Z.; Li, D.; Wang, Z.; Yan, P.; Ni, J. Energy Saving and Carbon Reduction Schemes for Hospital with Photovoltaic Power Generation and System Upgrading Technology. *Heliyon* **2023**, *9*, e21447. [CrossRef]
22. Motsi, L.; Chimbo, B. Success Factors for Evidence-Based Healthcare Practice Adoption. *S. Afr. J. Inf. Manag.* **2023**, *25*, a1622. [CrossRef]
23. *GB/T50378-2019*; Assessment Standard for Green Building. China Building Industry Press: Beijing, China, 2019.

24. Davis, F.D.; Bagozzi, R.P.; Warshaw, P.R. User Acceptance of Computer Technology: A Comparison of Two Theoretical Models. *Manag. Sci.* **1989**, *35*, 982–1003. [CrossRef]
25. Nagy, A.S.; Tumiwa, J.R.; Arie, F.V.; Erdey, L. An Exploratory Study of Artificial Intelligence Adoption in Higher Education. *Cogent Educ.* **2024**, *11*, 2386892. [CrossRef]
26. Helmiatin; Hidayat, A.; Kahar, M.R. Investigating the Adoption of AI in Higher Education: A Study of Public Universities in Indonesia. *Cogent Educ.* **2024**, *11*, 2380175. [CrossRef]
27. Magboul, I.; Jebreel, M.; Shawtri, F.; Qabajeh, M.; Shanti, A.; Alqudah, M.; Abu Huson, Y. Factors Influencing Accounting Information System Usage by Oil Companies & Performance Outcomes. *Cogent Bus. Manag.* **2024**, *11*, 2369211. [CrossRef]
28. Thi Hoa Sen, L.; Chou, P.; Dacuyan, F.B.; Nyberg, Y.; Wetterlind, J. Barriers and Enablers of Digital Extension Services' Adoption among Smallholder Farmers: The Case of Cambodia, the Philippines and Vietnam. *Int. J. Agric. Sustain.* **2024**, *22*, 2368351. [CrossRef]
29. Ozcan-Deniz, G.; Ozorhon, B.; Kaya, O.C. Building Information Modeling (BIM) Integration in Developing Countries: An in-Depth Examination of Adoption Factors from Public Clients' Perspectives. *Int. J. Archit. Comput.* **2024**, *23*, 481–491. [CrossRef]
30. Liu, Y.; Konduri, S.; Park, C. A Study on the Factors Influencing Willingness to Use Virtual Reality Systems for External Evaluation of Buildings. *Buildings* **2024**, *14*, 3714. [CrossRef]
31. Lee, S.-Y.; Chang, C.-Y. Factors Influencing Nurses' Satisfaction to Online Learning Approach: A Cross ICU Analysis. *BMC Nurs.* **2024**, *23*, 548. [CrossRef] [PubMed]
32. Drazin, R. The Processes of Technological Innovation. *Technol. Transf. Winter* **1990**, *16*, 45–46. [CrossRef]
33. N'Dri, A.B.; Su, Z. Successful Configurations of Technology–Organization–Environment Factors in Digital Transformation: Evidence from Exporting Small and Medium-Sized Enterprises in the Manufacturing Industry. *Inf. Manag.* **2024**, *61*, 104030. [CrossRef]
34. Ninčević Pašalić, I.; Ćukušić, M. Understanding E-Participation Adoption: Exploring Technological, Organizational, and Environmental Factors. *Technol. Forecast. Soc. Chang.* **2024**, *207*, 123633. [CrossRef]
35. Haripin; Warsono, S. Analysis of the Acceptance of E-Learning Platform (SIDEK-Edu) among High School Students with UTAUT2-TOE Approach: Implications for Digital Learning. *Int. J. Inf. Learn. Technol.* **2024**, *41–45*, 585–600. [CrossRef]
36. Xue, H.; Zhang, S.; Wu, Z.; Zhang, L. How to Improve the Smart Construction Technology Usage Behavior of Construction Enterprise Employees?—TOE Framework Based on Configuration Study. *Eng. Constr. Archit. Manag.* **2023**, *32*, 785–804. [CrossRef]
37. Sulaiman, T.T.; Mahomed, A.S.B.; Rahman, A.A.; Hassan, M. Understanding Antecedents of Learning Management System Usage among University Lecturers Using an Integrated TAM-TOE Model. *Sustainability* **2023**, *15*, 1885. [CrossRef]
38. Bach, M.P.; Meško, M.; Stjepić, A.M.; Khawaja, S.; Quershi, F.H. Understanding Determinants of Management Simulation Games Adoption in Higher Educational Institutions Using an Integrated Technology Acceptance Model/Technology–Organisation–Environment Model: Educator Perspective. *Information* **2025**, *16*, 45. [CrossRef]
39. Songera, D.; Young, R.; Davis, K.A. Social Architecture for Sustainable It Implementation in AEC/EPC. 2001. Available online: http://www.irbnet.de/daten/iconda/CIB20255.pdf (accessed on 1 July 2024).
40. Kimiagari, S.; Baei, F. Promoting E-Banking Actual Usage: Mix of Technology Acceptance Model and Technology-Organisation-Environment Framework. *Enterp. Inf. Syst.* **2022**, *16*, 8–9. [CrossRef]
41. Na, S.; Heo, S.; Han, S.; Shin, Y.; Roh, Y. Acceptance Model of Artificial Intelligence (AI)-Based Technologies in Construction Firms: Applying the Technology Acceptance Model (TAM) in Combination with the Technology–Organisation–Environment (TOE) Framework. *Buildings* **2022**, *12*, 90. [CrossRef]
42. Yu, X.; Sheng, G.; Sun, D.; He, R. Effect of Digital Multimedia on the Adoption of Agricultural Green Production Technology among Farmers in Liaoning Province, China. *Sci. Rep.* **2024**, *14*, 13092. [CrossRef]
43. Sanchís-Pedregosa, C.; Vizcarra-Aparicio, J.-M.; Leal-Rodríguez, A.L. BIM: A Technology Acceptance Model in Peru. *J. Inf. Technol. Constr.* **2020**, *25*, 99–108. [CrossRef]
44. Desmaryani, S.; Soleh, A.; Irmanelly; Wiarta, I. Integration of Technology Acceptance Models and Government Support to Improve Digital Literacy. *Heliyon* **2024**, *10*, e34086. [CrossRef] [PubMed]
45. Deng, M.; Guo, Y. A Study of Safety Acceptance and Behavioral Interventions for Autonomous Driving Technologies. *Sci. Rep.* **2022**, *12*, 17891. [CrossRef] [PubMed]
46. Kittipanya-ngam, P.; Tan, K.H.; Cavite, H.J. Future-Ready AI: A Framework for Ethical and Sustainable Adoption. *Technol. Soc.* **2025**, *83*, 102993. [CrossRef]
47. Jo, H.; Bang, Y. Understanding Continuance Intention of Enterprise Resource Planning (ERP): TOE, TAM, and IS Success Model. *Heliyon* **2023**, *9*, e21019. [CrossRef]
48. Katebi, A.; Homami, P.; Najmeddin, M. Acceptance Model of Precast Concrete Components in Building Construction Based on Technology Acceptance Model (TAM) and Technology, Organization, and Environment (TOE) Framework. *J. Build. Eng.* **2022**, *45*, 103518. [CrossRef]

49. Sampa, M.B.; Abdul Aziz, N.H.; Rahman, M.S.; Hasan, M.; Ab. Aziz, N.A. Factors Influencing the Adoption and Acceptance of eHealth in Malaysia: A Systematic Review. *Crit. Public Health* **2025**, *35*, 2519780. [CrossRef]
50. Zhao, Y.; Sun, Y.; Zhou, Q.; Cui, C.; Liu, Y. How A/E/C Professionals Accept BIM Technologies in China: A Technology Acceptance Model Perspective. *Eng. Constr. Archit. Manag.* **2023**, *30*, 4569–4589. [CrossRef]
51. Lee, M.; Chai, C.; Xiong, Y.; Gui, H. Technology Acceptance Model for Building Information Modelling Based Virtual Reality (BIM-VR) in Cost Estimation. *J. Inf. Technol. Constr.* **2022**, *27*, 914–925. [CrossRef]
52. Xiang, W.; Gao, J. Do Not Be Anticlimactic: Farmers' Behavior in the Sustainable Application of Green Agricultural Technology—A Perceived Value and Government Support Perspective. *Agriculture* **2023**, *13*, 247. [CrossRef]
53. Dai, Q.; Cheng, K. What Drives the Adoption of Agricultural Green Production Technologies? An Extension of TAM in Agriculture. *Sustainability* **2022**, *14*, 14457. [CrossRef]
54. Mousa, A.H.; Aljshamee, M.; Mohammed, A.H.; Mousa, S.H.; Obaid, H.A. A Systematic Literature Review: Measuring User Acceptance Models in e-Learning Adopted in Higher Education Institutes. The 8th International Conference on Applied Science and Technology (ICAST 2020). *AIP Conf. Proc.* **2020**, *2290*, 040021. [CrossRef]
55. Jo, H.; Bang, Y. Analyzing ChatGPT Adoption Drivers with the TOEK Framework. *Sci. Rep.* **2023**, *13*, 22606. [CrossRef]
56. Li, L.; Min, X.; Guo, J.; Wu, F. The Influence Mechanism Analysis on the Farmers' Intention to Adopt Internet of Things Based on UTAUT-TOE Model. *Sci. Rep.* **2024**, *14*, 15016. [CrossRef]
57. Davila Laguna, R.F.; Lizarzaburu Aguinaga, D.; Caceres Torres, D.E.; Larroche Cueto, B.A. Soft Skills and the Use of Industry 4.0 as Determinants of Professional Development in Engineering Graduates: A SEM Approach. *Sustain. Futures* **2025**, *10*, 100742. [CrossRef]
58. Wong, L.-W.; Leong, L.-Y.; Hew, J.-J.; Tan, G.W.-H.; Ooi, K.-B. Time to Seize the Digital Evolution: Adoption of Blockchain in Operations and Supply Chain Management among Malaysian SMEs. *Int. J. Inf. Manag.* **2020**, *52*, 101997. [CrossRef]
59. Ren, Z.; Zhong, K. Driving Mechanism of Subjective Cognition on Farmers' Adoption Behavior of Straw Returning Technology: Evidence from Rice and Wheat Producing Provinces in China. *Front. Psychol.* **2022**, *13*, 922889. [CrossRef]
60. Anderson, J.C.; Gerbing, D.W. Structural Equation Modeling in Practice: A Review and Recommended Two-Step Approach. *Psychol. Bull.* **1988**, *103*, 411–423. [CrossRef]
61. Zhou, Q.; Deng, X.; Hwang, B.-G.; Ji, W. Integrated Framework of Horizontal and Vertical Cross-Project Knowledge Transfer Mechanism within Project-Based Organizations. *J. Manag. Eng.* **2020**, *36*, 04020062. [CrossRef]
62. Huang, G.; Li, D.; Yu, L.; Yang, D.; Wang, Y. Factors Affecting Sustainability of Smart City Services in China: From the Perspective of Citizens' Sense of Gain. *Habitat Int.* **2022**, *128*, 102645. [CrossRef]
63. Abulail, R.N.; Badran, O.N.; Shkoukani, M.A.; Omeish, F. Exploring the Factors Influencing AI Adoption Intentions in Higher Education: An Integrated Model of DOI, TOE, and TAM. *Computers* **2025**, *14*, 230. [CrossRef]
64. Henao-Ramírez, A.M.; López-Zapata, E. Analysis of the Factors Influencing Adoption of 3D Design Digital Technologies in Colombian Firms. *J. Enterp. Inf. Manag.* **2022**, *35*, 429–454. [CrossRef]
65. Luo, G.; Serrão, C.; Liang, D.; Zhou, Y. A Relevance-Based Technology–Organisation–Environment Model of Critical Success Factors for Digital Procurement Adoption in Chinese Construction Companies. *Sustainability* **2023**, *15*, 12260. [CrossRef]
66. Terenzi, M.; Ogheri, C.; Locatelli, G. Understanding Homeowners' Behavioural Determinants: A Study on House Retrofitting and PV Adoption. *Appl. Energy* **2025**, *397*, 126319. [CrossRef]
67. Tang, X.; Li, J. Study of the Mechanism of Digitalization Boosting Urban Low-Carbon Transformation. *Front. Environ. Sci.* **2022**, *10*, 982864. [CrossRef]
68. Zhao, L.; Zhang, W.; Wang, W. BIM-Based Multi-Objective Optimization of Low-Carbon and Energy-Saving Buildings. *Sustainability* **2022**, *14*, 13064. [CrossRef]
69. Oguntuase, O.J.; Adu, O.B.; Obayori, O.S. How Do Individual-Level Factors Influence the Adoption of Low-Carbon Technology? Proposing and Validating the Bioeconomy Technology Acceptance Model in the Context of Africa. *Front. Sustain.* **2023**, *4*, 1148001. [CrossRef]
70. Do, M.-H.; Huang, Y.-F.; Hoang, T.-T. Blockchain Adoption in Green Supply Chains: Analyzing Key Drivers, Green Innovation, and Expected Benefits. *J. Theor. Appl. Electron. Commer. Res.* **2025**, *20*, 39. [CrossRef]
71. Sun, T.; Di, K.; Hu, J.; Shi, Q.; Irfan, M. Digitally Empowered Green Public Services in Environmentally Vulnerable Areas: Insights from SEM-ANN Analysis. *J. Retail. Consum. Serv.* **2025**, *84*, 104216. [CrossRef]
72. Huang, G.; Li, D.; Zhou, S.; Ng, S.T.; Wang, W.; Wang, L. Public Opinion on Smart Infrastructure in China: Evidence from Social Media. *Util. Policy* **2025**, *93*, 101886. [CrossRef]
73. Du, Q.; Pang, Q.; Bao, T.; Guo, X.; Deng, Y. Critical Factors Influencing Carbon Emissions of Prefabricated Building Supply Chains in China. *J. Clean. Prod.* **2021**, *280*, 124398. [CrossRef]
74. Ahankoob, A.; Abbasnejad, B.; Aranda-Mena, G. Building Information Modelling (BIM) Acceptance and Learning Experiences in Undergraduate Construction Education: A Technology Acceptance Model (TAM) Perspective—An Australian Case Study. *Buildings* **2025**, *15*, 1804. [CrossRef]

75. Rotimi, F.E.; Silva, C.; Ramanayaka, C.E.D.; Rotimi, J.O.B. Digital Technology Knowledge Transfer Enablers Amongst End-Users in Architecture, Engineering, and Construction Organisations: New Zealand Insights. *Buildings* **2025**, *15*, 772. [CrossRef]
76. Cudjoe, D.; Zhu, B.; Wang, H. The Role of Incentive Policies and Personal Innovativeness in Consumers' Carbon Footprint Tracking Apps Adoption in China. *J. Retail. Consum. Serv.* **2024**, *79*, 103861. [CrossRef]
77. Zhou, S.; Thomas Ng, S.; Huang, G.; Dao, J.; Li, D. Extracting Interrelated Information from Road-Related Social Media Data. *Adv. Eng. Inform.* **2022**, *54*, 101780. [CrossRef]

buildings

MDPI

Article

A Bibliometric Analysis of Research on Human Settlements Improvement Based on CNKI and Web of Science

Chenxi Li [1,2], Jinfeng Miao [3,*], Xuan Zhai [1] and Qiao Liu [1]

1 School of Public Administration, Xi'an University of Architecture and Technology, Xi'an 710055, China; lichenxi@xauat.edu.cn (C.L.)
2 Technology Innovation Center for Land Engineering and Human Settlements, Shaanxi Land Engineering Construction Group Co., Ltd., Xi'an Jiaotong University, Xi'an 712000, China
3 Department of Economic Management, Hebei Vocational University of Technology and Engineering, Xingtai 054000, China
* Correspondence: miaojinfeng@xpc.edu.cn

Abstract: In the context of rural revitalization, scientific and efficient improvement of the human settlement environment is of great practical significance to improve the quality of life of farmers as well as to promote modern rural construction. To understand the research status and trends of human settlement environment improvement, the paper visually analyzed the relevant literature from 2012 to 2022 based on CNKI and the Web of Science database with the help of CiteSpace software(developed by Chaomei Chen, Drexel University, Philadelphia, PA, USA) (v6.2.r2). The results show the following: (1) the research literature on human settlement environment improvement has undergone three stages of slow growth, steady growth, and explosive growth; (2) many countries and institutions have published research on the improvement of human settlement environment, forming a close cooperative relationship; (3) three research hotspots have been formed: the concept and classification of the human settlement environment, the improvement path of the human settlement environment, and the evaluation method of the human settlement environment; and (4) waste management and green space are the frontier issues in the field of human settlement environment improvement. The subsequent research on the improvement of human settlement environments should strengthen the breadth and depth of research, improve the evaluation index system, innovate research methods, and build interdisciplinary integration, to provide a perfect research paradigm for research on the improvement of human settlement environments in the future.

Keywords: human settlements; research progress; research frontier; bibliometrics

1. Introduction

The human settlement environment is a complex system with human beings as the core and villages, towns, cities, and other settlements, which encompass the most basic infrastructure needed for human survival, as well as the natural and social environment on which human life and production depend [1]. With the rapid development of the global economy, traffic congestion, environmental degradation, and housing shortages have emerged one after another, seriously affecting people's human settlement environment [2]. The increase in problems of the human settlement environment has become a major threat to socio-economic development and human health. Therefore, there is an urgent to carry out scientific and efficient work for improving the human settlement environment.

The study of the human settlement environment can be traced back to the 1950s, when Doxiadis first proposed the science of human settlement [3]. The "Vancouver Declaration on

https://doi.org/10.3390/buildings15111805

Human Settlements", issued at the Vancouver Conference on Human Settlements in 1976, and the "Adequate Shelter for All" and "Sustainable Human Settlements Development in an Urbanizing World" presented at the Second United Nations Conference on Human Settlements, held in Istanbul in 1996, have led to two major themes. "Sustainable human settlements development in an urbanizing world" made sustainable human settlement environment development a universal global agenda [4]. In 2002, the United Nations established the Living Agenda to promote the development of human settlements [5]. At the Third Plenary Session of the 16th Central Committee of the Communist Party of China (CPC) in 2003, the state emphasized rural development as a key issue in China's economic and social development and requested local governments and society to actively participate in rural construction planning practices. Since then, a series of rural planning practices have been developed, such as the "Modernization of New Rural Construction" in Jiangsu Province and the "Ten Thousand Villages Improvement and Thousand Villages Demonstration" in Zhejiang Province [6].

In 2008, the Urban and Rural Planning Law established the legal status of rural planning, the preparation system, and the relationship with urban and rural planning, and in 2010, the National No. 1 Document proposed to "improve the human settlement environment by improving the treatment of garbage and sewage". In 2012, the No. 1 Document of the Central Government again proposed to promote rural environmental improvement and improve the human settlement environment in rural areas. In 2014, the General Office of the State Council issued the "Guidance Opinions on Improving Rural Environment", proposing "to improve the rural environment as the focus, and comprehensively improve the human settlement environment". In 2018, in the context of China's rural development, the "Three-Year Action Plan for the Improvement of Human Settlement Environment" was issued, in which the requirement of "strengthening rural waste and gradually promoting urban sewage treatment" was put forward. In 2021, the General Office of the CPC Central Committee and the General Office of the State Council issued the "Five-Year Action Plan to Improve the Human Settlement Environment (2021–2025)", which further proposed to "promote the improvement of rural toilets, garbage treatment, sewage treatment, comprehensive river treatment, and improve the human settlement environment according to local conditions" as a key task. In 2022, the report of the 20th National Congress proposed to fundamentally combat soil pollution, improve infrastructure, and improve the rural human settlement environment. The report of the 20th National Congress in 2022 proposed to radically combat soil pollution, improve infrastructure, and improve the human settlement environment. However, a current specific human settlement environment improvement program is still not available. At the same time, the academic community has launched intense discussions and formed different opinions on the construction tasks, governance methods, and comprehensive evaluation of the human settlement environment [7].

Scholars are involved in a wide range of research fields on human settlement environment improvement, but there is a lack of research results based on scientometric analysis. Therefore, this paper uses the CiteSpace knowledge mapping visualization software (v6.2.r2) to systematically sort out the relevant literature of research on human settlement environment improvement, analyze research progress, and predict future research directions, aiming to help researchers gain a deeper understanding of the current situation and trends of human settlement environment improvement, summarize research hotspots and major research contents, and form a panoramic knowledge network structure. A comprehensive evaluation of the current research progress on human settlement environment improvement shows that the current academic-related research lacks research breadth and depth, research methods, evaluation index system, and multidisciplinary integration. The objectives of this study are as follows: (1) to understand the current status of research on the human settlement environment in terms of annual publication trends and cooperation between publishing countries and institutions; (2) to reveal research hotspots on the human settlement environment based on keyword co-occurrence; (3) to identify research frontiers on the human settlement environment based on the top 20 emergent keywords; and (4) to put forward the future outlook of human settlement environment improvement research at home and abroad.

This study includes five parts: introduction, data sources and research methodology, results and analysis, discussion, and conclusions. Firstly, this paper introduces the research background and raises the research questions. Secondly, this paper understands the research status through the changing trend of annual articles and the distribution of countries and institutions. On this basis, the hotspots and trends are summarized through high-frequency keywords and the sudden occurrence of keywords. Then, based on the above analysis, the paper proposes the shortcomings of the current research and further proposes future research directions. Finally, this paper presents a summary and conclusions.

2. Data Sources and Research Methods

2.1. Data Sources

In this paper, CiteSpace Knowledge Graph visualization software (v6.2.r2) was used to select the China Knowledge Network (CNKI) database and the Web of Science (WOS) database as data sources, and the relevant literature was searched. By using the advanced search function of CNKI and the basic search criteria of WOS, the paper selected academic journals with "topic" = "human settlement environment improvement", set the search period from 1 January 2012 to 31 December 2022, set the type of literature as journal papers, and retrieved 1878 Chinese papers and 2037 English papers in total. Then, the authors manually culled the non-academic literature and finally obtained 1828 Chinese and 1975 English papers on the improvement of the human settlement environment.

2.2. Research Methodology

CiteSpace software (v6.2.r2) mainly represents the relationship between nodes through connecting lines, tree diagrams, etc., and draws an information map to dynamically identify the current status of research, research hotspots, and future research directions [8,9]. Figure 1 shows the research process of this paper. The paper selects CNKI as the database of the Chinese literature and WOS as the database of the English literature. The topic is "human settlement environment improvement". Through searching, there are 1828 Chinese and 1975 English papers that meet the requirements. The research status includes "Trends in the annual production of literature" and "Distribution of research countries and institutions". Topics of high research interest include "Concept and Classification of human settlement environment", "Study on the path of human settlement environment improvement", and "Study on the evaluation method of human settlement environment". Research frontiers include "Waste management" and "Green space". Finally, the paper concludes that future research proposals are to "strengthen the breadth and depth of Research", "improve the evaluation index system", and "build multidisciplinary cross-fertilization".

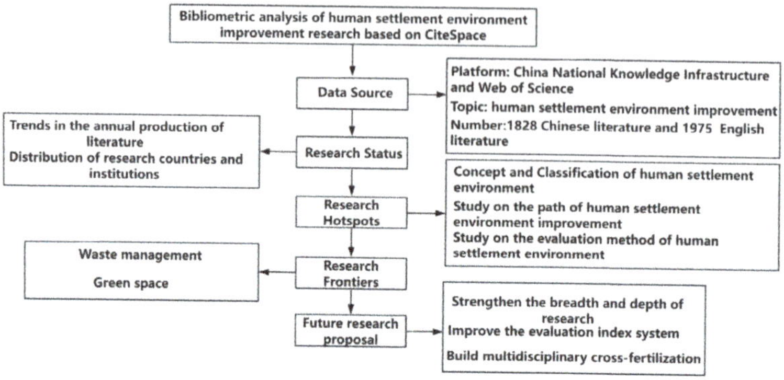

Figure 1. Research framework.

3. Results and Analysis

3.1. Research Status

3.1.1. Trends in the Annual Production of Literature

To understand more accurately the publication trend of the research literature on human settlement environment improvement, this paper draws a graph of the annual publication trend of research on human settlement environment improvement in the past ten years (Figure 2). Figure 2 shows that the amount of Chinese and international research on human settlement environment improvement has shown a trend of "slow growth—steady growth—explosive growth" in the past ten years. The average annual number of articles published in the Chinese literature was 15.5 during the slow growth phase (2012–2013), about 40 during the steady growth phase (2014–2017), and about 40 during the explosive growth phase (2018–2022). The number of articles published in 2020 and 2021 was 431 and 430, respectively, which is about 28 times that of the slow increase phase. A look at the international sample literature shows the following: in the slow growth phase (2012–2017), the number of articles on human settlement environment improvement rose to 3–11 per year; in the steady growth phase (2018–2019), the number of articles issued increased significantly compared with the previous period, reaching a maximum of 202. In the explosive growth phase (2020–2022), the number of articles increases rapidly, reaching 378 articles by 2022, with an average annual volume of about 319 articles. In summary, the literature related to improvement of the human settlement environment is increasing year by year, which indicates that it still has a high research intensity.

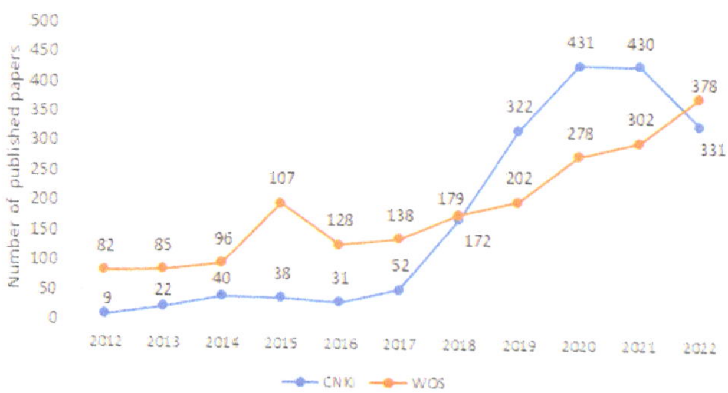

Figure 2. Change trend of human settlement environment improvement.

3.1.2. Distribution of Research Countries and Institutions

A study of the countries and institutions that have published research on human settlement environment improvement can provide an accurate understanding of the importance that scholars attach to this research topic. From the literature included in the WOS source journals, a total of 119 countries and 392 research institutions are included in the English research literature on human settlement environment improvement, and a co-occurrence map of foreign countries and research institutions is drawn in this paper (Figures 3 and 4). The size of the circles of nodes represents the number of publications, and it can be seen from Figure 3 that China and the United States are the top two countries in terms of the number of publications. As can be seen from Figure 4, the United States has the highest number of institutional publications, including the University of California (41), Harvard University (18), and Johns Hopkins University (18), in addition to the University of Chinese Academy of Sciences (39), and the rest are evenly distributed among the United Kingdom, France, and Australia. This shows that research on human settlement environment improvement has attracted widespread attention in various countries in the last decade.

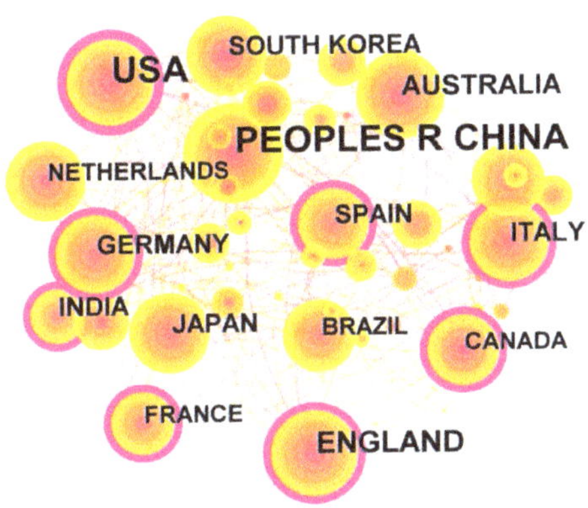

Figure 3. Mapping of the co-occurrence of countries publishing papers on international human settlement environment improvement studies.

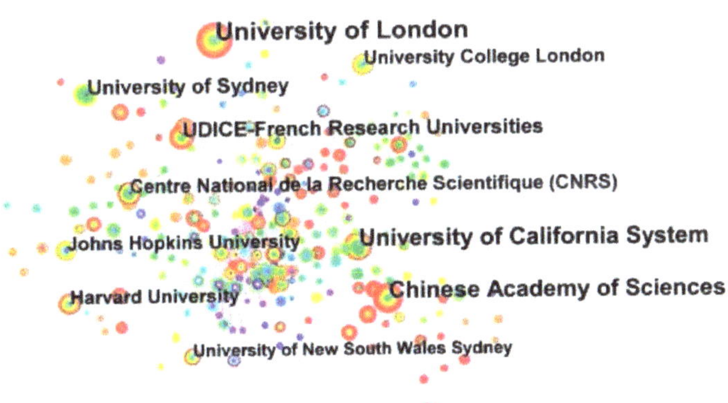

Figure 4. Mapping of international human settlement environment improvement research-publishing institutions' co-occurrence maps.

Table 1 shows the distribution of publishing institutions in China, with 288 research institutions involved in research on human settlement environment improvement, among which the Institute of Rural Development of the Chinese Academy of Social Sciences is in first place with 15 publications. It can be seen that various countries and research institutions attach importance to research on human settlement environment improvement. Nany Yuliastutil et al. proposed how local governments and communities should make decisions when building ecological villages with a good living environment [10]. Liaghat, M. explored and selected some evaluation indexes of coastal tourist destinations in the Dixon port area of Malaysia by using AHP alone [11]. Several factors influencing the environmental development of rural settlements were studied by E. Shcherbin et al. and a map of the potential of clustering in the Mogilev region of Russia was developed [12]. A significant number of scholars have also conducted research on the topic [13–18].

Table 1. Map of human settlement environment improvement research-issuing agencies in China.

Institution	Number of Articles Issued
Institute of Rural Development, Chinese Academy of Social Sciences	15
New Rural Office of Jilin Province	13
Department of Rural Social Enterprise Promotion, Ministry of Agriculture and Rural Affairs	9
Xiaokang—China Xiaokang Network	9
Xinhua News Agency	9
Ministry of Agriculture and Rural Affairs	8
Housing and Urban-Rural Development Department of Jiangsu Province	5
Yunnan Provincial General Station of Rural Energy Management	5
College of Marxism, South China Agricultural University	4
Contemporary Guizhou	4

3.2. Research Hotspots

Keywords are the core and essence of research, involving a high concentration and refinement of the research topic and research area. From the bibliometric point of view, the number of high-frequency co-cited keywords is directly proportional to the cluster size, and they can directly reflect the hotspots in the research field of a certain discipline [19]. Therefore, the frequency of keywords and related links provide a clearer understanding of the research hotspots in the field of human settlement environment improvement. In the CiteSpace keyword mapping, the larger the node represents the higher frequency of the keyword (Figure 5). The research literature on "human settlement environment, rural revitalization, environmental remediation, remediation work, and rural areas" appears more than 100 times in Chinese human settlement environment improvement (Table 2). International research on human settlement environment improvement mainly focuses on "health, physical, activity" and other keywords, which have become a research hotspot in the past decade (Table 2). According to the relevant literature on human settlement environment improvement in recent years, the research hotspots in the field of human settlement environment improvement can be summarized into the following three aspects.

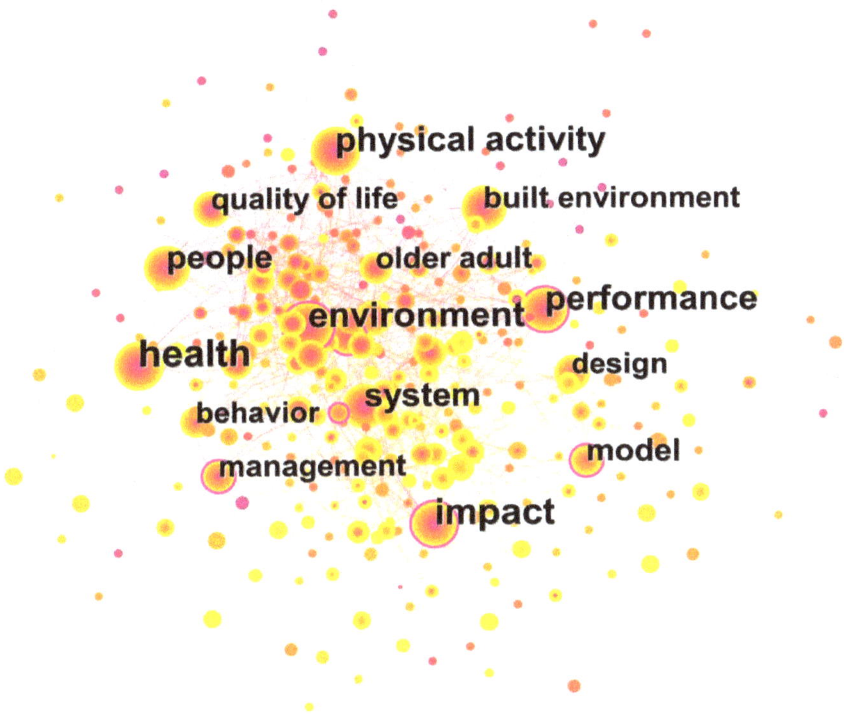

Figure 5. Keyword map of research on foreign human settlement environment improvement.

Table 2. Ranking of the top ten high-frequency keywords in human settlement improvement research.

Source	Sort	Keywords	Frequency
Chinese	1	Living environment	333
	2	Rural revitalization	305
	3	Environmental remediation	126
	4	Remediation work	112
	5	Rural	100
	6	Toilet revolution	59
	7	Remediation	57
	8	Comprehensive remediation	41
	9	Beautiful Country	41
	10	Village planning	40
International	1	Health	119
	2	Physical activity	106
	3	Performance	87
	4	Impact	86
	5	Environment	77
	6	Management	65
	7	Model	64
	8	People	63
	9	Quality of life	58
	10	Built environment	54

3.2.1. Concept and Classification of Human Settlement Environment

The concept of human settlement environment was initially put forth by China's human settlement environment science leader Wu Liang Yong [1], in the Doxiadis "human settlement" and other research based on the proposed human settlement environment, which states that the human settlement living space and survival base are an important place for people to use and transform nature, and are closely related to human production and development. He believed that the human settlement environment is a large system with rich content, including five subsystems—natural system, human system, housing system, social system, and support system—dividing living into five levels, namely, building, community (village), city, region, and global, concerning the type and scale of human settlement [1]. From the perspective of urban planning, Ren explains the human settlement environment as a sustainable environment that satisfies human living conditions while ensuring harmony between housing and nature [20]. From the perspective of architecture, Chen believes that the human settlement environment is the space where people gather and live, and it is not only the basis for human survival in nature, but also an important place for human transformation and use of nature [21]. When creating a building for humans to live in, the different requirements of living, working, and public services are taken into consideration. Many scholars define the human settlement environment from the perspective of different disciplines such as geography, ecology, and sociology [22–24].

The classification of human settlement environment has been greatly debated in academia. Li et al. defined the human settlement environment from two perspectives: narrow and broad [25]. The narrow sense of human settlement environment defines it as a space for human activities and focuses on the space or entity; the broad sense of human settlement environment is the sum of various material and immaterial elements surrounding the human subject. He believed that the object of scientific research on the human settlement environment is the broad sense of living. Tan et al. divided it into urban human settlement environment and rural human settlement environment in terms of geographical space [26]; Zhang divided it into macro, meso, and micro living from the perspective of spatial scale [27], while Ning and Zha pointed out that the hard environment is the carrier of the soft environment, and the livability of the soft environment is the value orientation of the hard environment [28].

In general, the human settlement environment is a complex and huge system in which the natural and artificial environments are interwoven and integrated [29].

3.2.2. Study on the Path of Human Settlement Environment Improvement

International human settlement environment improvement started earlier, and has gone through the stage of large-scale reconstruction and emphasis on quantity after the war to the stage of both quantity and quality, and then to the stage of emphasis on quality, especially environmental quality after the 1970s, and has made a series of progressive moves [30]. The UK focuses on the sustainable development of satellite cities, villages, and towns, involving various aspects such as resource consumption, environmental protection, social equity, public participation and decision-making, economic activities, and comprehensive evaluation, which are used to guide the practice of human settlement environment [31]. Germany has improved the human settlement environment through land consolidation, village renewal, nature conservation, urban–rural integration, and community creation [32]. In the Netherlands, rural areas are built from the persistence of traditional rural landscapes, the preservation of regional structural green areas, and the construction of rural recreational areas [33]. Japan is taking the opportunity of "one village, one product" to explore regional characteristics and advantages, promoting rural development through product economy and enhancing the internal vitality of rural areas [34]. Korea's "New Village Movement" aims to improve the quality of housing for rural households, and promotes rural development by using various organizations such as civil society organizations and agricultural technology, with government support and social follow-up as the development method [35]. In addition, Singapore, India, and other countries have also started to improve the quality of housing. In addition, Singapore and India have also started to improve their urban living, with Singapore focusing on housing, public services, and road construction [36].

Improvement of the human settlement environment in China has been further researched mainly based on foreign governance experience, and many scholars have mainly discussed it in macro and micro scales. From the macroscopic scale, the research on human settlement environment improvement mainly focuses on urban clusters and provincial levels. Taking six western provinces as examples, Su et al., based on collective action theory and planned behavior theory, proposed that various forms of rural human settlement environment publicity and education activities should be carried out to raise farmers' awareness of environmental protection [37]. Taking Henan Province as an example, Lu et al., by combing through the relevant literature, argued that environmental pollution in Henan Province is mainly caused by inappropriate emissions of rural household waste, rural farming manure, and agricultural production chemicals, and proposed governance initiatives such as further increasing financial investment in pollution control, strengthening infrastructure construction, enhancing farmers' environmental awareness, and strengthening the legal management of rural environmental protection [38].

From the microscopic scale, ethnic minority areas and villages become the key areas for human settlement environment improvement. Sun took the Tibetan ethnic region as the research object, conducting a household survey on 721 households, and using the generalized maximum entropy Logit analysis method, analyzed the willingness of rural households in ethnic minority regions to participate in rural human settlement environment improvement from the perspectives of household endowment and outworking; he proposed enhancing the willingness of Tibetan households to participate in the human settlement environment improvement in four aspects: environmental publicity, policy subsidies, technical training, and village rules and regulations [39]. Wang et al. used 296 villages in Chongqing as an example to investigate the situation of agricultural waste, infrastructure, farm buildings, and sanitary toilets and proposed that the organic combination of top–down decision-making by the government and bottom–up participation by social organizations and farmers should be promoted and that feasible paths should be taken to strengthen the sanitary facilities, promote the change of customs, and enhance environmental awareness at the farmer level [40].

3.2.3. Study on the Evaluation Method of Human Settlement Environment

Human settlement environment evaluation is an important way to understand the condition of human settlement environment and assess the effectiveness of management. The World Health Organization (WHO) was the first to propose the basic concept of healthy human settlement environment, including the four elements of safety, health, convenience, and comfort of living, and to conduct evaluation studies of the human settlement environment on this basis [5]. In 1964, the U.S. Commission on Housing and Health proposed the "Basic Principles of Healthy Human Settlement Environment", including six principles: conditions for disaster prevention and control, environmental conditions for human physiology, physiological requirements for living, conditions for disease prevention and infection prevention, spiritual satisfaction, and economic satisfaction. Since then, UN Living has focused on the areas of housing and social services, urban management, environment and infrastructure, and monitoring and information since 1989 [30]. Asami evaluated the human settlement environment in detail in terms of indicators of safety, health, convenience, comfort, and sustainability evaluation [41].

The evaluation of the human settlement environment in China is based on existing methods in foreign countries, mostly quantitative methods for in-depth research, which can be roughly divided into livability evaluation, sustainability evaluation, subjective and objective evaluation, satisfaction evaluation, and construction quality evaluation. Livability evaluation is mainly based on the human-oriented perspective of building a livable environment that is suitable for construction, production, and life. Zhu et al. constructed an index system from seven aspects of the urban landscape, environment, facilities, safety, transportation, economy, and management, and used PCA, GRA, and D-RSR to evaluate the livability of Xinyang City comprehensively. Sustainable development evaluation refers to the study of the human settlement environment from three aspects: production, living, and ecology [42]. Zhang et al. [43] established an index system for evaluating the sustainable development level of the rural human settlement environment from the perspective of natural and social resources, and used the development level (sustainable development potential) and coordination index (sustainable development trend) to construct a measurement model to explore the spatial variability and autocorrelation of human settlement environment. The comprehensive subjective and objective evaluation needs to consider the subjective and objective factors and combine the subjective feelings and objective situation to make a comprehensive evaluation of the rural human settlement environment. Xiao [44] put forward the four target layers of living facility environment, living ecological environment, living social environment, and living economic environment, and on this basis, five system layers of the rural human settlement environment evaluation index system are established. Satisfaction evaluation is to analyze the satisfaction degree of residents in the process of rural human settlement environment construction, and to evaluate the rural human settlement environment from the perspective of the main body of villagers. Ye et al. [45] used factor analysis to establish six index systems of living conditions, ecological environment, infrastructure, civilized countryside, effective governance, and affluent living, and made a comprehensive evaluation of villagers' satisfaction with the rural human settlement environment in the Chao Hu Lake area. Construction quality evaluation is a quantitative and qualitative description of the overall strengths and weaknesses of rural construction elements. Liu et al. constructed the index system from five dimensions of production, life, appearance, customs, and governance, using hierarchical analysis and entropy value method to calculate the weights, and used a fuzzy comprehensive evaluation method to establish the evaluation model [46].

3.3. Research Frontiers

Keyword co-occurrence refers to the terms that appear more frequently in certain years, and by analyzing the years in which the emergent keywords appear, it can reflect the development trend and frontier of a certain research field over a while. Using the CiteSpace emergent keyword's function, the top 20 emergent keywords in the field of human settlement environment improvement research in the past ten years were plotted (Tables 3 and 4). In the keyword emergent map, "Strength" represents the emergent strength of the keywords, and "Begin" and "End" represent the beginning and end of the emergent words' year.

Table 3. The top 20 emergent keywords of Chinese human settlement environment improvement research.

Keywords	Year	Strength	Begin	End
New Rural	2012	2.69	2012	2013
Village Improvement	2013	2.94	2013	2014
Construction	2014	1.8	2014	2016
Livable Countryside	2015	4.66	2015	2019
Ministry of Construction	2015	3.9	2015	2018
Comprehensive remediation	2016	4.24	2016	2018
Specialized actions	2017	1.87	2017	2018
State Council	2018	7.83	2018	2019
Solid promotion	2018	4.78	2018	2019
Jilin Province	2018	3.54	2018	2019
Beautiful China	2018	2.24	2018	2019
Informal	2018	2.24	2018	2019
Yongji County	2018	2.24	2018	2019
Yu Xinrong	2018	1.79	2018	2019
Constructed Village	2018	1.79	2018	2019
Chongqing	2018	1.79	2018	2019
Rural Ecology	2018	1.75	2018	2019
Territorial Tourism	2019	1.75	2019	2020
Harmless	2020	2.32	2020	2022
Waste Management	2020	1.9	2020	2022

Table 4. The top 20 emerging keywords in the study of international human settlement environment improvement.

Keywords	Year	Strength	Begin	End
Walking	2012	4.35	2012	2016
Care	2012	2.84	2012	2014
Associations	2013	4.74	2013	2016
Dementia	2013	2.95	2013	2015
Patterns	2012	3.48	2014	2015
Motion	2014	2.94	2014	2015
Activities of daily living	2015	3.66	2015	2017
Strategy	2016	4.51	2016	2018
Living cells	2013	3.43	2016	2018
Program	2017	4.4	2017	2019
Obesity	2015	4.37	2017	2019
Prevention	2013	3.96	2017	2019
Survival	2017	3.1	2017	2019
Model	2014	3.51	2018	2019
Mechanisms	2015	3.33	2019	2020
Pollution	2019	2.95	2019	2022
City	2020	4.74	2020	2022
Air pollution	2014	4.26	2020	2022
Quality	2012	3.31	2020	2022
Green space	2020	2.94	2020	2022

Among the top 20 emergent keywords of Chinese human settlement environment improvement research in CNKI, the emergent intensity of "solid promotion" is the largest, followed by "livable countryside", and "comprehensive"; the next keywords were "livable countryside", "comprehensive treatment", and "Ministry of Construction". In the keywords map of foreign human settlement environment improvement research from WOS, "city, associations" has the highest emergent intensity, followed by "strategy" and "obesity", "From 2020 to the present, "waste management, harmlessness, city, air pollution, quality, greenspace", which have become the frontier issues in the field of human settlement environment improvement in recent years.

3.3.1. Waste Management

Chinese waste management is an important part of improving the human settlement environment, and is of great significance in promoting green development in rural areas and realizing good ecological governance in villages [47]. According to the survey results of China's Ministry of Housing and Construction, 41 counties (cities and districts) nationwide have become "2020 demonstration counties for rural waste classification and resource utilization", and more than 90% of villages have achieved effective waste collection and disposal, but only 50% of administrative villages have carried out waste classification [48]. This shows that farmers' awareness of environmental protection is still relatively weak, the traditional thinking of rural residents is difficult to change, and there is still the phenomenon of throwing away garbage at will [49,50]. Therefore, in the future, the corresponding management units should enhance the awareness of environmental protection among villagers, carry out garbage classification publicity well, and improve the institutional mechanism of garbage classification, to improve the human settlement environment of rural residents [51–53].

3.3.2. Green Space

Promoting the green development of agriculture and rural areas is an inevitable requirement for implementing the new development concept and a major initiative to accelerate the modernization of agriculture and rural areas and promote sustainable development [54]. Since the implementation of the rural revitalization strategy, the special actions carried out around the country such as village appearance improvement and sewage treatment have achieved remarkable results in improving the human settlement environment and comprehensive management of outstanding environmental problems in rural areas, and have effectively promoted the development of the rural economy [55]. However, the process of rural environmental management is complex and complicated. However, the process of rural environmental management is complex and long-lasting, and the green transformation of agriculture and rural areas cannot be achieved overnight. It is necessary to actively promote green agricultural production methods, strengthen the protection and utilization of agricultural resources, continuously improve the rural human settlement environment, and promote green space for sustainable development [56,57].

4. Discussion

Through combing the hotspots and frontiers of research on human settlement environment improvement, there are still some problems and shortcomings in the current research on human settlement environment improvement, with insufficient research breadth and depth, incomplete research methods, unsound evaluation index system, and no cross-fertilization between disciplines. Based on the above analysis, the research potential of China's human settlement environment improvement should be further explored from the following aspects in the future, as given below.

4.1. Strengthen the Breadth and Depth of Research

As mentioned above, although the research on human settlements improvement involves multiple provinces and regions, there is still the problem that appropriate governance paths cannot be formulated according to local actual conditions (keyword co-occurrence). In addition, scholars mainly focus on specific issues such as rural garbage treatment and sewage treatment, while there is little research on new rural construction and green space (the top 20 emergent keywords). The research on human settlement environment is mainly carried out in urban and rural areas, and scholars mostly focus on rural human settlement environments, with many studies on waste classification, sewage treatment, toilet revolution, village appearance, etc., but fewer studies on new rural construction and management, and green development of rural environment [58]. There is less research on new rural construction and management and green development of the rural environment. At present, although the research on the human settlement environment involves several provinces and regions, there are still problems such as the inability to develop appropriate governance paths according to local conditions. Therefore, the research on the human settlement environment does not only stop at the governance work, but should focus on the organic connection of county and township territorial spatial planning in the new era and then form a new rural spatial governance system [59,60].

4.2. *Improve the Evaluation Index System*

In the current research on human settlement environment evaluation, researchers mostly focus on quantitative indicators and ignore qualitative indicators [61]. The concept boundary of indicators is relatively vague, which leads to a large difference in the weighting of the same indicator in different literature, resulting in controversial evaluation results. This has led to controversial assessment results. In the future, an evaluation system and model combining objective reality and subjective cognition should be adopted to improve the technical methods of human settlement environment evaluation, strengthen the definition and specification of evaluation indicators from the level of national regulations, improve the accuracy of research results and comparability of research results, and optimize the evaluation indicator system [62,63].

4.3. *Innovative Research Methods*

Most of the current research is limited to a single discipline, lacking interdisciplinary cooperation (keyword co-occurrence chart). The lack of knowledge integration among ecology, economics, sociology, and other fields limits the comprehensiveness and applicability of research results. So far, most of the studies on human settlement environment improvement have adopted a combination of case and policy research, and proposed corresponding governance paths through pilot studies of specific provinces, urban clusters, minority areas, and counties. Digital governance is one of the future development directions of the human settlement environment, and future research can try to use more modern research tools, such as 3S technology, big data analysis, remote sensing images, and other emerging technologies [64–66].

4.4. *Build Multidisciplinary Cross-Fertilization*

Human settlement environment improvement requires integrated analysis from different disciplinary fields such as geography, architecture, sociology, economics, psychology, pathology, and pharmacology. For example, geography's unique GIS geographic information science tools can provide strong technical support for architecture's living renewal practice, and big data platforms and mathematical and statistical methods can be an expansion of quantitative research tools [67]. Combining the macroscopic perspective of economics and management with the microscopic perspective of sociology and architecture, and the disciplinary knowledge of ecology and urban and rural planning will promote a qualitative leap in human settlement environment improvement [68]. Therefore, future research on the human settlement environment is yet to be realized. Therefore, future human settlement environment research needs to build cross-fertilization among disciplines, and provide new paths for future human settlement environment research by drawing on research methods, ideas, and technical paths of related disciplines while strengthening the characteristics of our disciplines.

5. Conclusions and Research Prospect

5.1. Conclusions

Based on the CiteSpace visualization tool and bibliometric methods, the paper analyzed the research progress of human settlement environment improvement. First, the number of publications in this field has been increasing, and various countries and issuing institutions have attached importance to the research on human settlement environment improvement. Secondly, according to the keyword co-occurrence analysis, it can be seen that the current research mainly focuses on three aspects: the concept and classification of the human settlement environment, the path of human settlement environment improvement, and the study of human settlement environment evaluation methods. Finally, the analysis of the first 20 emergent keywords shows that waste management and green space have gradually become the research frontiers in the field of human settlement environment.

Based on the findings of this study, future improvement of the human living environment should be further strengthened and expanded on the existing base. First, it is necessary to extend the breadth and depth of research, especially in the exploration of personalized solutions in specific areas such as rural areas and ethnic minority settlements. Second, the evaluation index systems should be improved, combining emerging technologies such as big data and artificial intelligence, and enhancing the scientificity and accuracy of the assessment methodology to adapt to the ever-changing needs. Furthermore, cross-disciplinary integration and cross-cooperation among engineering, social sciences, and ecology should be promoted and encouraged to provide a comprehensive perspective for solving complex human living environment problems. In addition, policy support and practical guidance should be strengthened. Multiple fields provide a comprehensive perspective for solving complex human settlement environment problems; furthermore, they strengthen policy support and practical guidance to ensure that theoretical research results can be effectively transformed into practical policy measures, and to promote the active participation of local governments and all sectors of society in the improvement of the human settlement environment. Through these efforts, the human living environment will not only meet people's basic needs, but also promote social equity and harmony as well as ecological protection. At the same time, continued attention to and integration of the latest scientific research results and technological advances around the world will help human society to better meet future challenges and create a better living environment.

This thesis also has certain limitations. This study mainly relies on the CNKI and WOF databases, which may have overlooked other important but unlisted resources, resulting in an incomplete understanding of the state of research on a global scale. As the period set is from 2012 to 2022, it fails to cover the latest research advances, especially those in rapidly developing emerging fields. Although CiteSpace software (v6.2.r2) provides powerful data analysis capabilities, it has limited ability to handle unstructured data, which may affect the validity of some deeper analyses. In summary, while this study provides valuable insights into understanding the current status and trends in the field of human settlement improvement, future work will need to overcome these limitations to more accurately capture the latest developments in the field and move it forward.

5.2. Research Prospect

In 2022, the Chinese government issued the "Five Year Action Plan for Improving Rural Human Settlements". Therefore, since 2022, the focus of China's human settlements research has been different from before, which is our main consideration in selecting the research period. In the future, bibliometric analysis of research on human settlements improvement will continue to conduct similar research starting from 2023. In terms of data acquisition, more source databases should be considered in the future to enhance the comprehensiveness of text retrieval.

Author Contributions: C.L.: Conceptualization and Writing—original draft; J.M.: Data curation; Project administration; Investigation; X.Z.: Methodology; Q.L.: Software; Writing—review and editing. All authors have read and agreed to the published version of the manuscript.

Funding: This work was supported by General Project of Humanities and Social Sciences Research of the Ministry of Education (23YJC630076), Shaanxi Natural Science Basic Research Program (grant number 2023-JC-QN-0803), Open Fund Funding Projects of Technology Innovation Center for Land Engineering and Human Settlements, Shaanxi Land Engineering Construction Group Co., Ltd. and Xi'an Jiaotong University (grant number 2024WHZ2056) and S&T Program of Baoding (No. 2363P022).

Data Availability Statement: The original contributions presented in this study are included in the article. Further inquiries can be directed to the corresponding author.

Acknowledgments: The authors sincerely thank the editor and anonymous reviewers for their valuable comments and suggestions to improve the quality of this paper.

Conflicts of Interest: Author Chenxi Li was employed by Shaanxi Land Engineering Construction Group Co., Ltd. The remaining author declares that the research was conducted in the absence of any commercial or financial relationships that could be construed as a potential conflict of interest.

References

1. Wu, L.Y. *Introduction to Sciences of Human Settlements*; China Architecture & Building Press: Beijing, China, 2001; pp. 97–112.
2. Zhang, Z.; Fang, B.; Li, X.; Wang, Y. Evaluation of Human Settlement Environment and Identification of Development Barriers Based on the Ecological Niche Theory: A Case Study of Northern Shaanxi, China. *Int. J. Environ. Res. Public Health* **2023**, *20*, 1772. [CrossRef] [PubMed]
3. Doxiadis, C.A. Action for human settlements. *Ekistics* **1975**, *40*, 405–448.
4. Carter, L.J. Dade County: The politics of managing urban growth. *Science* **1976**, *192*, 982–985. [CrossRef]
5. Mao, Q.Z. Theory and Practice of the Science of Human Settlements in China. *Urban Plan. Int.* **2019**, *34*, 54–63. (In Chinese) [CrossRef]
6. Zhou, L.; Yu, C.; He, P. Small village, big strategy: Practice in Jiangsu to promote integrated urban-rural development. *City Plan. Rev.* **2013**, *11*, 20–27.
7. Ye, G.; Luo, S.; Chen, Y.; Wang, Y. Study on Countermeasures of Rural Living Environment Improvement. In Proceedings of the 23rd International Symposium on Advancement of Construction Management and Real Estate (CRIOCM 2021), Beijing, China, 20–22 November 2021; Springer: Singapore, 2021; pp. 1145–1156.
8. Song, X.F.; Chi, P.J. Comparative study of the data analysis results by Vosviewer and Citespace. *Inf. Sci.* **2016**, *34*, 108–112, 146.
9. Zhang, X.; Su, N.; Yang, H.G.; Fang, X.K. Knowledge mapping of research on international E-government during 2000–2011 based on Citespace and VOSviewer. *J. Intell.* **2012**, *31*, 51–57.
10. Yuliastuti, N.; Wahyono, H.; Syafrudin, S.; Sariffuddin, S. Dimensions of community and local institutions' support: Towards an eco-village kelurahan in Indonesia. *Sustainability* **2017**, *9*, 245. [CrossRef]
11. Liaghat, M.; Shahabi, H.; Deilami, B.R.; Ardabili, F.S.; Seyedi, S.N.; Badri, H. A multi-criteria evaluation using the analytic hierarchy process technique to analyze coastal tourism sites. *Apcbee Procedia* **2013**, *5*, 479–485. [CrossRef]
12. Shcherbina, E.; Gorbenkova, E. Factors influencing the rural settlement development. *IFAC—PapersOnLine* **2019**, *52*, 231–235. [CrossRef]
13. Rigacci, L.N.; Giorgi, A.D.N.; Vilches, C.S.; Ossana, N.A.; Salibián, A. Effect of a reservoir in the water quality of the Reconquista River, Buenos Aires, Argentina. *Environ. Monit. Assess.* **2013**, *185*, 9161–9168. [CrossRef] [PubMed]
14. Bibri, S.E.; Krogstie, J. ICT of the new wave of computing for sustainable urban forms: Their big data and context-aware augmented typologies and design concepts. *Sustain. Cities Soc.* **2017**, *32*, 449–474. [CrossRef]
15. Trianni, G.; Angiuli, E.; Pesaresi, M. Statistical analysis of anisotropic rotation-invariant textural measurements of human settlements from multitemporal SAR data. In Proceedings of the 2011 Joint Urban Remote Sensing Event, Munich, Germany, 11–13 April 2011; IEEE: New York, NY, USA, 2011; pp. 117–120.
16. Harsányiová, M.; Prokop, P. Living condition, weight loss and cognitive decline among people with dementia. *Nurs. Open* **2018**, *5*, 275–284. [CrossRef] [PubMed]

17. Misinde, C. "An Intrinsic characteristics and Value of Poverty Indicators": A New Method for Deriving Child Living Condition Scores and Poverty, in Uganda. *Child. Indic. Res.* **2017**, *10*, 141–170. [CrossRef]
18. Anaraki, N.R.; Boostani, D. Living in and living out: A qualitative study of incarcerated mothers' narratives of their children's living condition. *Qual. Quant.* **2014**, *48*, 3093–3107. [CrossRef]
19. Nan, M.; Chen, J. Research Progress, Hotspots and Trends of Land Use under the Background of Ecological Civilization in China: Visual Analysis Based on the CNKI Database. *Sustainability* **2023**, *15*, 249. [CrossRef]
20. Ren, X.B. Exploration of residential building planning and design and human living environment. *Intell. City* **2018**, *4*, 20–21. (In Chinese)
21. Chen, D.E. Application of green building materials in habitat environment. *Inter. Archit. China* **2022**, 66–68. (In Chinese)
22. Zhang, Z.; Xiao, R.; Shortridge, A.; Wu, J. Spatial point pattern analysis of human settlements and geographical associations in eastern coastal China—A case study. *Int. J. Environ. Res. Public Health* **2014**, *11*, 2818–2833. [CrossRef]
23. Xie, T.; Liu, X.; Nie, P. Study on Spatial–Temporal Patterns and Factors Influencing Human Settlement Quality in Beijing. *Sustainability* **2022**, *14*, 3752. [CrossRef]
24. Yang, J.; Li, X.M.; Li, Y.H.; Sun, S.Z.; Wang, F. Assessment on spatial differences of human settlement environment in communities based on DPSIRM model: The case study of Dalian. *Geogr. Res-Aust.* **2012**, *31*, 135–143.
25. Li, H.; Xu, R.; Gao, Z.; Peng, B.Z. Quality Evaluation of Human settlements in a City Scale—A case study on Nanjing City. *Hum. Geogr.* **2005**, *20*, 1–5.
26. Tan, M.J.; Yan, L.J.; Li, H.B. Niche Fitness Model for the Quantitative Evaluation of the Environment Quality of the Urban Human Settlement. *Bull. Sci. Technol.* **2007**, *23*, 439–445. (In Chinese)
27. Zhang, R.K. Evaluation of the Current Situation of Urban Habitat in Changsha. *Urban Probl.* **2004**, 39–41. (In Chinese)
28. Ning, Y.M.; Zha, C.Q. The Study of Evaluation and Optimization for Human Settlement in the Metropolitan Areas: The case of Shanghai. *China City Plan. Rev.* **1999**, *23*, 15–20. (In Chinese)
29. Zhang, W. Study on intrinsic meanings of the livable city and the evaluation system of livable city. *Urban Plan. Forum* **2007**, *169*, 30–34. (In Chinese)
30. Deng, L.; Gu, J.T. Reviews and reflections on the index of residential environment. *J. Huaihua Univ.* **2011**, *30*, 35–38.
31. Minx, J.; Baiocchi, G.; Wiedmann, T.; Barrett, J.; Creutzig, F.; Feng, K.S.; Förster, M.; Pichler, P.; Weisz, H.; Hubacek, K. Carbon footprints of cities and other human settlements in the UK. *Environ. Res. Lett.* **2013**, *8*, 035039. [CrossRef]
32. Qian, L.; Gan, L.; Zhang, L. The functional restructuring and endogenous development of rural areas in Germany. *Urban Plan. Int.* **2020**, *35*, 6–13.
33. Wang, N. Modern Deduction of Traditional Farmland: The Construction Strategy of Randstad Rural Area in the Netherland. *Urban Plan. Int.* **2018**, *33*, 118–124. [CrossRef]
34. Zhou, W.H. The Transformation Practice of Japan's Rural Revitalization Path: An Examination of the Development Logic of the "One Village, One Product" Industrial Movement. *Jpn. Stud.* **2019**, 11–22. (In Chinese)
35. Nan, G.Y.; Pang, J.C. The New Village Movement and Life Changes in Korea. *Folk. Cult. Forum* **2019**, 26–37. (In Chinese)
36. Tan, P.Y.; bin Abdul Hamid, A.R. Urban ecological research in Singapore and its relevance to the advancement of urban ecology and sustainability. *Landsc. Urban Plan.* **2014**, *125*, 271–289. [CrossRef]
37. Su, Y.; Qiu, Y.; Xuan, Y.; Shu, Q.; Li, Z. A Configuration Study on Rural Residents' Willingness to Participate in Improving the Rural Living Environment in Less-Developed Areas—Evidence from Six Provinces of Western China. *Front. Environ. Sci-Switz.* **2002**, *10*, 2571. [CrossRef]
38. Lu, S.Q.; Liu, Q.; Zheng, F.; Zheng, X.R.; Liu, G. Rural Environmental Pollution and Countermeasures in the Construction of New Countryside: A Case of Henan Province. *J. Agric.* **2019**, *9*, 92. (In Chinese)
39. Sun, Q. Study on the influencing factors of the willingness of Tibetan farmer's households to participate in the improvement of rural living environment. *J. Ecol. Rural. Environ.* **2019**, *35*, 976–985.
40. Wang, Y.; Linna, S.; Zhu, L. Current situation and improvement framework of rural living environment in poverty-free areas: A county study from Chongqing City, China. *J. Agric. Resour. Environ.* **2022**, *39*, 417.
41. Asami, Y. Residential environment: Methods and theory for evaluation. *J. Womens Health* **2001**.
42. Zhu, W.; Cheng, Q.X.; Zhang, X.Y.; Xu, C.X.; Zhang, Z.K.; Yang, Y.L. Livability Evaluation of Xinyang City Based on PCA/GRA/D-RSR. *J. Xinyang Norm. Univ.* **2021**, *34*, 422–429.
43. Zhang, Y.N.; Zhu, H.Y.; OuYang, H.X.; Song, L.L. Sustainable Development Level and Spatial Pattern of Rural Living Environment In Yangtze River Economic Zone. *Resour. Ind.* **2022**, *24*, 42.
44. Xiao, Z.C. "Beautiful Countryside" Habitat Environment Comprehensive Value Index System Construction. *China Hous. Facil.* **2020**, 14–15. (In Chinese)
45. Ye, L.; Wu, Z.; Wang, T.; Ding, K.; Chen, Y. Villagers' Satisfaction Evaluation System of Rural Human Settlement Construction: Empirical Study of Suzhou in China's Rapid Urbanization Area. *Int. J. Environ. Res. Public Health* **2022**, *19*, 11472. [CrossRef]
46. Liu, B.T.; Zhang, J.R.; Yang, C.; Dong, J.H. Quality assessment on rural construction of human settlements in loess plateau. *J. Lanzhou Univ. Technol.* **2023**, *49*, 117.

47. Xu, D.; Qing, C.; Chen, Y.; He, J.; Zhang, F. Sustainable Development of Rural Human Settlements in the Information Age: Can Internet Use Drive Farmers to Participate in Garbage Classification? *Agriculture* **2023**, *13*, 846. [CrossRef]
48. Jiang, P.; Hu, R. Problems, causes and countermeasures of rural household waste separation. *Acad. Exch.* **2021**, *2*, 146–156. (In Chinese)
49. Li, X.; Bi, F.; Han, Z.; Qin, Y.; Wang, H.; Wu, W. Garbage source classification performance, impact factor, and management strategy in rural areas of China: A case study in Hangzhou. *Waste Manag.* **2019**, *89*, 313–321. [CrossRef]
50. Lu, S.; Zhou, Z.; Lu, Y. Rural Residents' Perceptions, Attitudes, and Environmentally Responsible Behaviors towards Garbage Exchange Supermarkets: An Example from Huangshan City in China. *Sustainability* **2022**, *14*, 8577. [CrossRef]
51. Shen, Y.; Wang, C. Optimisation of Garbage Bin Layout in Rural Infrastructure for Promoting the Renovation of Rural Human Settlements: Case Study of Yuding Village in China. *Int. J. Environ. Res. Public Health* **2021**, *18*, 11633. [CrossRef] [PubMed]
52. Cao, Y.Q.; Heng, X.; Zhang, X.N.; Qu, M. Influence of social capital on rural household garbage sorting and recycling behavior: The moderating effect of class identity. *Waste Manag.* **2023**, *158*, 84–92.
53. Chen, J.F.; Meng, C.G. Dilemma and Improvement Path of Rural Domestic Waste Treatment: A Case Study of Weiyuan County in Gansu Province. *J. Yunnan Agric. Univ.* **2022**, *16*, 71–77.
54. Chen, L.M.; Zheng, Q.Y.; Zhou, N. Green Development in China's Agriculture and Rural Areas: An Evaluation Based on TOPSIS and Hindrance Model. *J. Hebei Agric. Univ.* **2022**, *24*, 106–114. (In Chinese)
55. Zhong, S.; Li, X.; Zhang, Y.L. Research on the Connotation Interpretation, Realization Mechanism and Implementation Path of Rural Revitalization Led by the Concept of Green Development. *China Gard.* **2022**, *38*, 35–39. (In Chinese)
56. Zhang, J.; Feng, X.; Shi, W.; Cui, J.; Peng, J.; Lei, L.; Zhang, J.; Astell-Burt, T.; Jiang, Y.; Ma, J. Health promoting green infrastructure associated with green space visitation. *Urban Urban Green.* **2021**, *64*, 127237. [CrossRef]
57. Zhang, X.; Huang, X.; Li, J. The Evolution of Green Development, Spatial Differentiation Pattern and Its Influencing Factors in Characteristic Chinese Towns. *Sustainability* **2023**, *15*, 5079. [CrossRef]
58. Wei, Y.; Zhang, A.; Ma, Y. A Bibliometric Review of Rural Living Environment Improvement Research in China Based on CNKI Database: 1992–2022. *Sustainability* **2023**, *15*, 6561. [CrossRef]
59. Wang, Y.; Zhang, Y.; Yang, G.; Cheng, X.; Wang, J.; Xu, B. Knowledge Mapping Analysis of the Study of Rural Landscape Ecosystem Services. *Buildings* **2022**, *12*, 1517. [CrossRef]
60. Wang, Y.; Lu, Y.Q.; Zhu, Y.M.; Ding, Z.S. General Characteristics and Knowledge Maps Visualization Analysis of Human Settlement Research in China. *Trop. Geogr.* **2020**, *40*, 498–514. (In Chinese)
61. Li, J.; Jiang, C.; Xing, Z. Multi-dimensional influence measurement of urbanization on the quality of natural living environment in China. *Environ. Dev. Sustain.* **2021**, *23*, 12151–12168. [CrossRef]
62. Zhou, Q.J.; Chen, L.; Cui, P. A Habitat Evaluation Method from the Perspective of Residents' Experience. *Fujian Comput.* **2018**, *34*, 105. (In Chinese)
63. Wang, Q.B.; Xia, J.C.; Yu, L.Z.; Zhou, S.T.; Dong, R.J. The Objective Evaluation Index System of Rural Human Settlements' Environment Quality. *J. Agric.* **2020**, *10*, 71.
64. Chen, W.; Zhu, K.; Wu, Q.; Cai, Y.; Lu, Y.; Wei, J. Adaptability evaluation of human settlements in Chengdu based on 3S technology. *Environ. Sci. Pollut. Res.* **2022**, *29*, 5988–5999. [CrossRef]
65. Huang, G. Research on e-commerce security in data and cloud computing environment. In Proceedings of the 2022 International Conference on Artificial Intelligence in Everything (AIE), Lefkosa, Cyprus, 2–4 August 2022; IEEE: New York, NY, USA, 2022; pp. 482–487.
66. Chen, T.; Wu, W.B.; He, J.J.; Qiao, Y.X.; Liu, F.; Wen, Q. Urban human settlements monitoring model and its application based on multi-source spatial data fusion. *Acta Ecol. Sin.* **2019**, *39*, 1300–1308.
67. Zhu, M.; Wang, D.G. Comparison of human settlement research in geography and architecture from the perspective of a theory tree. *J. Geogr.* **2022**, *77*, 795–817. (In Chinese)
68. Ye, L.; Wu, Z.H.; Hu, X.M. A review of rural habitat evaluation research in China from a multidisciplinary perspective. *Huazhong Archit.* **2022**, *40*, 141–145. (In Chinese)

buildings

MDPI

Article

Classifying and Quantifying Resident Engagement in the Sponge-Style Old Community Renewal of China: An Analysis Using the Analytic Network Process–Preference Ranking Organization Method for Enrichment Evaluations II Method

Tiantian Gu [1,*], Muhan Xie [1], Enyang Hao [2], Yongchao Wang [1] and Shenghua Zhou [3]

1 School of Mechanics and Civil Engineering, China University of Mining and Technology, Xuzhou 221116, China; ts22030239p31@cumt.edu.cn (M.X.); ts22030232p31@cumt.edu.cn (Y.W.)
2 Department of Civil and Environmental Engineering, University of Auckland, Auckland 1010, New Zealand; ehao847@aucklanduni.ac.nz
3 School of Civil Engineering, Southeast University, Nanjing 211102, China; shenghua@connect.hku.hk
* Correspondence: 6072@cumt.edu.cn; Tel.: +86-15850762331

Abstract: Resident engagement is highlighted as a crucial aspect that directly impacts the residents' well-being of life in the process of sponge-style old community renewal (SOCR). However, there is a lack of in-depth study on exploring the classification of resident engagement clusters in the SOCR and developing a comprehensive evaluation system to measure this engagement. Therefore, this paper categorizes resident engagement into distinct clusters and develops an evaluation method for quantifying resident engagement in the SOCR. First, this study identified seven types of engagement clusters in the SOCR drawing from educational psychology and Arnstein's citizen engagement ladder theory. Second, an evaluation indicator system with 5 primary indicators and 21 secondary indicators was developed through a literature review and expert interviews. Third, a hybrid evaluation method combining the Analytic Network Process (ANP) and Preference Ranking Organization Method for Enrichment Evaluations II (PROMETHEE II) was established to effectively quantify resident engagement in the SOCR. Finally, 16 typical communities from five Chinese cities were selected as study areas to examine resident engagement in the SOCR. The results demonstrated that the full engagement cluster achieved the highest scores at the individual level, while the non-engagement cluster recorded the lowest scores. At the city level, Shanghai attained the highest overall engagement score, whereas Chizhou recorded the lowest. Several strategies were proposed to improve resident engagement in the SOCR. This study not only contributes to the body of knowledge on resident engagement in the SOCR but also provides guidance for fostering resident engagement in sponge city construction.

Keywords: resident engagement; sponge city; old community; evaluation

1. Introduction

Rapid global urbanization has significantly increased impervious surface areas in cities, encroaching on natural sponges like wetlands and watercourses [1,2]. This, coupled with low construction standards of drainage networks and incomplete drainage systems, has escalated urban waterlogging issues [3,4]. In response, various ecological rainwater management strategies have been proposed and adopted to address the challenge in countries [5,6], such as the "Low Impact Development" (LID) in the United States [7,8], "Sustainable Urban Drainage Systems" (SUDS) in the UK and "Water Sensitive Urban Design" (WSUD) in Australia [9,10]. Although these concepts may differ slightly based on local development and institutional context in different countries, their underpinning ideas share similarities, both prioritizing mitigating hydrological changes and improving water quality as fundamental principles [11]. In China, the "sponge city" concept was first proposed in 2012 to address urban water management challenges, and this national-level strategy was officially launched by the central government in 2014 [8]. Subsequently, China initiated the sponge city construction pilot projects in two batches in 2015 and 2016, encompassing a total of 30 cities. Notably, the construction of sponge cities is a critical urban planning strategy not just for the pilot cities but is deemed essential for all cities across the nation to implement. As the micro-unit of urban areas, old communities are particularly vulnerable to urban flooding, making them a focal point in sponge city initiatives. To enhance the flood resilience of these communities, the concept of sponge-style old community renewal (SOCR) was formally introduced in the "Guiding Opinions on Promoting the Construction of Sponge City" issued by the General Office of the State Council (GOSC) in October 2015 [12]. Following the issue of this policy document, the SOCR has gained widespread attention within the academic community, with its definition becoming increasingly precise: SOCR refers to initiatives in which low-impact development techniques are utilized to enhance the stormwater management and water environment [13–17]. In practice, several common SOCR techniques have been adopted in pilot cities to mitigate waterlogging and enhance sustainable management. These include the pavement of permeable bricks to improve water infiltration and reduce surface runoff, the construction of rain gardens to collect and retain rainwater for efficient utilization, and the installation of new underground drainage pipelines to ensure effective water discharge. Resident engagement is vital in determining the success or failure of the SOCR, given that they are the direct stakeholders and end-users [18–22]. For example, 32.8% of residents in Zhenjiang are dissatisfied with the maintenance of SOCR facilities, posing potential risks of protests and irreparable losses [15]. Consequently, resident engagement in the SOCR has gradually become the focus of attention, as the engagement level directly impacts the well-being of residents [23,24]. Therefore, an effective evaluation of resident engagement is essential to the promotion of SOCR.

Classifying engagement clusters has been promoted as a means to achieve long-lasting consensus and to foster engaged residents who actively support sustainable water management initiatives [15,25,26]. Resident engagement clusters were focused on categorizes engagement into different types based on the degree of participation, often drawing upon Arnstein's "A ladder of citizen engagement" [27]. This categorization has, to some extent, advanced the development of resident engagement clusters [28,29]. Regarding the field of the SOCR, while there has been a focus on the technical aspects of the SOCR to mitigate flooding, less attention has been paid to evaluating these efforts on SOCR from a management perspective [6]. Fortunately, the importance of evaluating public and resident engagement in SOCR has been increasingly recognized in China [6,30–32], aiming to deepen the understanding of various stakeholders and clarify key factors influencing engagement through various evaluation frameworks [33–35]. Promoting active resident engagement in such projects is crucial to be more responsive to residents' needs and interests, and foster community renewal and development [36–38]. Therefore, encouraging resident engagement in community renewal projects stands as a fundamental component of urban development strategies. However, there remains insufficient research on the clear classification of resident engagement clusters in the SOCR, with a lack of a comprehensive system for evaluating resident engagement in the SOCR.

To fill the aforementioned research gaps, this paper identifies distinct resident engagement clusters in the SOCR and establishes a comprehensive evaluation framework. The aims of this paper are to (1) classify resident engagement clusters in the SOCR; (2) develop an indicator system to evaluate the level of resident engagement using the ANP-PROMETHEE II method; and (3) propose actionable recommendations to enhance resident engagement. The innovative evaluation method not only enriches the knowledge domain of resident engagement in sponge city but also quantifies resident engagement at both individual and city levels, fostering broader community involvement and promoting sustainable development of sponge city.

2. State of the Art

2.1. Theoretical Background and Resident Engagement Framework

The origin and development of community participation can be traced back to 1947, when the UK's 'Urban and Rural Planning Act' required planning to seek opinions from affected community residents [39]. The World Health Organization (WHO) defines community engagement as a process that actively and genuinely involves people in shaping the issues that concern them. It enables them to make decisions about aspects that impact their lives, participate in policy formulation, and contribute to the planning, development, and delivery of services. Additionally, it empowers them to take actions that foster desired changes [40,41]. As one of many stakeholders in the community, residents play a central role in community engagement [42]. Voluntary engagement in community affairs is a crucial aspect of sustainable development [43], allowing residents to express the importance of various impacts from their perspective [44]. While scholars have varying perspectives on the concept of resident engagement, the underlying subject–method–object logic remains consistent, indicating that residents can influence public affairs and achieve cooperative governance with the government through specific measures [45–47]. Therefore, resident engagement in the SOCR is defined as the engagement of community residents in the entire process of the SOCR through specific engagement clusters. This engagement, driven by individual needs and collective welfare, includes decision-making, implementation, and the operation and maintenance of the SOCR.

Drawing from the field of educational psychology [48], and Arnstein's "A ladder of citizen engagement" [49], scholars have conducted extensive research on the resident engagement, which can be seen in Table 1. On this basis, this study established a framework for classifying the clusters of resident engagement, which can be seen in Figure 1 [15,49]. Specifically, the resident engagement clusters in the SOCR encompass not only behavioral engagement but also emotional engagement and cognitive engagement. Cognitive engagement refers to residents' comprehension of SOCR-related knowledge and their capacity to apply it effectively. Emotional engagement involves residents' attitude towards the SOCR and their willingness to participate in the SOCR. Behavioral engagement denotes the actions undertaken by residents throughout the SOCR process. This study classifies resident engagement clusters in the SOCR into seven types, including control engagement, informed engagement, non-engagement, passive attitude engagement, cooperative engagement, weak willing engagement, and full engagement. This categorization not only advances theoretical understanding but also promotes analyzing resident engagement in sponge city construction.

Table 1. Summary of studies of educational psychology and ladder of citizen participation.

Research Theory	Example	Main Elements	Clusters	Reference
Educational psychology	Residents engagement in water-related issues	Knowledge Environmental identity Attitudes Water-saving devices Water-saving behaviors Pollution-reduction behaviors	Disengaged Aware but in active Active but not engaged Engaged but cautious Engaged citizens	[25]
	Engagement in food-related issues	Health engagement Food involvement Motivation	Health-conscious Balanced Hedonist Careless	[50]
	Student engagement	Behavioral measures Cognitive measures Affective measures Social measures	Declining engagement profile Low engagement profile	[51]

Table 1. *Cont.*

Research Theory	Example	Main Elements	Clusters	Reference
Ladder of citizen participation	Community engagement	Engagement ability Engagement intention	Active-dominant Negative-coping Self-development Equity-appealing	[52]
	Calgary's engage program	Engage policy Engagement practice Citizen power	Listen and learn Consult Collaborate Empower.	[53]
	Resident engagement in tourism destination governance	Residents perception Residents attitude	Activists Disenchanted Favorers Opposers	[54]

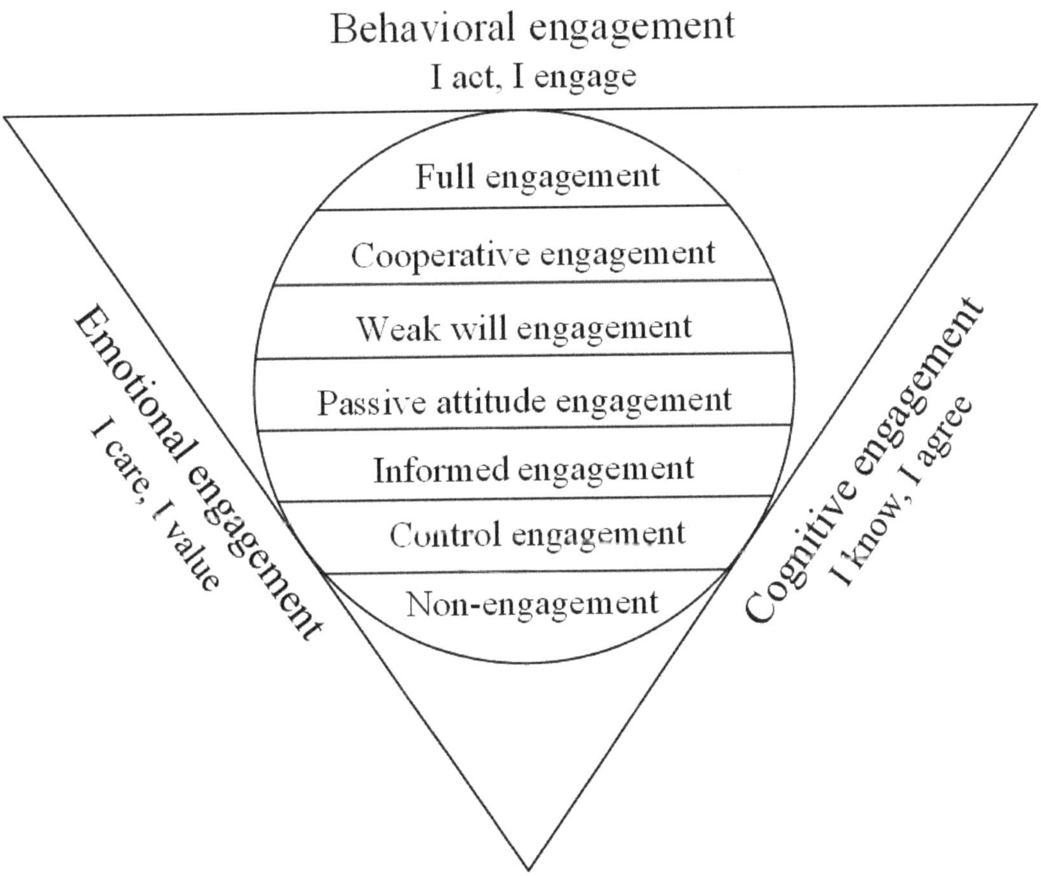

Figure 1. Framework for classifying resident engagement clusters in the SOCR.

2.2. *Clusters of Resident Engagement in the SOCR*

2.2.1. Control Engagement

According to Arnstein's citizen participation ladder, the therapy cluster is categorized as a "non-participation" level of participation. In this cluster, residents are subjected to therapy and believe that they are part of the planning process, while in reality, they have no impact on the planning itself. In the SOCR, the engagement cluster in which residents are given opportunities to participate but are almost not empowered actually is classified as a control engagement cluster. The typical characteristics of this cluster include residents' substantial understanding of common technologies, the significance, and participation methods associated with the SOCR. However, these residents demonstrate minimal endorsement and a lack of emotional investment in the initiative [15].

2.2.2. Informed Engagement

Based on Arnstein's citizen participation ladder, the informing cluster is the participation of tokenism, which denotes a form of participation where citizens possess an awareness of their rights and choices yet have minimal opportunities to influence plans formulated for their benefit and responsibilities. In this type of participation, information solely flows unilaterally from officials to citizens, devoid of feedback channels or negotiating power. In the SOCR, the engagement cluster where residents only have partial understanding, with no action and right is called informed participation in clustering in the SOCR is classified as informed engagement. These residents maintain a pessimistic disposition towards the SOCR and exhibit limited enthusiasm for participating in its renewal, consequently leading to infrequent involvement [55,56].

2.2.3. Non-Engagement

The cluster called manipulation, as rung 1 in Arnstein's citizen participation ladder, is not true citizen participation but rather distorted into a public relations tool by those in power. In the name of citizen participation, individuals are placed on rubberstamp advisory committees or advisory boards with the sole intent of "educating" them or engineering their support. In the SOCR, the engagement cluster where residents have hardly engaged in the process of the SOCR is classified as non-engagement. The typical characteristics of this cluster are that residents possess relatively one-sided understandings of the SOCR, demonstrating minimal support for it and exhibiting reluctance to engage emotionally. Furthermore, they display an unwillingness to take action and engage in the entire process of the SOCR [55].

2.2.4. Passive Attitude Engagement

Even though the placation cluster is situated at the highest rung of the tokenism level within Arnstein's citizen participation ladder, it embodies a scenario where individuals possess a degree of influence but still lack the decision-making ability or substantial power, thereby unable to exert a significant impact on ongoing planning processes. In the SOCR, the engagement cluster where residents cannot engage in the key content of the SOCR, though they have a certain power, is classified as passive attitude engagement. The typical characteristics of this cluster are that residents demonstrate a comprehensive understanding of commonly utilized technologies, the significance of the SOCR, and various participation channels. They are willing to participate and take action in the entire process of the SOCR. However, due to persisting issues and ineffective avenues for complaints, they maintain a pessimistic outlook towards the SOCR [57].

2.2.5. Cooperative Engagement

The consultation cluster represents a form of participation that solely focuses on gathering opinions or attitudes from residents, without actively involving them in specific behaviors or actions in accordance with Arnstein's citizen participation ladder. In the SOCR, the engagement cluster where residents have active emotional engagement without specific behavioral engagement is classified as cooperative engagement. The typical characteristics of this cluster are that residents possess a comprehensive understanding of commonly used technologies, the significance of the SOCR, and various participation channels. They maintain a positive attitude towards the SOCR and are willing to participate in its entirety. However, their ultimate participation may be hindered by external environmental factors [58].

2.2.6. Weak Willing Engagement

In Arnstein's citizen participation ladder, both the partnership and representation clusters signify complete citizen participation. In these clusters, citizens hold decision-making power regarding planning and affairs, although the extent of their power varies. In the SOCR, the engagement cluster where residents have certain power and participate in the SOCR actively is classified as weak willing engagement. Residents in this cluster demonstrate a thorough understanding of commonly used technologies, the significance of the SOCR (SOCR), and available participation channels. They maintain a positive attitude towards the SOCR, viewing it as a solution to waterlogging issues, and actively engage in the initiative. However, their willingness to participate may be influenced by external policy measures [58].

2.2.7. Full Engagement

The citizen control cluster is at the highest rung of Arnstein's citizen participation ladder, in which residents have a certain degree of power (or control) to ensure that they are able to manage a plan or institution, fully responsible for policy and management aspects, and have conditions for negotiation and change. In the SOCR, the engagement cluster, in which residents not only have high power but also above-average cognitive engagement, emotional engagement, and behavioral engagement, is classified as full engagement. These residents not only understand the common technologies, significance, and participation channels of the SOCR but also exhibit a positive attitude toward it. They demonstrate a willingness to participate and actively engage in the entire process of the SOCR by taking specific actions [56].

3. Methods and Materials

This paper employs a mixed-methods approach, integrating qualitative methods including a systematic literature review (SLR) and expert interviews with the quantitative ANP-PROMETHEE method. The evaluation process is shown in Figure 2. Specifically, Step 1 is to identify the evaluation indicators for resident engagement in the SOCR qualitatively using the SLR and expert interviews. Step 2 is to quantify resident engagement in the SOCR at the individual level using the ANP-PROMETHEE II method. Expert-related data were processed to calculate the weights of each indicator through the ANP and evaluate resident engagement in the SOCR at the individual level through PROMETHEE II. This step highlights the significance of individual engagement in the construction process. Step 3 is to quantitatively evaluate the performance of resident engagement in the SOCR at the city level. This step uses resident-related data to comprehensively evaluate how the individual engagement of residents, as a fundamental element of the city, contributes to promoting resident engagement at the city level. Following the establishment of the evaluation model, data collection, including expert-related data collection and resident-related data collection, was conducted across typical communities to enhance the evaluation model and provide practical guidance.

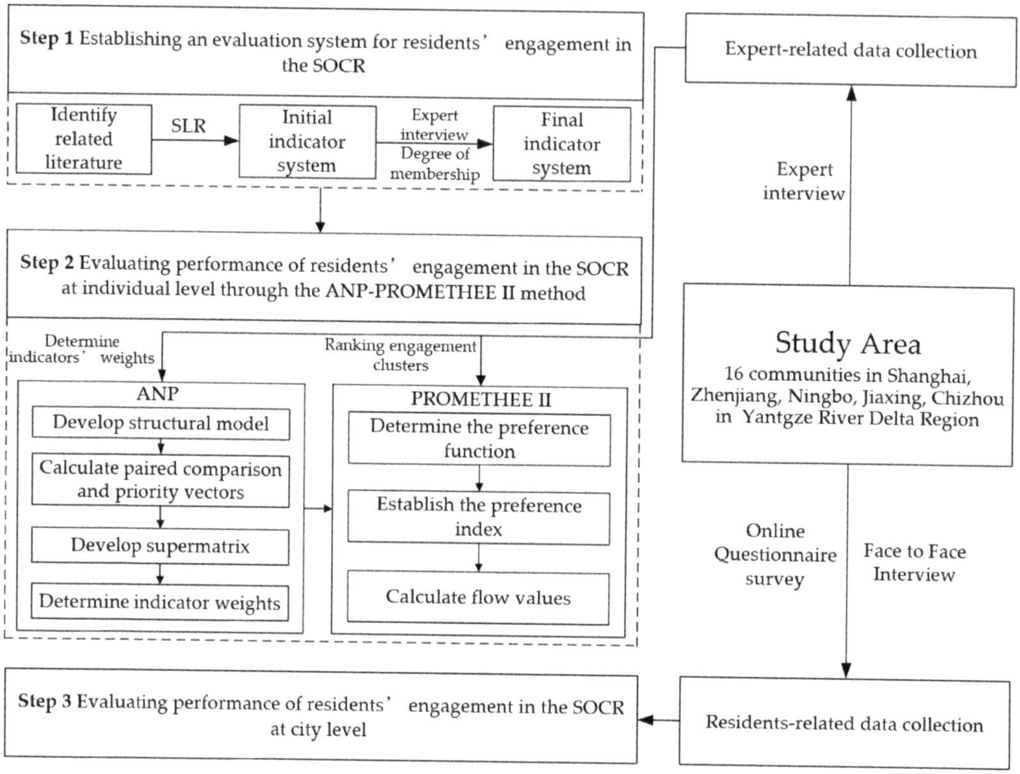

Figure 2. The flow chart of the evaluation process of resident engagement in the SOCR.

3.1. Identifying Indicators for the Evaluation of Resident Engagement in the SOCR

Identifying evaluation indicators is the basis for evaluating the level of resident engagement with the SOCR. Many scholars have adopted the method of SLR to sort out evaluation indicators [59]. Therefore, this study systematically and accurately identified corresponding indicators based on the SLR method and participatory governance theory [60], combined with the resident engagement clusters in the SOCR, to establish a preliminary evaluation indicator system. The initial indicator system includes 5 primary indicators and 35 secondary indicators. All selected indicators are from the literature and evaluation guidelines related to resident engagement in the SOCR.

To select the evaluation indicators, in-depth interviews with experts are conducted in the fields of "sponge city implementation", "old community renewal", and "city governance". The interview mainly focused on the rationality of evaluation indicators of resident engagement in the SOCR. During the interview, experts provided professional opinions on these indicators, eliminating those that were redundant or showed low correlation. On this basis, experts were invited to score the importance of the evaluation indicators from 1 (very unimportant) to 5 (very important). The select process adhered to the following criteria: (1) Indicators should be aligned with the concept of a sponge city and encapsulate the principles of ecological rainwater management and sustainable urban development. (2) Indicators should reflect the current state of implementation within both pilot cities and nationwide sponge city projects in China, ensuring practical relevance and adaptability to real-world conditions. The outline of the expert interviews is provided in Supplementary Material A. Detailed information on the experts interviewed is provided in Supplementary Material B. After obtaining 38 questionnaires, the membership degree of 21 evaluation indicators was calculated using the membership degree formula [61] and finally formed an evaluation indicator system for resident engagement in the SOCR. The indicators at all levels and their sources are shown in Table 2.

Table 2. Final evaluation indicator system of level of resident engagement in the SOCR.

Primary Indicators	Code	Secondary Indicators	Source
Empowerment	EP1	The government departments at the community level provide residents with certain power to participate in decision making	[62–66]
	EP2	Relevant government departments at the community level build follow-up tracking mechanisms to solve the problem of resident engagement	
	EP3	The government departments at the community level provide sufficient economic resources for residents to participate, such as providing funds for holding pre-consultation meetings	
	EP4	The government departments at the community level allocate sufficient human resources to help residents engage, such as setting up specialized governance institutions	
Engagement	EG1	Residents have the opportunity to participate in the decision-making stage of the SOCR and put forward suggestions	[62,64–67]
	EG2	Residents have the opportunity to participate in the implementation stage of the SOCR and give feedback	
	EG3	Residents have the opportunity to participate in the maintenance phase of the SOCR and supervise the follow-up process	
Cooperation	CP1	There are specialized government departments responsible for coordinating problems of resident engagement at the community level	[62,64,65,68,69]
	CP2	Third-party organizations have specialized agencies at the community level responsible for coordinating problems of resident engagement	
	CP3	The construction party has special personnel responsible for coordinating problems of resident engagement	
	CP4	The resident representatives communicate with the coordinating unit about the problems in the engagement process	
Network	NW1	Government departments actively guide residents to engage	[62,64–66,68]
	NW2	Third-party organizations actively guide residents to engage	
	NW3	Construction party actively guides residents to engage	
Validity	EF1	The resident engagement helps to deepen the understanding of the importance of the SOCR	[63,65,70,71]
	EF2	The resident engagement helps to deepen the understanding of the technologies and targets of the SOCR	
	EF3	The resident engagement helps to improve residents' satisfaction with the SOCR	
	EF4	The resident engagement helps to enhance residents' trust in the government	
	EF5	The resident engagement has effectively accelerated the progress of project implementation and improved efficiency	
	EF6	The resident engagement has effectively reduced the complaints of residents and conflicts with government departments	
	EF7	The resident engagement has effectively improved the community environment	

3.2. The Evaluation Model for Resident Engagement in the SOCR

Resident engagement in SOCR at individual and city levels is evaluated in this part. The individual-level evaluation focuses on the personal engagement indicators and resident engagement clusters, while the city-level evaluation aggregates these individual contributions to determine overall urban engagement performance. The interplay between these levels is integral, as individual engagement, cumulatively, serves as foundational inputs that shape the broader city-level engagement outcomes.

Combining different evaluation methods to achieve favorable outcomes has been widely adopted in engineering management and environment management domains. For example, the hybrid CRITIC-TOPSIS method has been used to evaluate smart community governance, and the CRITIC-PROMETHEE II method has been employed to quantify residents' perceptions of smart community construction [72,73]. Inspired by this research, this study integrates ANP and PROMETHEE II to quantitatively evaluate the level of resident engagement in the SOCR. Specifically, the ANP is particularly suited for calculating the indicators' weights of resident engagement in the SOCR. It adeptly captures the subjective weighting of interrelated indicators, which is crucial for accurately reflecting the diverse levels of engagement among residents in our selected communities. Additionally, it facilitates the modeling of interdependencies among criteria and sub-criteria. On this basis, an approach is required to evaluate the level of resident engagement in the SOCR. The PROMETHEE II, as a multi-criteria decision-making (MCDM) tool, is ideal for ranking alternatives (in this case, engagement clusters) based on the weights derived from ANP. It ensures effective comparison and prioritization of engagement types, which is critical for renewal projects. Notably, the integration of ANP's subjective indicator weighting with PROMETHEE II's effective ranking capabilities can fully take into account the interdependence of indicators between the upper and lower levels [74], which can better reflect the situation of the alternatives to be compared without complete compensation problem [75]. The ANP-PROMETHEE II method, which has been successfully applied in the field of environment management [76–78], is capable of handling complex decision-making environments with interdependent criteria. This capability aligns with the multi-dimensional nature of resident engagement in the SOCR. Therefore, this method was developed as the preferred evaluation method to improve the reliability and robustness of the evaluation results.

3.2.1. Evaluating Performance of Resident Engagement in the SOCR at Individual Level Through the ANP-PROMETHEE II Method

(1) Determining weights of indicators in the SOCR through the ANP

ANP is a decision-making method proposed by Saaty in 1996 that adapts to a non-independent hierarchical structure. The ANP fundamentally involves identifying the key elements within a system, constructing a relational network among these elements, and assessing their relative importance through expert judgment [79,80]. The process for calculating the indicators' weight of resident engagement in the SOCR in the ANP is as follows:

(a) Development of the model and structure of decisions.

The network structure of the ANP approach is usually divided into two parts: the control layer and the network layer [79]. The objectives and criteria are contained in the control layer, where the criteria are independent of each other and dominated by the target elements. The network layer is composed of elements dominated by the control layer, and they interact with each other to form the network structure [79]. Suppose that there are criteria $P_1, P_2, \ldots, P_n$ in the control layer of the ANP structure. The network layer contains element groups $C_1, C_2, \ldots, C_m$ and C_i contains n_i elements $P_{i1}, P_{i2}, \ldots, P_{in_i}$ $(i = 1, 2, 3, \ldots, m)$.

(b) Calculation of paired comparison and priority vectors.

Compare the importance of each element in C_i with that in C_j. P_{jl} $(j = 1, 2, 3, \ldots, m; l = 1, 2, 3, \ldots, n)$, that is, under the control criterion P_s $(s = 1, 2, 3, \ldots, N)$, consider P_{jl} as a sub-criterion; and construct a judgment matrix. The importance of criteria or sub-criteria is determined and given values between 1 and 5 using experts' opinions. In total, 38 experts in the research field were interviewed to determine the weight of each indicator. Thereafter, the extent of the inconsistency of the judgments is measured by the consistency ratio (CR). If this ratio is smaller than 0.1, the consistency in the judgment is acceptable; otherwise, the judgment should be revised. The index and rate of consistency can be calculated in Equations (1) and (2).

$$CI = \frac{\lambda_{max} - n}{n - 1} \tag{1}$$

$$CR = \frac{RI}{CI} \tag{2}$$

where CR is the consistency ratio, CI is the compatibility index pair-wise comparison matrix, λ_{max} is the maximum eigenvalue of the judgment matrix, RI is the random index, and n is the number of compared components in the matrix.

Once the consistency of the judgments is ensured, the significant coefficients of the criteria should be determined. For this, a method called "eigenvector" is used.

$$AW = \lambda_{max} W \tag{3}$$

where A is the pair-wise comparison matrix of the criteria, W represents the eigenvector, and λ_{max} is the maximum eigenvalue of the judgment matrix.

According to the eigenvector method, the eigenvector of the judgment matrix is obtained and is represented in the form of a matrix; that is, an unweighted super matrix under criterion P_s is obtained, as shown in Equation (4) [73]:

$$W_{ij} = \begin{bmatrix} W_{i1}^{j1} & \cdots & W_{i1}^{jn} \\ \vdots & \ddots & \vdots \\ W_{in}^{j1} & \cdots & W_{in}^{jn} \end{bmatrix} \tag{4}$$

(c) Development of the supermatrix.

The supermatrix is used to represent the way in which a cluster affects other clusters (considering external connections), or the effect of the elements present inside the cluster itself (considering internal connections) [81]. If the two element groups are independent of each other, then $W_{ij}= 0$. In this study, the supermatrix W has five groups recorded as C_n ($i = 1, 2, 3, \ldots, 5$), where C_1 contains the four indicators $EP_1 - EP_4$, C_2 contains the three indicators $EG_1 - EG_3$, C_3 contains the four indicators $CP_1 - CP_4$, C_4 contains the three indicators $NW_1 - NW_3$, and C_5 contains the seven indicators $EF_1 - EF_7$. Similarly, by comparing the elements in other element groups, the supermatrix of the influence of each element on the element group C_i under the criterion is obtained in Equation (5):

$$W = \begin{matrix} C_1 \\ \vdots \\ C_n \end{matrix} \overset{\begin{matrix} C_1 & \cdots & C_n \end{matrix}}{\begin{bmatrix} W_{11} & \cdots & W_{1n} \\ \vdots & \ddots & \vdots \\ W_{n1} & \cdots & W_{nm} \end{bmatrix}} \tag{5}$$

To develop a supermatrix in systems with dependent variables and calculate the final priorities of the elements, all the primary priority vectors obtained from paired comparison matrices are entered into a column matrix. The result of this process is the development of an unweighted super matrix. Next, through multiplication of the data in the cluster matrix by the unweighted super matrix and then normalizing the calculated matrix, the weighted supermatrix $\overline{W}$ is obtained.

The limited supermatrix is formed thereafter. For this, according to Equation (6), the weighted super matrix reaches the power limit to converge.

$$\overline{W}^{\infty} = \lim_{k \to \infty} \overline{W}^{2k+1} \tag{6}$$

where $\overline{W}^{\infty}$ is the limit super matrix, $\overline{W}$ is the weighted super matrix, and k is the exponent determined by iteration.

As the calculation process of ANP is relatively complex, this paper uses Super Decision (SD) V3.2 software, a software that implements ANP, to calculate the weights of all indicators.

(2) Ranking resident engagement clusters in the SOCR through the PROMETHEE II

PROMETHEE is a comprehensive multiple-criteria decision-making (MCDM) method initially developed by Brans [82] and subsequently extended by Brans and Vinke [83,84]. Its basic principle is based on the pairwise comparison of alternatives across chosen criteria, making PROMETHEE II a promising tool for integrating multiple rankings of alternatives. Furthermore, PROMETHEE II can take into account diverse decision criteria with varying dimensions and scales, allowing for the thorough and efficient consideration of information contained within the decision matrix when reaching a final decision [85]. The steps for PROMETHEE II are as follows.

(a) Establishment of the matrix.

The PROMETHEE II approach requires the establishment of an evaluation matrix, which includes the performance of M alternative solutions (7 clusters in this research) on a set of N evaluation criteria (21 indicators in this research). During data collection, experts are invited to assess the level of resident engagement clusters on a scale of "1–5" values.

(b) Determination of the preference function.

The preference function P represents the relative preference of alternative a over alternative b according to criterion j. In general, the Gaussian rule is the most commonly adopted in practical applications [83]. This is primarily because its preference calculations are more sensitive to changes in distance, thereby yielding more accurate preference results. Furthermore, it lacks any abrupt discontinuities, offering outcomes characterized by increased robustness and stability [86]. Since the major criteria in this study contain continuing data, the Gaussian criterion was selected for evaluation, and the preference function Equation (7) is as follows:

$$P(d) = \begin{cases} 0 & d \le 0 \\ 1 - e^{-\frac{d^2}{2s^2}} & d > 0 \end{cases} \tag{7}$$

where P(d) is the preference function and s is a parameter that needs to be determined by experts.

(c) Calculation of preference indexes.

Aggregated preference index $\pi(a,\ b)$ expresses the degree to which resident engagement cluster a prefers all the criteria to resident engagement cluster b. The aggregated preference indexes are given in Equation (8):

$$\pi(a,\ b) = \sum_{j=1}^{n} P_j(a,b)w_j \tag{8}$$

(d) Calculate flow values.

The positive outranking flow expresses how an alternative a is outranking all the others. The positive outranking flow is of Equation (9). Higher $\varnothing^{+}(a)$ gives a better alternative. On the other hand, the negative outranking flow expresses how an alternative a is outranked by all the others. The negative outranking flow is of Equation (10). The lower $\varnothing^{-}(a)$ gives a better alternative. The net outranking flow is applied and is in Equation (11). The higher $\varnothing(a)$ specifies the final better alternative. Then, rank the clusters according to the net flow value to obtain the ranking of the modes of level of resident engagement of the SOCR

$$\varnothing^{+}(a) = \frac{1}{M-1}\sum_{b=1}^{M}\pi(a, b) \tag{9}$$

$$\varnothing^{-}(a) = \frac{1}{M-1}\sum_{b=1}^{M}\pi(b, a) \tag{10}$$

$$\varnothing(a) = \varnothing^{+}(a) - \varnothing^{-}(a) \tag{11}$$

3.2.2. Evaluating Performance of Resident Engagement in the SOCR at City Level

The preceding section delves into the individual performance of resident engagement in the SOCR. Nevertheless, for a city, it is crucial to compute the comprehensive performance of resident engagement. Throughout the SOCR process, it can be considered that each resident falls into an engagement cluster, which is ascertained through the aggregation of behavioral, cognitive, and emotional engagement. This cluster assignment is unique to each resident, resulting in distinct engagement performance outcomes. Consequently, to obtain the comprehensive performance of resident engagement in the SOCR, it is necessary to aggregate their individual performance. According to the above, resident engagement clusters were divided into seven categories. Assuming that there are n residents in an old community of a city and the engagement cluster y_k adopted by the resident i in the sponge-style renewal belongs to the cluster A_k ($k = 1, 2, \ldots, 7$), where A_k ($k = 1, 2, \ldots, 7$) represents, respectively, the following:

A1 = {Control engagement}
A2 = {Informed engagement}
A3 = {Non-engagement}
A4 = {Passive attitude engagement}
A5 = {Cooperative engagement}
A6 = {Weak willing engagement}
A7 = {Full engagement}

If there are b_k residents, adopt the engagement cluster y_k, and the engagement performance corresponding to the cluster y_k of type k ($k = 1, 2, \ldots, 7$) resident engagement performance is Y_k; then, the equation of the level of comprehensive resident engagement performance in the SOCR is defined as Equation (12):

$$I = \sum_{k=1}^{7}\frac{b_k Y_k}{n} \tag{12}$$

This study designed a questionnaire and distributed it to residents in selected communities. The questionnaire includes a brief introduction to the research objectives, demographic and socio-economic information of the respondents, and detailed information about resident engagement clusters. The questionnaires can be seen in Supplementary Material C. Residents are invited to choose the resident engagement cluster they belong to according to the questionnaire. Afterward, the collected questionnaire results were inputted into Equation (12) for calculation.

3.3. Study Area

Five cities in the Yangtze River Delta of China, namely Shanghai, Ningbo, Jiaxing, Zhenjiang, and Chizhou, were selected as investigation cities (Figure 3) in order to gather resident-related data for the following reasons. First, these five cities confront severe challenges including rapid urbanization, water scarcity, and severe waterlogging issues, making them pertinent for implementing the SOCR. Second, these five cities are all the pilot cities of sponge cities in China, which provides a favorable knowledge base and practical environment for resident engagement, thereby facilitating the progress of this study [87,88]. Third, these cities exhibit variations in terms of economic and social levels, enabling comparative analysis.

Within these five cities, 16 communities were selected for evaluation. These communities were chosen because they have implemented the SOCR, resulting in enhanced water conservation and significant environmental improvements. For example, Yaojianghuayuan, a community in Ningbo, was selected for the SOCR due to severe issues with waterlogging, limited parking spaces, and outdated infrastructure. The renewal measures included constructing rain gardens, paving permeable bricks, and upgrading drainage systems, which have substantially increased the community's water retention capacity. Thus, they represent a suitable diagram for evaluating the level of resident engagement in the SOCR, providing valuable insights into the applicability of the evaluation models and the enhancement of resident engagement. Figure 3 illustrates the locations and photos of chosen communities.

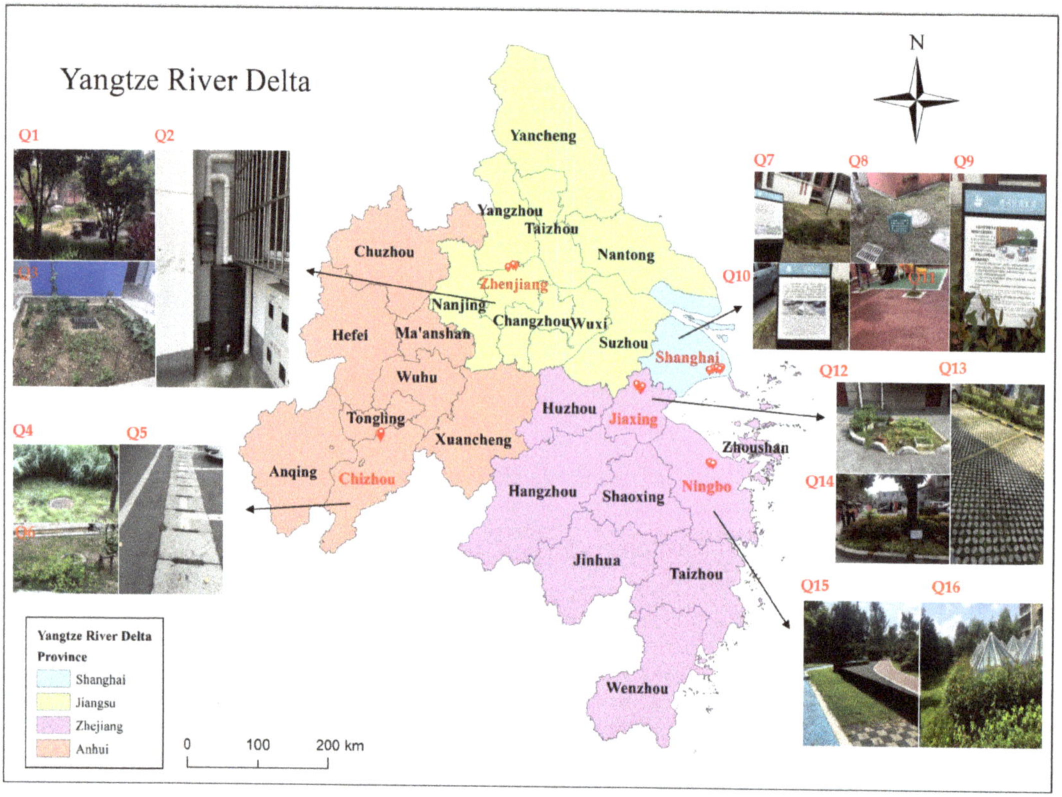

Figure 3. The location of the surveyed cities in the Yangtze River Delta Region. Note: (1) The red fonts highlight the location of the surveyed cities. (2) The name and code of 16 communities—Q1: Sanmaogong community; Q2: Jiangbinxincun Second community; Q3: Huarunxincun community; Q4: Yijingyuan community; Q5: Qingxinjiayuan community; Q6: Huijing community; Q7: Xinluyuan A community; Q8: Xinluyuan F community; Q9: Haishangmingyue community; Q10: Hailuhuiming community; Q11: Hailuyuehua community; Q12: Yanyu community; Q13: Lingxiangfang community; Q14: Zhenhe community; Q15: Yaojianghuayuan community; Q16: Sanhejiayuan community.

3.4. Data Collection

3.4.1. Expert-Related Data Collection

Expert interviews were utilized to collect weighted data pertaining to each evaluation indicator of resident engagement in the SOCR, as well as to collect scoring data on the seven types of engagement clusters. The expert-related data for this study were obtained through both Wenjuanxing and VooV meetings. Wenjuanxing is an online survey collection platform

(The official website is www.sojump.com, accessed on 20 April 2022) that allows for the transformation of designed questionnaires into online questionnaires with hyperlinks or QR codes. This platform enables respondents to conveniently answer questions using their mobile devices, which has been widely utilized in previous studies [89,90]. In order to incentivize active participation and careful completion of the online survey, respondents who completed the survey were rewarded via the internet. The online survey questionnaire was distributed via e-mail to 38 experts from May to June 2022. The specific questionnaire can be found in Supplementary Material D. These experts are mainly from the field of sponge city and community engagement. This guaranteed that the experts possessed abundant experience and knowledge regarding resident engagement in the SOCR, thus ensuring the high reliability of the collected data. Subsequent to the survey, each expert participated in a detailed virtual discussion via VOOV meeting, lasting at least one hour. The interview was pivotal, concentrating on evaluating the relative importance and specific values of the indicators, particularly in relation to the distinctive characteristics of five selected cities within the Yangtze River Delta Region. A total of 32 valid questionnaires were collected, with an effective response rate of 84.21%.

3.4.2. Resident-Related Data Collection

From May to June 2022, our research team distributed questionnaires in each community to collect residents' related data. To determine the sample size for large populations, this study used the sample size calculation formula proposed by Cochran, which is expressed as [91]:

$$n = \frac{z^2 p(1 - p)}{e^2}$$

where n is the required sample size, p represents the percentage of a specific characteristic within the total population (when the proportion of participating residents is unknown, a value of 50% is commonly used for p to maximize variance and produce the largest sample size [92]), e is the percentage maximum error required, and z corresponds to the confidence level value. It is common practice to use a 95% confidence interval (corresponding z = 1.96) and a 5% margin of error (corresponding e = 5%) in relevance research [93]. Consequently, the optimal sample size for this study has been calculated as 385 residents per city, using a z-value of 1.96, a p-value of 0.5, and an e-value of 5%. Considering potential errors in the distribution of the questionnaires, 2000 questionnaires were distributed in total, with 400 to each city.

The data were obtained through face-to-face interviews and an online survey platform conducted via the Wenjuanxing platform. During the face-to-face interviews, simple random sampling was employed, with questionnaires distributed in densely populated areas of the communities, ensuring equal chances of selection for all units in the population [94,95]. Questionnaires were filled out under the guidance of our research team members on-site and collected immediately after completion, targeting the authenticity and reliability of the data. Additionally, online questionnaires were distributed to community managers in the selected communities. These managers further distributed these questionnaires to residents through various promotional strategies, including QR codes, posters, residents' WeChat group shares, and posts on the community's official WeChat account. To encourage residents to participate actively in the online questionnaire survey and provide thoughtful responses, respondents who completed the survey were rewarded through the Internet. In total, 1657 valid responses were received, which corresponds to an effective response rate of 82.85%. Among the five cities, Shanghai had the highest number of collected questionnaires, with a total of 391, while Ningbo, Jiaxing, Zhenjiang, and Chizhou collected 309, 306, 312, and 339 questionnaires, respectively. The basic information about all respondents is presented in Table 3.

Table 3. Basic characteristics of respondents.

Variable	Option	Quantity	Proportion	Variable	Option	Quantity	Proportion
Gender	Male	785	47.37%	Living alone or not	No	962	58.06%
	Female	872	52.63%		Yes	695	41.94%
Age	Younger than 20 years old	132	7.97%	Rent or not	No	1181	71.27%
	20–34 years old	736	44.42%		Yes	476	28.73%
	35–49 years old	419	25.29%	Working condition	No job or looking for a job	205	12.37%
	50–64 years old	267	16.11%		In work	1342	80.99%
	Older than 65 years old	103	6.22%		Retirement	110	6.64%
Education level	Primary school and below	84	5.07%	Income	Less than 2000 yuan	170	10.26%
	Middle school	495	29.87%		2000–3999 yuan	338	20.40%
	Polytechnic school or high school	460	27.76%		4000–5999 yuan	415	25.05%
	Junior college education	326	19.67%		6000–7999 yuan	422	25.47%
	Bachelor degree or above	292	17.62%		More than 8000 yuan	312	18.83%
Length of residence	Less than 1 year	529	31.93%	City	Shanghai	391	23.60%
	2–5 years	716	43.21%		Ningbo	309	18.65%
	6–10 years	344	20.76%		Jiaxing	306	18.47%
	More than 10 years	68	4.10%		Zhenjiang	312	18.83%
					Chizhou	339	20.46%

4. Findings

4.1. The Performance of Resident Engagement in the SOCR at the Individual Level

Each evaluation indicator was scored by expert interview, and the weights of evaluation indicators were computed by equations of the ANP. In this paper, the second part of the questionnaire on the relationship and importance of evaluation indicators for the level of resident engagement focuses on the mutual influence between the evaluation indicators of the level of resident engagement in the SOCR. Experts were asked to indicate whether the indicators of a given row affect the indicators of a given column by assigning a value of 1 for yes and 0 for no. Based on the survey findings, indicators that more than 50% of experts (i.e., more than 16 experts) agreed to be interdependent were retained. The detailed outcomes are presented in Supplementary Material E.

After sorting out the mutual influence relationship among the evaluation indicators for the level of resident engagement, the SD software was employed to calculate the weights of indicators for the evaluation of the level of resident engagement in the SOCR. Data from 32 experts were input into the SD software. However, 11 data sets failed to pass the conformity test, leaving 21 data sets for calculating the indicator weights. The final weights of primary and secondary indicators were derived by averaging these weights, as illustrated in Table 4.

The findings showed significant differences in the weights of indicators for evaluating the level of resident engagement in the SOCR. The weights of these evaluation indicators reflected their significance and contribution to the evaluation framework. As shown in Table 4, "empowerment" held the highest weight at the primary level, reaching 0.4284, followed by "validity" (0.2840), "engagement" (0.1600), and "network" (0.0907). The willingness of local authorities to share power is deemed an important factor contributing to the high levels of resident participation [44]. Additionally, validity is considered one of the criteria for good governance, and promoting public participation is regarded as an essential prerequisite for strengthening governance effectiveness [96,97]. Notably, "collaboration" held the lowest weight (0.0370). Collaboration is a process that becomes even more complex when multiple levels are involved and requires effective coordination [98].

Table 4. Evaluation indicator weight of level of resident engagement of the SOCR based on ANP.

Primary Indicators		Secondary Indicators		Comprehensive Weights	Ranking
Name	W_i	Code	W_{ij}	$W = W_i \times W_{ij}$	
Empowerment	0.4284	EP1	0.6574	0.2816	1
		EP2	0.2027	0.0868	3
		EP3	0.0942	0.0404	9
		EP4	0.0457	0.0196	12
Engagement	0.1600	EG1	0.2897	0.0464	8
		EG2	0.6554	0.1048	2
		EG3	0.0549	0.0088	17
Cooperation	0.0370	CP1	0.5267	0.0195	13
		CP2	0.0630	0.0023	21
		CP3	0.1098	0.0041	20
		CP4	0.3005	0.0111	16
Network	0.0907	NW1	0.7049	0.0639	6
		NW2	0.0841	0.0076	19
		NW3	0.2109	0.0191	14
Validity	0.2840	EF1	0.0698	0.0198	11
		EF2	0.0411	0.0117	15
		EF3	0.2809	0.0798	4
		EF4	0.2809	0.0798	4
		EF5	0.1801	0.0512	7
		EF6	0.1183	0.0336	10
		EF7	0.0290	0.0082	18

Similarly, in the secondary indicators, the weight of different indicators can also provide insights into the various aspects of resident engagement. EP1 exhibited the highest contribution, with an indicator weight of 0.2816. The ability of residents to express their opinions and actively partake in decision-making processes is fundamental to the concept of participation [99]. The followings are EG2 (0.1048), EP2 (0.0868), EF3 (0.0798), EF4 (0.0798). Conversely, CP2 exhibited the lowest indicator contribution, at 0.0023. This may be attributed to difficulties in communicating problems to a trusted third party, potentially causing significant delays in resolving issues [100].

Following the steps of the PROMETHEE II, the decision matrix and parameter values were determined based on scoring data provided by 21 experts. Subsequently, the preference function, and evaluation differences between resident engagement clusters, positive flow, and negative flow were calculated, as delineated in Table 5. Due to the complexity of the PROMETHEE II calculation process, Visual PROMETHEE 1.4 software was employed for data processing.

Table 5. Flow score of resident engagement cluster.

Type	Resident Engagement Cluster	Positive Flow	Negative Flow	Average Net Flow Φ
Type 1	Control engagement	0.0279	0.0801	−0.0522
Type 2	Informed engagement	0.0238	0.0744	−0.0505
Type 3	Non-engagement	0.0005	0.2122	−0.2117
Type 4	Passive attitude engagement	0.0732	0.0259	0.0474
Type 5	Cooperative engagement	0.0467	0.0692	−0.0226
Type 6	Weak will engagement	0.0934	0.0137	0.0797
Type 7	Full engagement	0.2103	0.0004	0.2099

It can be seen from the table that the positive flow scores of the resident engagement cluster are full engagement, weak will engagement, passive attitude engagement, cooperative engagement, control engagement, informed engagement, and non-engagement from high to low. The negative flow scores of the resident engagement cluster are full engagement, weak will engagement, passive attitude engagement, cooperative engagement, informed engagement, control engagement, and non-engagement from low to high. The average net flow is obtained by subtracting the negative flow from the positive flow. The average net flow of control engagement, informed engagement, non-engagement, passive attitude engagement, cooperative engagement, weak will engagement, and full engagement are −0.05222, −0.0505, −0.2117, 0.0474, −0.0226, 0.0797, 0.2099, respectively. It can be seen that full engagement has the highest score, followed by weak will engagement, passive attitude engagement, cooperative engagement, informed engagement, and control engagement, while non-engagement has the lowest score.

4.2. The Performance of Resident Engagement in the SOCR at the City Level

Upon scoring the resident engagement clusters of the SOCR and considering the distribution of residents across different clusters in cities such as Shanghai, Ningbo, Jiaxing, Zhenjiang, and Chizhou, a comprehensive evaluation of resident engagement can be derived. The detailed proportion of each cluster and the comprehensive score of each city can be seen in Figure 4. Analysis of the figure reveals that among the five cities, Shanghai has the highest number of residents in clusters 1, 6, and 7, while Chizhou City leads in cluster 3. With the exception of Ningbo and Zhenjiang, the population in cluster 4 exceeds 60 in other cities. Shanghai attains the highest overall score for resident engagement within the SOCR, with a score of 5.6189, securing the top position among the five cities. It is followed by Ningbo (5.4401), Jiaxing (5.2516), and Zhenjiang (4.9904). Conversely, Chizhou records the lowest overall score at 4.6814.

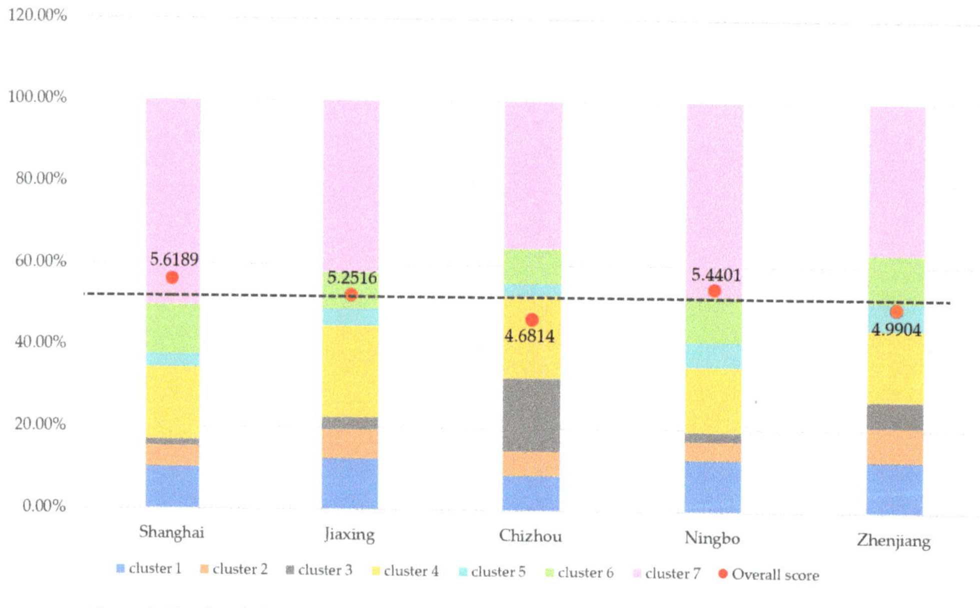

Figure 4. The detailed proportion of each cluster and the comprehensive score of each city. Note: The dashed line represents the average score of the five cities. The average score is 0.5196.

5. Discussion

5.1. Differences in the Performance of Resident Engagement in the SOCR at the Individual Level

The Action Profile function in Visual PROMETHEE is employed to analyze the characteristics of each resident engagement cluster, facilitating a detailed discussion of the differences in evaluation indicators across various clusters.

For the control engagement cluster, the scores across various evaluation indicators are presented in Figure 5. From the figure, it is evident that this engagement cluster exhibits higher scores on EP3, EG1, NW1, and NW3, yet lower scores on EP1, EG2, CP4, EF1, EF2, EF3, and EF6. Sufficient economic resources within the community provide residents with the opportunity to engage in the decision-making phase of the SOCR. Furthermore, the active guidance of government and construction companies significantly enhances resident engagement [101]. However, poor coordination between government and residents could lead to communication gaps [102], resulting in the overall ineffectiveness of renewal efforts.

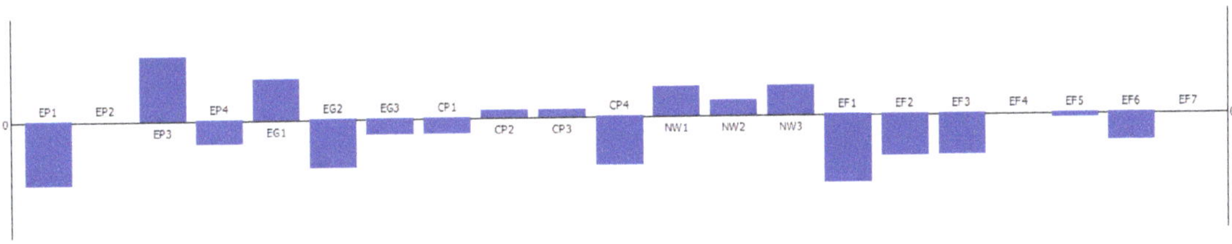

Figure 5. Scores of control engagement cluster on each evaluation indicator.

For the informed engagement cluster, the scores across various evaluation indicators are depicted in Figure 6. Analysis reveals higher scores on EG1 and EF1, juxtaposed with lower scores on EP2, EP3, EG2, CP1, CP2, CP4, NW3, EF3, EF4, and EF7. Increasing opportunities for residents to participate in decision making can enhance their comprehension of the significance of the SOCR [103]. Nevertheless, insufficient tracking mechanisms by the government and limited financial resources may diminish resident engagement in the implementation phase. Similarly, inadequate collaboration among communities, third-party organizations, construction parties, and residents may undermine residents' satisfaction with the renewal initiative and erode their trust in governmental authorities [104].

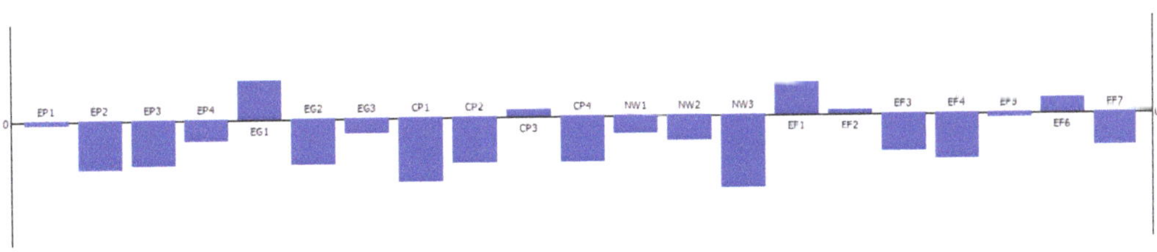

Figure 6. Scores of informed engagement cluster on each evaluation indicator.

For the non-engagement cluster, the scores on various evaluation indicators are illustrated in Figure 7, which exhibits notably low scores across the five dimensions, particularly in EP1 for empowerment and EG1 for participation [105]. The network, as a tool for acquiring external information, becomes more influential and informative with a denser and larger network size. A diverse array of information channels equips residents with the necessary tools to identify opportunities for engagement, enhances their participatory capabilities, and consequently augments the validity of their engagement [106,107].

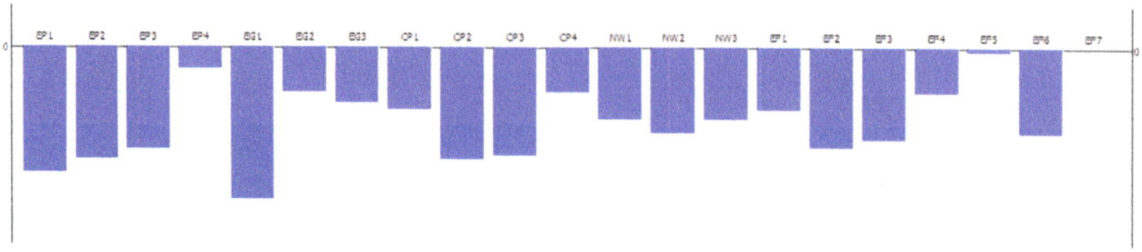

Figure 7. Scores of non-engagement cluster on each evaluation indicator.

For the passive attitude engagement cluster, the scores across various evaluation indicators are depicted in Figure 8. This cluster demonstrates higher scores on EP1, EG1, CP1, CP2, CP4, EF4, and EF5, juxtaposed with a lower score on EF3. The empowerment of residents to participate in decision-making processes fosters a heightened sense of ownership and engenders positive attitudes, thereby cultivating robust collaborative relationships among government departments, third-party organizations, and the residents themselves [1,23]. This enhancement of collaborative dynamics significantly bolsters residents' trust in government, which, in turn, facilitates the expedited implementation of projects [27,108].

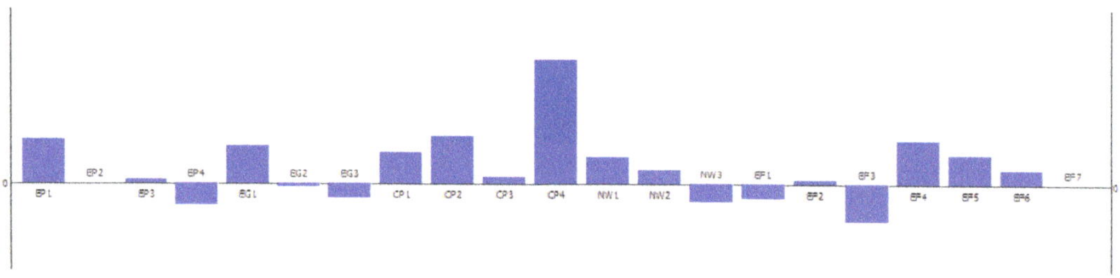

Figure 8. Scores of passive attitude engagement cluster on each evaluation indicator.

For the cooperation engagement cluster, the scores on various evaluation indicators are displayed in Figure 9. Analysis reveals higher scores on EP2, EF3, and EF6, while lower scores are evident for EG1, EG2, CP4, NW1, EF4, and EF5. The provision of a tracking mechanism by the government to address residents' issues plays a significant role in augmenting residents' satisfaction with the SOCR and effectively reducing complaints. However, to further bolster residents' trust in the government, it is essential for the government to offer guidance and provide opportunities for residents to participate actively in both the decision-making and implementation stages of the SOCR. Such measures are crucial for advancing the progress of project implementation. A lack of trust could hinder the success of resident engagement in the SOCR [57,109].

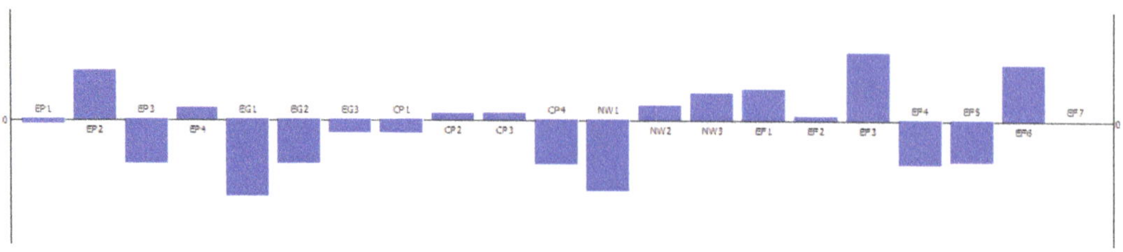

Figure 9. Scores of cooperation engagement cluster on each evaluation indicator.

The weak will engagement cluster scores on various evaluation indicators are shown in Figure 10. Although residents were afforded adequate rights to participate and a robust collaborative network among all parties was established, leading to higher satisfaction with the SOCR, the effectiveness of resident participation in enhancing perceptions of the SOCR was limited [105]. This limitation persisted in the absence of a stronger willingness among residents to engage actively. The government can develop different engagement strategies for residents of different ages and educational backgrounds, continuously increasing their willingness to engage [110].

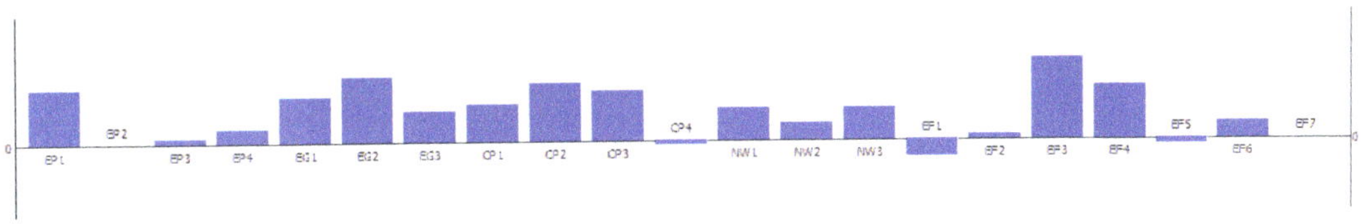

Figure 10. Scores of weak will engagement cluster on each evaluation indicator.

The full engagement cluster scores on various evaluation indicators are depicted in Figure 11. Notably, this cluster demonstrates high scores across the five dimensions of empowerment, participation, collaboration, network, and validity, particularly excelling in the indicators of empowerment and participation. It indicates that residents within this cluster possess ample participation rights, enjoy greater opportunities to engage in the SOCR, and maintain a positive collaborative relationship with all stakeholders. Consequently, a robust resident engagement network has been established, leading to highly effective resident engagement overall.

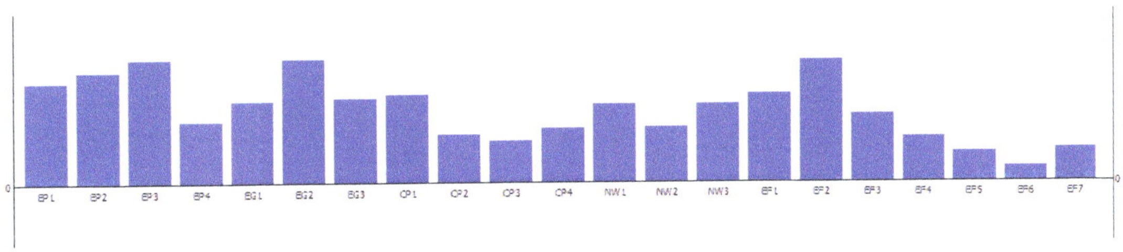

Figure 11. Scores of full engagement cluster on each evaluation indicator.

5.2. Differences in the Performance of Resident Engagement in the SOCR at City Level

The analysis above reveals a disparity in the levels of resident engagement among urban populations, which can be attributed to differences in resident engagement clusters. Large discrepancies in the built environment, community organization, property management, and other factors among different cities and communities give rise to diverse avenues for community development, subsequently leading to the formation of distinct resident engagement clusters across various urban areas [3,111]. Furthermore, this phenomenon may also be attributed to a combination of individual conditions, community capital, and environmental factors.

Regarding individual conditions, there is a positive correlation between residents' educational levels and their willingness to participate, suggesting that residents in areas with higher educational attainment exhibit a stronger inclination towards engagement [112]. Similarly, residents' cognitive levels also play a role in their willingness to engage in the community, as individuals with higher cognitive abilities are more inclined to partake in such activities [16]. The Q7, Q8, Q9, Q10, and Q11 communities in Shanghai possess comparatively superior educational resources among the five cities. This region not only hosts multiple high-quality schools but also actively enhances its educational resources by establishing strategic collaborations with prestigious universities such as Shanghai Jiao Tong University, East China Normal University, and Shanghai Normal University, aiming to co-develop several foundational education institutions. Consequently, the residents in these communities exhibit a higher overall cognitive level, and their willingness and cognitive engagement are better. Conversely, the Q4, Q5, and Q6 communities in the Guichi District of Chizhou City, despite recent vigorous efforts to advance educational development, still lag behind in educational attainment when compared to the aforementioned communities in Shanghai. This discrepancy highlights a need for ongoing educational enhancements in Chizhou to foster greater community engagement and participation.

With regard to community capital, the level of economic development significantly influences resident engagement in the SOCR, as higher economic development corresponds to a higher level of resident engagement with SOCR-related issues [29,113]. Economic growth drives local governments to prioritize resident engagement, and regions with advanced economic development have implemented more policies, such as social and volunteer activities, to facilitate resident engagement, thereby stimulating residents' enthusiasm for engagement [29,112,114]. For instance, Shanghai has a better level of economic development among the five cities and is capable of bearing higher project costs. The government has renovated public facilities such as roofs, corridors, and garages in residential areas, while also actively providing opportunities for residents to participate. Residents in Shanghai have shown a strong willingness to engage and openness toward the SOCR. In addition to Shanghai, Ningbo exhibits a relatively robust level of economic development, ranking among the top two cities in Zhejiang Province. Consequently, the Q15 community in Ningbo has benefited from substantial economic backing throughout the SOCR process, culminating in Q15 becoming the first community in Ningbo to successfully complete its SOCR initiative. This financial support has positively influenced resident participation.

As for the filed and environment, the spatial layout and facilities within a community significantly impact resident engagement, as inconvenient environments may limit engagement levels [115]. Person-friendly environments, which facilitate access to essential services and foster engagement in community activities, have been shown to positively affect residents' participation [116]. For example, residents in the Q3 community of Zhenjiang reported several environmental issues, including malodors emanating from the renovated overflow wells and irregularities in the road surfaces. Despite proactive efforts by the community to address these concerns, the pace of construction remained slow, which adversely impacted the daily lives and mobility of the residents, resulting in a diminished attitude towards participation and a reduced willingness to engage. Taking these factors into account, Shanghai's overall progress has been quite successful, which is reflected in the high proportion of full engagement clusters in Shanghai, thus leading to the city's top overall ranking.

5.3. *Implications for Improving Resident Engagement in the SOCR*

5.3.1. Theoretical Implications

According to the findings of this study, several theoretical implications emerge. First, this study extends the application of educational psychology and the ladder of citizen participation theory to the domain of resident engagement in the SOCR. Unlike previous research that focused on cognitive load theory and motivation within urban planning [117–119], this work utilizes these theories to develop a framework for classifying resident engagement clusters in SOCR. It not only expands the application scope of both theories but also enhances their relevance, providing a robust framework for future research on resident engagement in similar community renewal projects.

Second, this study develops a hybrid evaluation method by combining ANP and PROMETHEE II to quantify resident engagement in the SOCR. This method is particularly advantageous in handling interdependencies among multiple evaluation indicators and in providing a robust ranking of resident engagement levels [120,121]. In contrast to traditional evaluation methods, this method enhances reliability by accounting for complex, multidimensional interactions within the evaluation framework [76]. Such enhancements are critical as they offer practical insights for the decision-makers, enabling them to tailor strategies more effectively to improve resident engagement in the SOCR.

Third, this study highlights the vital link between individual engagement and the sustainable development of sponge city initiatives. Engaged residents not only enhance urban resilience to flooding and bolster public trust in government-led projects but also are crucial to the success of such initiatives [21]. By demonstrating how localized actions collectively advance macro-level goals, this research emphasizes the essential role of promoting residents' participation to ensure the effective implementation and success of sponge city.

5.3.2. Practical Implications

This study quantitatively evaluated the level of resident engagement level in the SOCR at both individual and city levels through the ANP-PROMETHEE II method. Considering the weight of each indicator in the indicator system and the impact of different indicators on resident engagement clusters, as well as the differences in resident engagement among cities, some targeted suggestions have been proposed to enhance resident engagement.

First, strengthening the popularization of knowledge related to the SOCR could improve resident engagement in the SOCR. Inadequate publicity mechanisms and residents' limited understanding of participation-related matters contribute to a lack of awareness and comprehension of the SOCR, ultimately impacting the level of resident engagement [16]. The extent of residents' understanding of renewal initiatives plays a pivotal role in shaping their attitudes and capabilities when engaging in community affairs [3,122]. If residents lack a fundamental understanding of the SOCR, then the policy that is based on social acceptance may be biased. Stated differently, if residents do not comprehend the policies related to SOCR, their implementation may face substantial impediments. Hence, it is imperative for the government to strengthen residents' awareness of the significance of SOCR and related information through various channels such as media and other communication tools [123,124].

Second, fostering communication among residents can enhance levels of resident engagement. In contrast to previous community renewal efforts, the SOCR, centered on resident engagement, advocates for a human-centric society wherein a transparent, equitable, and inclusive platform for resident engagement in community affairs management is essential [125], aiming to promote a sense of community as well as stronger community connections and community support [126]. More connections and frequent interactions with neighbors can contribute to a higher expectation of engagement and cooperation in the SOCR [8]. Only through fostering greater social ties within the neighborhood can residents perceive attitudes towards community renewal more widely and be susceptible to the influence of others' opinions [8]. Residents engaging in the SOCR initiatives not only foster enhanced public spirit, mutual understanding and civic pride but also increase positive feelings about the local community [124]. Specifically, optimizing a community feedback channel for information dissemination can assist participants in tracking the progress of SOCR initiatives. Moreover, maintaining effective communication platforms within these communities is crucial. This can involve establishing a public participation platform within the community, setting up online and offline workshops, and providing residents with avenues for open communication. Decision-makers can gain insight into residents' preferences through surveys and other feedback mechanisms [98]. Additionally, enhancing the operational capabilities of the owners' committee to address disputes and encourage resident communication is vital. As the owners' committee's effectiveness improves, resident satisfaction with the SOCR is likely to increase, further strengthening their willingness to actively engage in the process of the SOCR.

Third, boosting funding is crucial for enhancing the level of resident engagement in the SOCR. Government funding and subsidies play pivotal roles in influencing residents' satisfaction with the SOCR and serve as fundamental drivers for promoting their engagement [122]. Increasing government funding can profoundly impact resident engagement in a positive and proactive manner [127]. Providing ample material and spiritual incentives to residents in old communities may incentivize greater investment of their time, energy, and resources into participation [16]. To ensure financial support for the development of the SOCR, modern financial models such as public–private partnerships are recommended for private enterprises to invest in the implementation of smart community initiatives [58].

Furthermore, the actual situation of the SOCR varies from place to place, necessitating tailored approaches to resident engagement based on local conditions. Consequently, understanding the focal points of the SOCR in the local community is of importance.

6. Conclusions

Classifying and evaluating the engagement of residents in the SOCR is an effective way to facilitate resident engagement in the SOCR. To address the critical question of how to evaluate resident engagement, this study classified resident engagement clusters in the SOCR, proposed an integrated evaluation model combining the ANP and PROMETHEE II approaches to scientifically quantify the level of resident engagement in the SOCR and applied it in five selected cities in the Yangtze River Delta of China. Specifically, resident engagement in the SOCR was classified into 7 types drawing from the field of educational psychology and Arnstein's "a ladder of citizen participation". Then, the SLR and expert interviews were used to identify indicators for evaluating resident engagement, resulting in an evaluation indicator system with 5 primary indicators and 21 secondary indicators. Then, the ANP-PROMETHEE II method was adopted to evaluate the level of resident engagement.

There are several key findings in this study. At the individual level, the evaluation results indicate that "empowerment" has the highest weight among the primary indicators, while "collaboration" has the lowest weight. Among the engagement clusters, the full engagement cluster represented the highest level, while the non-engagement cluster indicated the lowest. Moreover, regarding performance at the city level, notable disparities were observed, with Shanghai exhibiting the highest level of resident engagement, followed by Ningbo, Jiaxing, Zhenjiang, and Chizhou. These findings not only enrich the theory of educational psychology and the ladder of citizen participation but also develop and expand the application of the ANP-PROMETHEE II method. Furthermore, this study bridges micro-level resident engagement with macro-level sustainability objectives in sponge city construction. Practically, the developed evaluation model can effectively evaluate the level of resident engagement in the SOCR, enabling the implementation of targeted improvement measures to enhance future resident engagement and promote sponge city initiatives.

However, this research has two limitations: the small size of expert-related data collected and the static nature of the evaluation model used, which does not reflect the dynamic process of sponge city implementation. Future research should consider collecting more resident-related data to evaluate the level of resident engagement in the SOCR. Given that the level of resident engagement may vary over time, it is advisable to monitor and dynamically evaluate the data to facilitate ongoing improvement and enable longitudinal comparisons.

Supplementary Materials: The following supporting information can be downloaded at: https://www.mdpi.com/article/10.3390/buildings14123773/s1. Supplementary Material A: Outline of expert interview on the evaluation system of resident engagement level in the SOCR; Supplementary Material B: Basic information statistics of experts; Supplementary Material C: Questionnaire about Evaluation of Resident Engagement clusters in the SOCR; Supplementary Material D: Questionnaire about the relationship and importance of evaluation indicators for resident engagement level in the SOCR; Supplementary Material E: Interaction relationship of resident engagement level in the SOCR.

Author Contributions: Conceptualization, T.G.; methodology, M.X.; software, Y.W., E.H. and Y.W.; validation, T.G., M.X., E.H. and Y.W.; formal analysis, T.G. and S.Z.; writing—original draft, M.X; writing—review and editing, T.G., E.H. and Y.W.; visualization, Y.W.; supervision, T.G.; project administration, T.G.; funding acquisition, T.G. All authors have read and agreed to the published version of the manuscript.

Funding: This study was supported by the National Natural Science Foundation of China (Grant No. 72104233) and the China Postdoctoral Science Foundation (Grant No. 2023M743767).

Data Availability Statement: The original contributions presented in the study are included in the article/Supplementary Material, further inquiries can be directed to the corresponding author.

Acknowledgments: The authors hereby express their special gratitude to all the respondents who presented the needed data with great patience, as well as the surveyors and interviewers who did their best in terms of data collection.

Conflicts of Interest: The authors declare that they have no conflict of interest.

References

1. Li, W.; Guo, J.; Yuan, J.; Liu, H.J.; Edwards, D.J. Exploring the key indicators of social impact assessment for sponge city PPPs: A sustainable development perspective. *Buildings* **2022**, *12*, 1329. [CrossRef]
2. Vitale, C.; Meijerink, S.; Moccia, F.D. Urban flood resilience, a multi-level institutional analysis of planning practices in the Metropolitan City of Naples. *J. Environ. Plan. Manag.* **2023**, *66*, 813–835. [CrossRef]
3. Li, Y.; Li, H.X.; Huang, J.; Liu, C. An approximation method for evaluating flash flooding mitigation of sponge city strategies—A case study of Central Geelong. *J. Clean. Prod.* **2020**, *257*, 120525. [CrossRef]
4. Myers, B.R.; Pezzaniti, D. Chapter 6-flood and peak flow management using WSUD systems. In *Approaches to Water Sensitive Urban Design*; Sharma, A.K., Gardner, T., Begbie, D., Eds.; Woodhead Publishing: Cambridge, UK, 2019. [CrossRef]
5. Radcliffe, J.C. History of water sensitive urban design/low impact development adoption in Australia and internationally. In *Approaches to Water Sensitive Urban Design*; Woodhead Publishing: Cambridge, UK, 2019; pp. 1–24. [CrossRef]
6. Yang, M.; Chen, D.; Shi, L.; He, J.; Liu, L.; Shang, X. How do construct a sponge city that can improve residents' satisfaction? Evidence from a suburb of Huizhou City, China. *Ecol. Indic.* **2022**, *142*, 109238. [CrossRef]
7. Mao, X.; Jia, H.; Shaw, L.Y. Assessing the ecological benefits of aggregate LID-BMPs through modelling. *Ecol. Model.* **2017**, *353*, 139–149. [CrossRef]
8. Huang, Y.; Lin, J.; He, X.; Lin, Z.; Wu, Z.; Zhang, X. Assessing the scale effect of urban vertical patterns on urban waterlogging: An empirical study in Shenzhen. *Environ. Impact Assess. Rev.* **2024**, *106*, 107486. [CrossRef]
9. Ellis, J.B.; Deutsch, J.C.; Mouchel, J.M.; Scholes, L.; Revitt, M.D. Multicriteria decision approaches to support sustainable drainage options for the treatment of highway and urban runoff. *Sci. Total Environ.* **2004**, *334*, 251–260. [CrossRef]
10. Xiong, H.; Sun, Y.; Ren, X. Comprehensive Assessment of Water Sensitive Urban Design Practices based on Multi-criteria Decision Analysis via a Case Study of the University of Melbourne, Australia. *Water* **2020**, *12*, 2885. [CrossRef]
11. Fletcher, T.D.; Shuster, W.; Hunt, W.F.; Ashley, R.; Butler, D.; Arthur, S.; Trowsdale, S.; Barraud, S.; Semadeni-Davies, A.; Bertrand-Krajewski, J.-L.; et al. SUDS, LID, BMPs, WSUD and more—The evolution and application of terminology surrounding urban drainage. *Urban Water J.* **2015**, *12*, 525–542. [CrossRef]
12. Liu, J.; Zhou, W. Impact of the Chinese Sponge City and underground utility tunnel construction on the infrastructure development in developing countries. In Proceedings of the ICCREM, Guangzhou, China, 10–12 November 2017; pp. 288–297. [CrossRef]
13. Dai, L.; van Rijswick HF, M.W.; Driessen PP, J.; Keessen, A.M. Governance of the Sponge City Programme in China with Wuhan as a case study. In *Urban Resilience to Droughts and Floods*; Routledge: London, UK, 2020; pp. 106–124.
14. Li, H.; Ding, L.; Ren, M.; Li, C.; Wang, H. Sponge city construction in China: A survey of the challenges and opportunities. *Water* **2017**, *9*, 594. [CrossRef]
15. Gu, T.; Li, D.; Zhu, S.; Wang, Y. Does sponge-style old community renewal lead to a satisfying life for residents? An investigation in Zhenjiang, China. *Habitat Int.* **2019**, *90*, 102004. [CrossRef]
16. Gu, T.; Hao, E.; Ma, L.; Liu, X.; Wang, L. Exploring the determinants of residents' behavior towards participating in the sponge-style old community renewal of China: Extending the theory of planned behavior. *Land* **2022**, *11*, 1160. [CrossRef]
17. Jiang, Y.; Qiu, L.; Gao, T.; Zhang, S. Systematic application of sponge city facilities at community scale based on SWMM. *Water* **2022**, *14*, 591. [CrossRef]
18. Wang, X.; Hawkins, C.; Berman, E. Financing sustainability and stakeholder engagement: Evidence from US cities. *Urban Aff. Rev.* **2014**, *50*, 806–834. [CrossRef]
19. Fu, X.; Qian, Q.K.; Liu, G.; Zhuang, T.; Visscher, H.J.; Huang, R. Overcoming inertia for sustainable urban development: Understanding the role of stimuli in shaping residents' participation behaviors in neighborhood regeneration projects in China. *Environ. Impact Assess. Rev.* **2023**, *103*, 107252. [CrossRef]
20. Zheng, S.; Fu, X.; Zhuang, T.; Wu, W. Exit, voice, loyalty, and neglect framework of residents' responses to urban neighborhood regeneration: The case of Shanghai, China. *Environ. Impact Assess. Rev.* **2023**, *100*, 107087. [CrossRef]
21. Xie, H.; Zhang, L.; Cui, P.; Yuan, J.; Li, Q. Exploring the Evolution Mechanisms of Social Risks Associated with Urban Renewal from the Perspective of Stakeholders. *Buildings* **2024**, *14*, 1470. [CrossRef]
22. Liu, G.; Huang, R.; Li, K.; Shrestha, A.; Wang, H.; Cai, M. Exploring the dilemma of enterprises participating in the old community renewal: Perspective of managers. *Cities* **2024**, *150*, 105073. [CrossRef]

23. Szczepańska, A.; Zagroba, M.; Pietrzyk, K. Participatory budgeting as a method for improving public spaces in major Polish cities. *Soc. Indic. Res.* **2022**, *162*, 231–252. [CrossRef]
24. Sun, C.; Liu, J.; Chen, Y. Impact of empowerment on public participation awareness in community renewal projects: Mediating role of trust and community identity. *Journal of Asian Architecture and Building Engineering*, 2024; Advanced online publication. 1–14. [CrossRef]
25. Dean, A.J.; Lindsay, J.; Fielding, K.S.; Smith, L.D. Fostering water sensitive citizenship–Community profiles of engagement in water-related issues. *Environ. Sci. Policy* **2016**, *55*, 238–247. [CrossRef]
26. Hall, D.M.; Gilbertz, S.J.; Anderson, M.B.; Ward, L.C. Beyond "buy-in": Designing citizen participation in water planning as research. *J. Clean. Prod.* **2016**, *133*, 725–734. [CrossRef]
27. Serrao-Neumann, S.; Harman, B.; Leitch, A.; Low Choy, D. Public engagement and climate adaptation: Insights from three local governments in Australia. *J. Environ. Plan. Manag.* **2015**, *58*, 1196–1216. [CrossRef]
28. Hurlbert, M.; Gupta, J. The split ladder of participation: A diagnostic, strategic, and evaluation tool to assess when participation is necessary. *Environ. Sci. Policy* **2015**, *50*, 100–113. [CrossRef]
29. Ye, K.; Liang, Y.; Shi, J. Evaluation and classification of public participation in EIA for transportation infrastructure megaprojects in China. *Environ. Impact Assess. Rev.* **2023**, *101*, 107138. [CrossRef]
30. Ding, L.; Ren, X.; Gu, R.; Che, Y. Implementation of the "sponge city" development plan in China: An evaluation of public willingness to pay for the life-cycle maintenance of its facilities. *Cities* **2019**, *93*, 13–30. [CrossRef]
31. Zhao, Y. The Difficulties and Explorations of Old Community Retrofitting Based on the Sponge City Concept-The case of 240 community in the old urban area of Xi'an. *Highlights Bus. Econ. Manag.* **2024**, *26*, 124–129. [CrossRef]
32. Zhang, F.; Chan, A.P.; Chen, L.; Li, D.; Cui, P. Exploring and measuring the health resilience of urban buildings against the pandemic: A case study of Hong Kong public housing during COVID-19. *Int. J. Disaster Risk Reduct.* **2024**, *103*, 104343. [CrossRef]
33. Díez, M.A.; Etxano, I.; Garmendia, E. Evaluating participatory processes in conservation policy and governance: Lessons from a Natura 2000 pilot case study. *Environ. Policy Gov.* **2015**, *25*, 125–138. [CrossRef]
34. Kozová, M.; Dobšinská, Z.; Pauditšová, E.; Tomčíková, I.; Rakytová, I. Network and participatory governance in urban forestry: An assessment of examples from selected Slovakian cities. *For. Policy Econ.* **2018**, *89*, 31–41. [CrossRef]
35. Huo, X.; Xue, H.; Xu, X.; Hao, T.; Jiao, L. A risk sharing model for old community renewal project based on bargaining game model. *Sci. Rep.* **2024**, *14*, 24316. [CrossRef]
36. Chen, Y.; Qu, L. Emerging participative approaches for urban regeneration in Chinese megacities. *J. Urban Plan. Dev.* **2020**, *146*, 04019029. [CrossRef]
37. Kim, G.; Newman, G.; Jiang, B. Urban regeneration: Community engagement process for vacant land in declining cities. *Cities* **2020**, *102*, 102730. [CrossRef] [PubMed]
38. Chi, C.S.; Ruuska, I.; Xu, J. Environmental impact assessment of infrastructure projects: A governance perspective. *J. Environ. Plan. Manag.* **2016**, *59*, 393–413. [CrossRef]
39. Cullingworth, B.; Nadin, V.; Hart, T.; Davoudi, S.; Pendlebury, J.; Vigar, G.; Webb, D.; Townshend, T. *Town and Country Planning in the UK*; Routledge: London, UK, 2006.
40. Mthembu, Z.; Mogaka, J.J.; Chimbari, M.J. Community engagement processes in low-and middle-income countries health research settings: A systematic review of the literature. *BMC Health Serv. Res.* **2023**, *23*, 457. [CrossRef] [PubMed]
41. Gautam, V.; Bhalla, S. Why residents exhibit environmentally responsible behavior? *J. Clean. Prod.* **2023**, *427*, 139253. [CrossRef]
42. Manturuk, K.; Lindblad, M.; Quercia, R. Homeownership and civic engagement in low-income urban neighborhoods: A longitudinal analysis. *Urban Aff. Rev.* **2012**, *48*, 731–760. [CrossRef]
43. Zhu, Y. Toward community engagement: Can the built environment help? Grassroots participation and communal space in Chinese urban communities. *Habitat Int.* **2015**, *46*, 44–53. [CrossRef]
44. Khaledi Koure, F.; Hajjarian, M.; Hossein Zadeh, O.; Alijanpour, A.; Mosadeghi, R. Ecotourism development strategies and the importance of local community engagement. *Environ. Dev. Sustain.* **2023**, *25*, 6849–6877. [CrossRef]
45. Flora, C.B. Social aspects of small water systems. *J. Contemp. Water Res. Educ.* **2004**, *128*, 6–12. [CrossRef]
46. Cousins, J.J. Infrastructure and institutions: Stakeholder perspectives of stormwater governance in Chicago. *Cities* **2017**, *66*, 44–52. [CrossRef]
47. Liu, Y. Research on the Dilemma and Path of Resident Participation in Urban Community Governance in China. *Soc. Public Welf.* **2020**, *11*, 44–47. (In Chinese)
48. Fredricks, J.A.; Blumenfeld, P.C.; Paris, A.H. School engagement: Potential of the concept, state of the evidence. *Rev. Educ. Res.* **2004**, *74*, 59–109. [CrossRef]
49. Arnstein, S.R. A ladder of citizen participation. *J. Am. Plan. Assoc.* **2019**, *85*, 24–34. [CrossRef]
50. Palamenghi, L.; Usta, D.; Leone, S.; Graffigna, G. Food-related behavioral patterns in patients with inflammatory bowel diseases: The role of food involvement and health engagement. *Nutrients* **2024**, *16*, 1185. [CrossRef] [PubMed]
51. Chen, F.; Li, S.; Lin, L.; Huang, X. Identifying temporal changes in student engagement in social annotation during online collaborative reading. *Educ. Inf. Technol.* **2024**, *29*, 16101–16124. [CrossRef]
52. Xu, L.; Yang, F. Multi-level analysis of community participation: A two dimensional perspective based on subject's intention and ability. *J. Beijing Adm. Coll.* **2016**, *6*, 92–99. (In Chinese) [CrossRef]

53. Levenda, A.M.; Keough, N.; Rock, M.; Miller, B. Rethinking public participation in the smart city. *Can. Geogr./Géographe Can.* **2020**, *64*, 344–358. [CrossRef]
54. Presenza, A.; Del Chiappa, G.; Sheehan, L. Residents' engagement and local tourism governance in maturing beach destinations. Evidence from an Italian case study. *J. Destin. Mark. Manag.* **2013**, *2*, 22–30. [CrossRef]
55. Fung, A. Putting the public back into governance: The challenges of citizen participation and its future. *Public Adm. Rev.* **2015**, *75*, 513–522. [CrossRef]
56. Karanth, K.K.; Nepal, S.K. Local residents perception of benefits and losses from protected areas in India and Nepal. *Environ. Manag.* **2012**, *49*, 372–386. [CrossRef]
57. Liu, H.; Gao, H.; Huang, Q. Better government, happier residents? Quality of government and life satisfaction in China. *Soc. Indic. Res.* **2020**, *147*, 971–990. [CrossRef]
58. Luetke Lanfer, H.; Brew-Sam, N.; Rossmann, C. An analysis of power dynamics affecting handwashing interventions in Sierra Leone: Findings from a qualitative participatory study. *JMIR Form. Res.* **2023**, *7*, e39226. [CrossRef]
59. Li, Z.; Zhang, H.; O'Brien, L.; Cai, R.; Flint, S. On evaluating commercial cloud services: A systematic review. *J. Syst. Softw.* **2013**, *86*, 2371–2393. [CrossRef]
60. Speer, J. Participatory governance reform: A good strategy for increasing government responsiveness and improving public services? *World Dev.* **2012**, *40*, 2379–2398. [CrossRef]
61. Bas, E.; Yolcu, U.; Egrioglu, E. Picture fuzzy regression functions approach for financial time series based on ridge regression and genetic algorithm. *J. Comput. Appl. Math.* **2020**, *370*, 112656. [CrossRef]
62. Zhao, G. *Government Reform: Institutional Innovation and Participatory Governance: A Hangzhou Empirical Study of Local Government Governance Reform*; Zhejiang University Press: Hangzhou, China, 2013. (In Chinese)
63. Lin, J.; Deng, A.; Zhao, X.; Kuang, F.; Lin, W. Analysis on influence of public participation on satisfaction of land consolidation plan. *China Land Sci.* **2018**, *32*, 54–60. (In Chinese)
64. Johnson, A.L. *Engaging Queenslanders: Evaluating Community Engagement*; Department Communities: Brisbane, QLD, Australia, 2004.
65. Yu, X.; Zhao, Y. *A Study on Participatory Governance at the Local Level*; Zhejiang University Press: Hangzhou, China, 2009. (In Chinese)
66. Wang, J. *Participatory Governance: A Positive Study on Urban Community Building in China*; China Social Science Press: Beijing, China, 2006. (In Chinese)
67. March, M.P.I.H. *Guide to Evaluating Participatory Processes*; Department of Governance and Institutional Relations: Barcelona, Spain, 2013.
68. Nadeem, O.; Fischer, T.B. An evaluation framework for effective public participation in EIA in Pakistan. *Environ. Impact Assess. Rev.* **2011**, *31*, 36–47. [CrossRef]
69. Zhao, N.N.; Liu, Y.T.; Liu, Z. Co-wisdom, co-strategy, and co-benefit: Urban regeneration and community governance in Guangzhou City. *Urban Stud.* **2019**, *26*, 117–124. (In Chinese)
70. Zhao, G. The practice, impact variables, and application limits of participatory governance. *J. Gansu Adm. Inst.* **2015**, *2*, 43–51. (In Chinese)
71. Zhang, L. The study on perceived organization support on members' satisfaction of cooperative: Based on the regulating of members' participation. *Stat. Inf. Forum* **2014**, *29*, 84–91. (In Chinese)
72. Yin, J.; Wang, J.; Wang, C.; Wang, L.; Chang, Z. CRITIC-TOPSIS based evaluation of smart community governance: A case study in China. *Sustainability* **2023**, *15*, 1923. [CrossRef]
73. Gu, T.; Wang, Y.; Wang, L.; Hao, E.; Wang, C.; Xie, M. Quantifying Residents' Perceptions of Smart Community Construction through a Hybrid EWM-PROMETHEE II Method: A Case Study of Shenzhen, China. *Land* **2024**, *13*, 1197. [CrossRef]
74. Si, W.; Niu, L. Enhancing Human Reliability Prediction in Smart Tower Crane Interfaces: A Refined Approach Using Simplified Plant Analysis Risk-Human Reliability Assessment and the Decision Making Trial and Evaluation Laboratory-Analytic Network Process. *Buildings* **2024**, *14*, 1083. [CrossRef]
75. Santos, D.S.; Primo, R.G.B.; de Araújo Lima, A.P.H.G.; Schramm, V.B.; Rodrigues, Y.V.S.; Belderrain, M.C.N.; Pessoa, F.L.P.; de Araújo Kalid, R.; Callefi, M.H.B.M. Evaluation of the social impacts of small-and medium-sized biorefineries in the Southern Coast Territory of Bahia considering the selection of technologies for bioactives: An MCDA model. *Environ. Dev. Sustain.* **2024**, *26*, 13117–13137. [CrossRef]
76. He, Z.; Chen, H.; Yan, H.; Yin, Y.; Qiu, Q.; Wang, T. Scenario-based comprehensive assessment for community resilience adapted to fire following an earthquake, implementing the analytic network process and preference ranking organization method for enriched evaluation II techniques. *Buildings* **2021**, *11*, 523. [CrossRef]
77. Hamzeh, M.; Ali Abbaspour, R.; Davalou, R. Raster-based outranking method: A new approach for municipal solid waste landfill (MSW) siting. *Environ. Sci. Pollut. Res.* **2015**, *22*, 12511–12524. [CrossRef]
78. Shuowei, B.; Huisingh, D.; Wang, Q.; Qiu, H.; Elwert, T.; Jia, S.; Wang, Y.; Sun, Y. Green multiple-criteria decision-making method of selection of diamond cutting tool for promoting cleaner production in the natural stone industry. *J. Clean. Prod.* **2023**, *415*, 137580. [CrossRef]
79. Saaty, T.L. *Theory and Applications of the Analytic Network Process: Decision Making with Benefits, Opportunities, Costs, and Risks*; RWS Publications: Bangalore, India, 2005.

80. Wan, X.; Liu, X.; Du, Z.; Du, Y. A novel model used for assessing supply chain sustainability integrating the ANP and ER approaches and its application in marine ranching. *J. Clean. Prod.* **2021**, *279*, 123500. [CrossRef]
81. Saaty, T.L.; Vargas, L.G. *Decision Making with the Analytic Network Process: Economic, Political, Social and Technological Applications with Benefits, Opportunities, Costs and Risks*; Springer Science and Business Media: Dordrecht, The Netherlands, 2013; Volume 195.
82. Brans, J.P. L'ingénierie de la décision. Elaboration d'instruments d'aide à la décision. Méthode PROMETHEE. In *L'aide à la Décision: Nature, Instruments et Perspectives D'avenir*; Nadeau, R., Landry, M., Eds.; Presses de L'université Laval: Quebec City, QC, Canada, 1982; pp. 183–214.
83. Brans, J.P.; Vinke, P. A preference ranking organization method (the PROMETHEE method for multiple criteria decision making). *Manag. Sci.* **1985**, *31*, 647–656. [CrossRef]
84. Vavatsikos, A.P.; Demesouka, O.E.; Anagnostopoulos, K.P. GIS-based suitability analysis using fuzzy PROMETHEE. *J. Environ. Plan. Manag.* **2020**, *63*, 604–628. [CrossRef]
85. Abedi, M.; Torabi, S.A.; Norouzi, G.-H.; Hamzeh, M.; Elyasi, G.-R. PROMETHEE II: A knowledge-driven method for copper exploration. *Comput. Geosci.* **2012**, *46*, 255–263. [CrossRef]
86. Brans, J.P.; Mareschal, B. Promethee methods. In *Multiple Criteria Decision Analysis: State of the Art Surveys*; Springer: Berlin/Heidelberg, Germany, 2015; pp. 163–186.
87. Jia, H.; Wang, Z.; Zhen, X.; Clar, M.; Yu, S.L. China's sponge city construction: A discussion on technical approaches. *Front. Environ. Sci. Eng.* **2017**, *11*, 18. [CrossRef]
88. Wang, H.; Mei, C.; Liu, J.; Shao, W. A new strategy for integrated urban water management in China: Sponge city. *Sci. China Technol. Sci.* **2018**, *61*, 317–329. [CrossRef]
89. Xiáng, Y.; Wáng, J.; Li, Y.T. AME survey-003 A2: On the attractiveness of an medicine career in current China with a survey of 7508 medical professionals and 443 non-medical professionals. *Quant. Imaging Med. Surg.* **2016**, *6*, 84. [CrossRef]
90. Liao, C.; Zhao, D.; Zhang, S. Psychological and conditional factors influencing staff's takeaway waste separation intention: An application of the extended theory of planned behavior. *Sustain. Cities Soc.* **2018**, *41*, 186–194. [CrossRef]
91. Cochran, W.G. *Sampling Techniques*, 2nd ed.; John Wiley and Sons: New York, NY, USA, 1963.
92. Kotrlik, J.; Higgins, C. Organizational research: Determining appropriate sample size in survey research appropriate sample size in survey research. *Inf. Technol. Learn. Perform. J.* **2001**, *19*, 43.
93. Taherdoost, H. Sampling methods in research methodology; how to choose a sampling technique for research. *Int. J. Acad. Res. Manag.* **2016**, *5*, 18–27. [CrossRef]
94. Imai, K. Survey Sampling. *Curr. Sociol.* **1998**, *46*, 75–87. [CrossRef]
95. Singh, S. Simple Random Sampling. In *Advanced Sampling Theory with Applications*; Springer: Berlin/Heidelberg, Germany, 2003. [CrossRef]
96. Xing, Q.; Yao, W. Digital governance and its benchmarking college talent training under the rural revitalization in China—A case study of Yixian County (China). *Front. Public Health* **2022**, *10*, 984427. [CrossRef]
97. Doorn, N. Governance experiments in water management: From interests to building blocks. *Sci. Eng. Ethics* **2016**, *22*, 755–774. [CrossRef]
98. Mahlangu, P.; Goudge, J.; Vearey, J. Towards a framework for multisector and multilevel collaboration: Case of HIV and AIDS governance in South Africa. *Glob. Health Action* **2019**, *12*, 1617393. [CrossRef] [PubMed]
99. González, A.; Therivel, R.; Lara, A.; Lennon, M. Empowering the public in environmental assessment: Advances or enduring challenges? *Environ. Impact Assess. Rev.* **2023**, *101*, 107142. [CrossRef]
100. Heselmans, A.; Delvaux, N.; Laenen, A.; Van de Velde, S.; Ramaekers, D.; Kunnamo, I.; Aertgeerts, B. Computerized clinical decision support system for diabetes in primary care does not improve quality of care: A cluster-randomized controlled trial. *Implement. Sci.* **2020**, *15*, 5. [CrossRef] [PubMed]
101. Kurtenbach, S.; Küchler, A.; Zick, A. Prevention of Neighborhood Effects on the Susceptibility to Radicalization: Results of a Comparative Study in Germany. *Urban Aff. Rev.* **2024**, *advanced online publication*, 10780874241259423. [CrossRef]
102. Akinola, G.A.; Fagbenle, O.I.; Ogunde, A.O. Exploring stakeholders' perspectives on the barriers to the application of cost-reduction techniques in public higher educational building delivery. *Buildings* **2024**, *14*, 2551. [CrossRef]
103. Zwick, A.; Spicer, Z. Examining the Smart City Generational Model: Conceptualizations, Implementations, and Infrastructure Canada's Smart City Challenge. *Urban Aff. Rev.* **2024**, *60*, 1229–1253. [CrossRef]
104. Wang, Y.; Cai, J.; Zuo, J.; Bartsch, K.; Huang, M. Conflict or consensus? Stakeholders' willingness to participate in China's Sponge City program. *Sci. Total Environ.* **2021**, *769*, 145250. [CrossRef]
105. Rich, M.J.; Stoker, R.P. Rethinking empowerment: Evidence from local empowerment zone programs. *Urban Aff. Rev.* **2010**, *45*, 775–796. [CrossRef]
106. Yang, J.; Zhang, T.; Zhang, L. Research on the influence of digital penetration on the entrepreneurial behavior tendency of rural residents in tourism. *Environ. Dev. Sustain.* **2024**, *26*, 25549–25567. [CrossRef]
107. Cui, P.; You, Z.; Shi, Q.; Feng, L. Research on the factors influencing the epidemic resilience of urban communities in China in the post-epidemic era. *Buildings* **2024**, *14*, 2838. [CrossRef]
108. Di Napoli, I.; Dolce, P.; Arcidiacono, C. Community trust: A social indicator related to community engagement. *Soc. Indic. Res.* **2019**, *145*, 551–579. [CrossRef]

109. Khan, M.Z.; Shah, S.M.; Khan, A.A. Assessing conservation attitudes of mountain communities under different resource management regimes in northern Pakistan. *Environ. Dev. Sustain.* **2023**, *25*, 2550–2570. [CrossRef]
110. Yao, W.; Zhang, T.; Fukuda, H.; Wang, M.; Yang, J. Assessing Utilization and Comfort in a Village Scenic Park: Implications for Rural Revitalization. *Buildings* **2024**, *14*, 1538. [CrossRef]
111. Zeng, C.; Deng, X.; Yang, J.; Cheng, Y.; Zhao, T.; Wang, P. The exploration of residents' perception of eco-urbanization at community and driving factors in China. *Cities* **2020**, *122*, 103513. [CrossRef]
112. Lu, J.; Liu, L.; Wang, Y.; Zhou, Z. Social Engagement and Urban–Rural Disparity in Self-Management Behaviors: Study of Middle-Aged and Older Chinese Hypertension Patients. *Front. Public Health* **2022**, *9*, 801307. [CrossRef]
113. Zhang, X.; Jennings, E.T.; Zhao, K. Determinants of environmental public participation in China: An aggregate level study based on political opportunity theory and post-materialist values theory. *Policy Stud.* **2018**, *39*, 498–514. [CrossRef]
114. Ye, K.; Guo, Z.; Zhang, W.; Liang, Y. Heterogeneous environmental policy tools for expressway construction projects: A crossregional analysis in China. *Environ. Impact Assess. Rev.* **2022**, *97*, 106907. [CrossRef]
115. Noring, L.; Ohler, L.P.; Struthers, D. City government capacity and patterns in urban development project governance. *Urban Aff. Rev.* **2021**, *57*, 1343–1371. [CrossRef]
116. Ripat, J.D.; Redmond, J.D.; Grabowecky, B.R. The winter walkability project: Occupational therapists' role in promoting citizen engagement. *Can. J. Occup. Ther.* **2010**, *77*, 7–14. [CrossRef]
117. Hassan, W.; Martella, A.M.; Robinson, D.H. Identifying the Most Cited Articles and Authors in Educational Psychology Journals from 1988 to 2023. *Educ. Psychol. Rev.* **2024**, *36*, 94. [CrossRef]
118. Puskás, N.; Abunnasr, Y.; Naalbandian, S. Assessing deeper levels of participation in nature-based solutions in urban landscapes–A literature review of real-world cases. *Landsc. Urban Plan.* **2021**, *210*, 104065. [CrossRef]
119. Rahmayanti, K.P.; Rukmana, D. Scoping review of the depth of participation and barriers to community participation in post-disaster housing reconstruction. *Int. J. Disaster Risk Reduct.* **2024**, *104*, 104375. [CrossRef]
120. Xu, J.; Li, L.; Ren, M. A hybrid ANP method for evaluation of government data sustainability. *Sustainability* **2022**, *14*, 884. [CrossRef]
121. Wątróbski, J. Temporal PROMETHEE II—New multi-criteria approach to sustainable management of alternative fuels consumption. *J. Clean. Prod.* **2023**, *413*, 137445. [CrossRef]
122. Gao, H.; Wang, T.; Gu, S. A study of resident satisfaction and factors that influence old community renewal based on community governance in Hangzhou: An empirical analysis. *Land* **2022**, *11*, 1421. [CrossRef]
123. Sun, C.; Min, J.; Li, J.; Cai, W. Public participation and policy evaluation in China's smog governance. *Environ. Impact Assess. Rev.* **2023**, *100*, 107052. [CrossRef]
124. Selim, G.; Farhan, S.L. Reactivating voices of the youth in safeguarding cultural heritage in Iraq: The challenges and tools. *J. Soc. Archaeol.* **2024**, *24*, 58–78. [CrossRef]
125. Duan, X.; Xu, T.M.; Duan, C.H. Research on renewal of old communities based on public participation—A case study of Gejiaying. *IOP Conf. Ser. Earth Environ. Sci.* **2020**, *510*, 062007. [CrossRef]
126. Dickinson, P.R.; Morrison, P.S. Aversion to local wellbeing inequality is moderated by social engagement and sense of community. *Soc. Indic. Res.* **2022**, *159*, 907–926. [CrossRef]
127. Habibah, U.; Hassan, I.; Iqbal, M.S.; Naintara. Household behavior in practicing mental budgeting based on the theory of planned behavior. *Financ. Innov.* **2018**, *4*, 28. [CrossRef]

buildings

MDPI

Article

Exploring the Evolution Mechanisms of Social Risks Associated with Urban Renewal from the Perspective of Stakeholders

Hong Xie [1], Lei Zhang [2], Peng Cui [3], Jingfeng Yuan [1] and Qiming Li [1,*]

1 Department of Construction and Real Estate, School of Civil Engineering, Southeast University, Nanjing 211189, China
2 Research Center of Smart City, Nanjing Tech University, No. 30 Puzhu Rd., Nanjing 211816, China
3 Department of Engineering Management, Nanjing Forestry University, Nanjing 210037, China
* Correspondence: seu_liqm@163.com; Tel.: +86-135-0518-6838

Abstract: The urban renewal (UR) process involves various stakeholders and related activities, and the various risks arising from this endeavor can affect these stakeholders. Additionally, the impact of adverse factors such as policy discontinuity and inequitable distribution of benefits among stakeholders can easily result in collective tensions or conflicts, as well as the gradual emergence of potential social frictions and confrontations. These social risks (SRs) not only impede the smooth execution of urban renewal projects but also pose challenges to social harmony and stability. Hence, to mitigate and control the SRs in the UR process (URSRs) and ensure the successful implementation of effective and sustainable UR projects, it is of paramount importance to gain a comprehensive understanding of the occurrence and evolution mechanisms of these SRs. Although existing studies have touched upon the influence of stakeholder conflicts on URSRs, there remains a lack of systematic examination of the evolution mechanisms of these risks from the perspective of stakeholder theory. The resulting fragmented and specialized comprehension of URSRs has hindered the effectiveness of risk governance strategies. This study adopts stakeholder theory to analyze the potential sources of risk throughout the entire UR process. By considering the conflicts of interests among stakeholders, a systematic analysis of the evolution mechanisms of URSRs is explored and targeted governance recommendations for URSRs are proposed.

Keywords: urban renewal; stakeholders; social risks; structural equation modelling; evolution mechanism

1. Introduction

Urban renewal (UR) involves the transformation of urban regions, particularly those grappling with urban decay, achieved through measures such as functional alterations, demolition, reconstruction, and comprehensive management [1,2]. This process serves as a vital undertaking to equip cities to cope with emerging development trends and address the challenges posed by modernization and multifaceted disasters [3]. As a pivotal instrument of government in urban development, UR appreciably impacts the socioeconomic, cultural, and ecological fabric of specific regions, thereby bearing direct relevance to the well-being of the general populace [4,5]. By the end of 2022, China's urbanization rate soared to an impressive 65.22%, signifying the commencement of the stock development phase [6]. During this period, megacities and megalopolises continued to exert a persistent allure for residence and employment, becoming focal points of intensive construction activities [7]. As the urbanization journey enters its middle and later stages in China, numerous cities are expected to embrace a dual approach, equally prioritizing new construction and renovation of existing buildings, thus intertwining the processes of urbanization and UR [8]. UR not only constitutes a crucial mechanism for transforming urban development and construction patterns, as well as innovating land utilization methods, but also represents a key driver for advancing urban governance and elevating urban vitality [9].

UR goes beyond the mere construction and sale stages of real estate projects; it entails processes such as relocation, repurchasing, transformation, reconstruction, and resettlement of residents [5]. As secondary development projects with established residential neighborhoods nearby, areas earmarked for renewal often suffer from disarrayed layouts, inadequate supporting infrastructure, deterioration of sanitation conditions, disaster preparedness, and environmental quality, and a high-density population—comprising a significant number of migrants—with low educational attainment, limited adaptability to change, and a reduced capacity to cope with environmental shifts [9,10]. These unfavorable conditions could easily ignite dissatisfaction among neighboring residents. Mishandling such issues can act as catalysts for social risks (SRs), potentially escalating into significant social disruptions, thereby posing a substantial threat to both the progress of UR projects and the social stability within affected areas [11]. Uncertainties raised by SRs stemming from human practices or social factors (such as uneven distribution of benefits, discontent with government policies, and a rising population of unemployed individuals) can jeopardize social stability and order, further leading to social conflicts [12]. When the uncertainties accumulate to a critical point, SRs can transform into full-fledged social crises, threatening social stability, balance, and sustainable development [13,14].

The UR process involves various stakeholders and related activities, and the various risks arising from this endeavor can influence these stakeholders. The key stakeholders in UR are the public, government, and developers [15]. The public constitutes the most direct stakeholders, whereas the government assumes a pivotal role as the primary public entity and leader [16,17]. On the other hand, developers function as entities driven by commercial interests [8]. Throughout the UR process, the dynamics among these three major stakeholders impact social stability. Although UR brings forth substantial economic and societal benefits, it also gives rise to a range of social problems that directly influence public interests and rights [8,15]. These issues encompass land expropriation, relocation, and the transformation of living and ecological environments, leading certain stakeholders to experience objective or subjective repercussions from their interests being adversely impacted, prompting them to take action [2,18]. Particularly noteworthy is the potential lack of public involvement in decision-making mechanisms during the renewal process, which can result in the genuine demands of the majority not being adequately reflected in the planning stages and their exclusion from active participation in the actual decision-making process [15,19]. These consequences may, in turn, foster or even exacerbate social alienation [20]. Additionally, the impact of adverse factors such as policy discontinuity and inequitable distribution of benefits among stakeholders can easily result in collective tensions or conflicts, as well as the gradual emergence of potential social frictions and confrontations [8,21]. These SRs not only impede the smooth execution of UR projects but also pose challenges to social harmony and stability [22]. As large-scale UR persists in the future, it is inevitable that various social issues will arise, with those affecting social stability or leading to collective incidents of resistance taking center stage [2,22].

Hence, to mitigate and control the SRs in the UR process (hereinafter referred to as URSRs) and ensure the successful implementation of effective and sustainable UR projects, it is of paramount importance to gain a comprehensive understanding of the occurrence and evolution mechanisms of these SRs [23]. To date, prior studies have examined URSRs. For example, Liu et al. [20] employed inclusive development theory to elucidate the extent to which various social exclusion risks arising from UR impacted the original residents. Bai et al. [15] proposed that the success of UR projects depends on the effective mediation of conflicts among stakeholders. Mai et al. [11] investigated different stakeholders involved in UR projects and their interactions, offering management strategies for addressing SRs based on stakeholders' actions. Although existing research has touched upon the influence of stakeholder conflicts on URSRs, a systematic examination of the evolution mechanisms of these risks from the perspective of stakeholder theory is still necessary. The fragmented and specialized comprehension of URSRs needs to be overcome to improve the effectiveness of risk governance strategies.

To address these shortcomings, this study adopts stakeholder theory to analyze the potential sources of risk throughout the entire UR process. By considering the conflicts of interest among stakeholders, a systematic analysis of the evolution mechanisms of URSRs is conducted. Moreover, the study employs structural equation modeling (SEM) to validate the URSR evolution paths, ultimately leading to the establishment of a model depicting the evolution and evolution of these risks. Based on this model, critical SR factors are identified and targeted governance strategies for URSRs are proposed. This study not only strengthens stakeholders' understanding and awareness of URSRs but also optimizes the knowledge framework concerning the evolution of SRs at the urban level. Furthermore, it provides stakeholders with a scientific and rational basis for setting risk monitoring points throughout the entire UR process, thus facilitating resource allocation.

2. Literature Review

2.1. URSR Factor Idenfication

URSRs, similar to general risks, encompass SR factors, SR events, and SR consequences [21,24]. SR factors represent the underlying sources of URSRs, such as issues related to expropriation and demolition [1], management [8], the environment [25], policy [1,25], and cultural heritage preservation [1,8]. SR events are typically classified into two categories: individual extreme events and collective events [26]. Collective events refer to scenarios where a specific number of individuals, forming a coincidental group with shared interests, gather together, either legally or illegally, to express demands or claims, which can have a profound impact on social order and stability [26]. Based on the three dimensions of UR, namely, the project, the public, and the government, the consequences of URSRs can be categorized into several outcomes: the UR project may either continue, be delayed or suspended, or be terminated following coordination efforts; the scale and casualties of public incidents; and whether the local government is held accountable by higher authorities [11,26]. URSRs originate from SR factors, facilitated by various risk carriers such as human resources, finances, and the environment. They propagate to different risk trigger points, where risks continuously interpenetrate, intertwine, and accumulate with both internal and external risk flows, leading to a continuous buildup of risk energy [11,26]. Upon reaching a critical threshold, the risk energy can be triggered by specific conditions and surpass the critical threshold, leading to the emergence of individual risk events [26,27]. If individual risk events are not effectively addressed, they may escalate, culminating in individual risk events or collective risk events, ultimately causing losses at project, public, and governmental levels [28].

URSR factors can be divided into internal and external SR factors [2]. Internal SRs correspond to internal stakeholders and encompass a collection of uncertain factors that are directly related to the project during the site selection, land expropriation, and demolition phases, and they have the potential to trigger social instability and disorder [11,27]. On the other hand, external SRs are linked to external stakeholders and pertain to the social instabilities that emerge as a result of internal risk factors, such as potential environmental pollution in the area during the operational phase of the project [11,29]. Considering the implementation period of UR projects, the SR factors encompass land expropriation, building demolition, financial compensation, environmental impacts, relocation arrangements, and government policies concerning public infrastructure [2,28].

UR typically takes place in inner-city areas with high land and property prices in the vicinity [1,2]. Additionally, the involvement of developers in UR projects often leads to compensation and demolition approaches that differ from those used in traditional government-led initiatives [3,5]. Consequently, property owners' expectations regarding compensation for their properties tend to be higher [6,9]. Throughout the expropriation and demolition process, many issues, including inadequate compensation packages, forced evictions, insufficient resettlement arrangements, and rising rents, can evoke feelings of discontent among property owners and trigger collective events [11]. Thus, there is an inherent risk associated with expropriation and demolition (EDR) during the UR process [11].

The well-being of residents is closely interlinked with environmental factors, which primarily comprise the natural and cultural environments [30,31]. Within the natural environment, pollution arises from various sources in the project's environment, such as potential noise, water, dust, and radiation hazards, as well as the generation of construction waste [15]. Additionally, these projects can lead to disruptions in the existing road and transportation networks, resulting in traffic congestion [5]. On the other hand, older city areas have a long-standing presence, embodying the city's historical development and holding sentimental value for its inhabitants [3,22]. Therefore, it is crucial to approach the transformation of these areas with due respect for local culture and the viewpoints of residents, while also considering the uncertainties that cultural differences may bring [5]. The risk factors associated with the cultural environment involve the potential removal of or damage to cultural landmarks within the renewal area [32]. Moreover, these projects may conflict with the cultural beliefs and values of ethnic minorities residing in the area, as well as pose threats to the area's original geomancy [5,32].

UR projects encompass not only reinforcing and preserving existing buildings and historical relics but also strengthening municipal infrastructure to cater to residents' basic needs and enhance urban aesthetics [6]. However, this approach may introduce technical uncertainties during the UR process, thereby giving rise to potential technical risk (TR). For example, UR projects may encounter risks arising from challenging construction conditions due to unfavorable natural circumstances [33]. Additionally, there is potential technical specification risk linked to the absence of guidelines for novel technologies or instances where construction practices fail to meet established standards [15,25]. Furthermore, insufficient safety measures can give rise to the risk of construction-related accidents [33]. The occurrence of TR during the UR process can engender highly adverse social consequences [4,32].

Despite the ongoing development of UR models, the government's prominent position among numerous stakeholders remains unaltered, which continues to hold the reins in policy formulation, safeguarding public interests, and driving the overall process [1,4]. Given the government's special role, it is imperative to ensure an unhindered channel for public expression throughout the UR process, adhere to standardized workflows, and maintain complete transparency in sharing information [34,35]. Such measures are vital to prevent public skepticism stemming from any work-related errors [36]. UR projects often entail intricate ownership structures, necessitating a delicate balance between the interests of property owners, developers, local governments, and the public [8,15]. Erroneous decision making in this context can lead to improper distribution of benefits and project setbacks and even trigger social conflicts or tensions, leading to social unrest or instability [29,37]. As a consequence, UR endeavors are susceptible to organizational management risk (OMR) [38].

The government's ideology regarding UR is undergoing a shift from the conventional practices of "large-scale demolition and construction", "renovation and restoration", and "government-led initiatives" to more diversified approaches such as "comprehensive renovation", "functional replacement", and "incremental renewal" [4,18,36]. However, on the one hand, due to the prolonged timelines of UR projects, the change in overall policy direction may bring challenges during implementation [3,36]. Some initiatives may experience delays or even come to a halt [3,38]. On the other hand, as UR transitions from traditional extensive development to a more refined and meticulous mode, there may be a delay in the development of policy frameworks, hindering the smooth progress of UR efforts [38,39]. Consequently, policy change risk (PCR) that emerges during UR is evident in areas such as ambiguous responsibility allocation, insufficient safeguard measures, policy environment instability, and a lack of alignment with urban planning [3,36].

Based on the existing research, the UR process entails several potential sources of risk, primarily revolving around social factors. These include expropriation and demolition risk (EDR), external environmental risk (EER), TR, OMR, and PCR.

2.2. Stakeholders and Their Actions in UR

An organization's stakeholders comprise not only those directly impacted by its objectives but also external groups with direct or indirect affiliations to the organization, which means any organization or individual with a vested interest in the organization can potentially influence the realization of its objectives [40]. UR is primarily developed by the government, which acquires land through methods such as purchase, repurchase, exchange, and expropriation based on urban planning [8,15]. After ensuring the proper resettlement of the original property owners, the government then proceeds to publicly auction the land and invites third-party investors for development [6]. In UR, developers typically communicate with the original property owners, obtain consent from the majority of them, and then apply to the government for project approval. Once the project is approved, developers sign contracts with the original property owners to complete property transfers and advance UR projects. Throughout the entire process, the government only assumes a role of macro-guidance and coordination [6,41]. Additionally, the representatives of the original property owners, often led by village committees or neighborhood committees, play a vital role in maximizing collective interests and ensuring social stability, in which the interests of affected groups, such as tenants, also should not be disregarded [3,6]. As a result, the primary stakeholders in UR encompass the government, which holds management rights over the projects; investors responsible for funding the UR efforts; the original property owners; and the affected original residents [8,15].

During the UR process, several adverse factors, such as environmental pollution and disruptions in geomancy due to land expropriation and demolition, can impair stakeholders' interests [42]. This specifically refers to situations where the interests of stakeholders cannot be adequately reconciled during the project, leading to losses [43,44]. The convergence of multiple potential risk factors in UR contributes to the emergence of URSRs, increasing the likelihood of SR events that lead to social instability and disorder [11,26]. In the planning and execution of UR projects, the anticipated or actual impacts on the surrounding environment and economy are acquired and perceived by stakeholders through various communication channels [27]. Subsequently, different parties become involved in the UR process, heightening the risk of conflict escalation [15,29].

In UR activities, when the interests of stakeholders suffer detrimental impacts, they often engage in what is known as "expression of demands" to defend their interests [11,27]. However, relationships among stakeholders are intricate, with distinct differences in their interests and channels of expression. Government, developers, and the public encounter numerous obstacles in communication and coordination, such as information asymmetry and uneven distribution of interests, which may exacerbate conflicts of interest and affect the ultimate realization of societal value [11,34,35]. For instance, when residents in old urban areas face the possibility of their homes being demolished, they may organize delegations to attend planning meetings, presenting their needs and concerns and hoping to be adequately considered in the demolition plans [8,11]. However, if these conventional channels of expression are not effective or if stakeholders feel their appeals are not being adequately addressed, they may resort to unconventional means of expression [21]. For example, some residents may organize protest marches, hold public gatherings, or engage in petition-signing activities to draw attention from the public and media, thereby increasing the effectiveness of pressure exertion [21,29].

At this juncture, the government's response becomes particularly crucial [24]. If the government can promptly and effectively address these demands, such as through engaging in dialog with displaced residents, adjusting relocation plans, or providing more detailed information on the progress of UR projects, public trust will be bolstered, and project advancement will proceed more smoothly [24,45]. However, if the government lacks timely responsiveness or takes inappropriate actions that lead to dissatisfaction among displaced residents, it may trigger public distrust and dissatisfaction. Such sentiments of distrust often spread through the media, fueling further resistance actions by more stakeholders and potentially even culminating in large-scale SR events [24,46].

Government trust is a crucial reflection of the interaction between the public and the government, representing the public's confidence in government actions and institutions [24,45]. However, in the UR process, information asymmetry can lead to uncertainty in public trust in the government [11,32]. For instance, the lack of transparency or information regarding demolition and resettlement policies, compensation standards, and other critical aspects may leave the public feeling confused and anxious. In such cases, the public often seeks information through various channels [47,48]. If the information gathered fails to meet their expectations, it can lead to suspicions about the fairness of UR projects [49,50], resulting in a crisis of trust in government institutions and activities, ultimately leading to SRs [47,51].

Protest actions are actions undertaken by stakeholders to actively or passively defend their rights and interests when they perceive violations [21,24]. The type of protest action is closely related to the characteristics of the stakeholder groups [11,26]. In line with stakeholder theory, internal stakeholders are often individuals or groups whose interests are directly impaired [52,53]. Internal stakeholders are the primary initiators of protest actions, although they may not always take on a leading position [15,21]. This group's protest actions are characterized by their involvement when their interests suffer harm, their benefits outweighing the costs of participation, and having a strong motivation to participate [8,54]. On the other hand, external stakeholders are individuals or groups whose interests are indirectly affected as a result of their cooperative or trust-based relationships with the directly affected parties [40,44]. The protest actions of external stakeholders are characterized by the benefits being smaller than the costs of participation, and their actions may exhibit short-term irrational traits [37,55].

3. Hypotheses

As mentioned in Section 2.1, this study identified the primary SR factors based on the potential sources of risk in the entire UR process, as follows: EDR (expropriation and demolition risk), EER (external environmental risk), TR (technical risk), OMR (organizational management risk), and PCR (policy change risk). These five categories of SR factors are interconnected and can influence each other. The interplay of these potential risk factors in UR leads to the occurrence of URSRs, which in turn can give rise to SR events that lead to social instability and disorder [29]. Throughout the planning and implementation of UR projects, the anticipated or actual impacts on the surrounding environment and economy are acquired and perceived by stakeholders through various communication channels [11,15]. As different stakeholders become involved in the UR process, the risk of intensified conflict emerges

Given their subjective agency, stakeholders often seek channels to express their demands and advocate for their rights when they experience discontent [29,37]. In most cases, stakeholders first utilize conventional channels to voice their concerns and make appeals [8,44]. However, if these channels are obstructed or their expressed demands are not addressed or are met with dissatisfaction, stakeholders may resort to unconventional channels for expressing demands, leading to individual or localized acts of rebellion [56].

However, if the government fails to fulfill its responsibilities, it can lead to a crisis of public trust [28,51]. The spread of mistrust through the media can encourage more groups of stakeholders to take action, potentially culminating in large-scale SR events [29,45].

Therefore, through the theoretical analysis above, the evolution of SRs in the UR process can be conceptualized within the framework illustrated in Figure 1. This framework highlights the interplay among the five categories of SR factors: EDR, EER, TR, OMR, and PCR. As these factors interact and intertwine, they can trigger negative impacts on stakeholders' interests. In response, stakeholders may resort to expression and protest while the government intervenes and adjusts its approach, leading to URSRs.

Based on the above analysis, hypothesized relationships among the five categories of SR factors, adverse impacts on stakeholders' interests, stakeholders' protest actions, government actions, and SRs can be proposed.

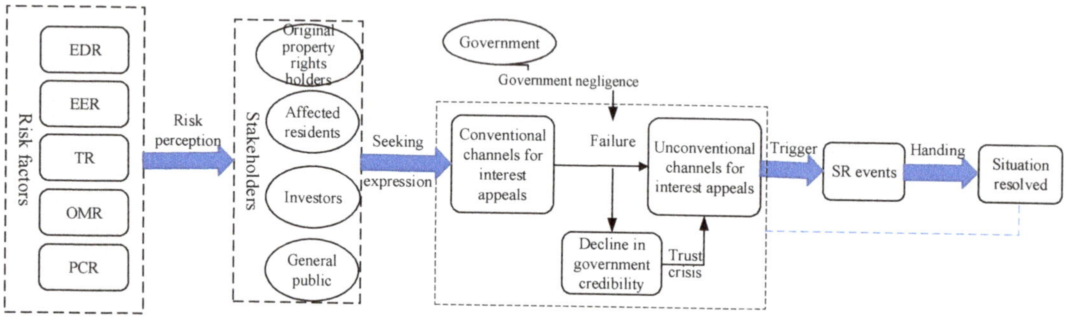

Figure 1. Theoretical framework for the evolution of URSRs.

3.1. Hypotheses Regarding the Interrelationships among the Five Categories of SR Factors

3.1.1. Interrelationship between PCR and EDR, EER, and OMR

In China, there are appreciable disparities among different regions regarding key policies related to overall planning, land, funding, fiscal matters, and demolition [21]. Specific policies tailored to various types of UR projects are lacking (PCR) [57]. The uncertainty of policies creates a fragmented and localized nature of UR regulations, limiting their applicability [58]. Consequently, issues such as inadequate compensation schemes, forced demolition, and insufficient resettlement of residents emerge during the expropriation and demolition process, giving rise to EDR in the UR process [57,58].

Moreover, the practical outcomes of UR have shown that its implementation can bring about significant positive externalities, leading to improved convenience facilities in surrounding areas [2]. However, as UR transitions from traditional extensive development to more refined and focused approaches, there is a lag in policy system construction [1,2]. The urban ecological and environmental issues arising from UR activities have garnered widespread attention [5,9]. For example, some of the environmental concerns surrounding UR include congestion in the existing road transportation network, as well as the removal of or damage to culturally significant sites, cultural conflicts among ethnic minorities, and disruption of the original geomancy in the area [11,32]. It is evident that PCR can trigger environmental challenges during the UR process.

The government's dominant position among various stakeholders remains unchanged as it continues to wield prevailing influence in policy formulation, safeguarding public interests, and driving the process [11,36]. However, the allocation of interests has emerged as a core challenge in UR. Balancing the interests of property rights owners, developers, management operators, and laborers, safeguarding the often-neglected public interests, and formulating UR policies that carefully consider the rights of landowners and market interests while also addressing public concerns and ensuring social equity present a critical test for the government's governance capabilities [8,15]. Hence, it can be concluded that PR can lead to the corresponding OMR.

Therefore, this study proposes the following hypotheses:

Hypotheses 1 (H1). *PCR (policy change risk) leads to EDR (expropriation and demolition risk).*

Hypotheses 2 (H2). *PCR (policy change risk) leads to EER (external environmental risk).*

Hypotheses 3 (H3). *PCR (policy change risk) leads to OMR (organizational management risk).*

3.1.2. Interrelationship between OMR and EDR, EER

Land expropriation and demolition inherently involve a complex interplay of interests, wherein the government often assumes the roles of both "player" and "referee", necessitating a balance between efficiency and fairness [42,59]. An imbalance between these factors can easily trigger conflicts, leading to OMR. The experience in Shenzhen highlights the significance of establishing a well-structured compensation policy system and a supervisory incentive mechanism, all within a stable organizational and institutional framework [57,58]. Complete transparency at the institutional and societal levels is crucial, enabling open dialog and negotiation between the government and the affected parties under the rule of law [50,60]. Such an approach effectively regulates the government's exercise of administrative power and safeguards the rights of affected parties, allowing them to be informed and exercise their oversight over administrative actions [45,49]. Consequently, it is evident that OMR associated with UR significantly impacts EDR.

During the UR process, the government and developers often prioritize investment returns, leading to excessive development in economically promising areas [8,21]. This concentration of population and industries results in increased industrial and domestic waste, reduced green spaces, and deterioration of the city's natural environment [5,42]. Furthermore, improper expropriation and demolition can lead to the destruction of architectural and cultural heritage and loss of the city's traditional character, causing a decline in the overall cultural environment [5]. To address these challenges, a positive combination of government leadership and public participation is crucial, particularly in the process of identifying and protecting cultural relics. Ensuring public participation rights and the right to seek accountability for the results of the identification and protection of cultural relics are essential steps [27]. Consequently, it becomes evident that OMR factors, such as unregulated workflow and inadequate project planning, can further exacerbate EER in the UR process [5,22]. Conversely, fostering a collaborative relationship based on negotiation can protect the cultural environment of local residents during UR.

Therefore, the following hypotheses are proposed:

Hypotheses 4 (H4). *OMR (organizational management risk) leads to EDR (expropriation and demolition risk).*

Hypotheses 5 (H5). *OMR (organizational management risk) leads to EER (external environmental risk).*

3.1.3. Interrelationships between TR and EER

Technical issues in UR can influence the city's environment. For example, the repeated construction of deep pile foundations may lead to the compression of urban underground spaces, whereas the implementation of rainwater harvesting techniques can address urban waterlogging problems in older city areas [5,9]. Consequently, this study posits the following hypothesis:

Hypotheses 6 (H6). *TR (technical risk) leads to EER (external environmental risk).*

3.1.4. Interrelationship between the Five Categories of SR Factors and Adversely Impacted Stakeholders' Interests

Since the initiation of reform and opening-up, China has undergone rapid urbanization. However, during this process, certain issues have arisen concerning the management system, local governance, and distribution mechanisms [22,25]. As a result, some segments of society (such as low-income individuals, residents of old urban areas, the elderly, etc.) have been adversely affected, becoming vulnerable groups and contributing to social instability [11,22]. Consequently, specific SR factors in the context of UR can endanger the interests of stakeholders. Urbanization has facilitated the construction of numerous new buildings, expanding the geographical boundaries of cities [4]. Simultaneously, it has triggered large-scale transformation, demolition, and reconstruction, exemplified by UR initiatives. The stakeholders in UR are highly diverse, each holding different interest preferences, and interest compromises among specific groups are unavoidable [15,61]. Based on the definition of URSRs in this study, these risk factors are closely tied to negative impacts on stakeholders' interests. This study approaches these potential SR factors from the perspectives of land expropriation and demolition, the environment, technical aspects, organization, and policy. All of these risk factors have the potential to significantly impact the interests of the diverse stakeholders involved in UR. The corresponding hypotheses are as follows:

Hypotheses 7 (H7). *EDR (expropriation and demolition risk) has an adverse impact on stakeholders' interests.*

Hypotheses 8 (H8). *EER (external environmental risk) has an adverse impact on stakeholders' interests.*

Hypotheses 9 (H9). *TR (technical risk) has an adverse impact on stakeholders' interests.*

Hypotheses 10 (H10). *OMR (organizational management risk) has an adverse impact on stakeholders' interests.*

Hypotheses 11 (H11). *PCR (policy change risk) has an adverse impact on stakeholders' interests.*

3.2. Interrelationships between Adversely Impacted Stakeholders' Interests and Stakeholders' Protest Actions, Government Actions, and SRs

During the process of UR, grievances over damaged interests often become the core expression of demands [62]. When stakeholders, especially vulnerable groups such as displaced families, affected residents, and tenants, perceive social inequality or negative impacts on their interests, they commonly resort to actions expressing their interests [10]. For instance, if families displaced by UR activities do not receive adequate compensation, they may file lawsuits against the government to voice their demands [63]. Subsequently, if these demands remain unmet, it signifies a certain degree of failure in the institutionalized channels for expressing grievances [64]. In such situations, some stakeholders may resort to more confrontational, violent, or disruptive irrational methods of protest, such as blocking traffic or damaging public property [26]. Therefore, the present study puts forth the following hypotheses:

Hypotheses 12 (H12). *Adverse impacts on stakeholders' interests drive them to actively seek channels to express their demands.*

Hypotheses 13 (H13). *The expression of demands by stakeholders further motivates them to engage in protest actions.*

In the process of the development and diffusion of SRs associated with UR, the public's expression of demands can influence their attitudes toward the government. The specific impact and direction of this influence may vary depending on the particular situation [47,51]. Prior studies indicate that when stakeholders take action to express their demands and directly engage with the government, they form subjective judgments about the government's administrative capabilities, credibility, and level of concern in the UR process [34,49]. Such interactions can play a role in elevating the level of trust in government. Thus, this paper posits the following hypothesis:

Hypotheses 14 (H14). *Stakeholders' expression of demands positively correlates with an increase in their trust in the government.*

Public trust in various levels of government exhibits a differentiated pattern, commonly characterized as high trust in the central government and low trust in local governments [45]. The level of trust that the public has in the government can influence their consciousness and actions of protest [45,65]. For example, when rural migrant workers display lower trust in local governments but higher trust in the central government (thus creating a considerable trust disparity between the two levels), their inclination toward collective protest consciousness might become more pronounced [45]. In this paper, the term "government" primarily refers to the local level, as local governments are the ones initiating and implementing UR projects [49,60]. As the government's public credibility strengthens, residents demonstrate higher levels of cooperation driven by their trust in the government [65,66]. They actively respond to government appeals and willingly reduce protest actions [45,67]. Thus, the following hypothesis is proposed:

Hypotheses 15 (H15). *The trust of stakeholders in the government mitigates the occurrence of protest actions.*

The expression of demands by stakeholders and the government response are closely interconnected [65,67]. Improving the government's responsiveness to demands can yield two significant outcomes. Firstly, it helps to prevent a shift in the channels used by the public to express their interests; secondly, it fosters higher levels of trust in the government's governance capabilities, thereby elevating the government's authority [24,45]. As a result, this study proposes the following hypothesis:

Hypotheses 16 (H16). *Government response plays a moderating role in the path from the expression of demands by stakeholders to their adoption of protest actions.*

The social combustion theory posits that SR factors, under the influence of accelerants, continuously intensify, resulting in negative effects on society [68]. Once these negative impacts reach a critical threshold, they become the "ignition temperature" that triggers social unrest [26,68]. During the development of URSRs, vulnerable groups whose interests are impaired and who face barriers in expressing their concerns are motivated by their interests to gradually unite into negative groups of a certain scale [64]. Subsequently, they engage in a series of destructive protest actions, leading to adverse consequences across the entire social system [69]. As the cumulative negative impact of stakeholders' protest actions reaches a certain level, it may surpass the ignition temperature threshold of SRs, ultimately leading to a comprehensive outbreak of URSRs [27,68]. Hence, this study proposes the following hypothesis:

Hypotheses 17 (H17). *Protest actions lead to SRs.*

By integrating the aforementioned hypothesized relationships, this paper constructed a theoretical model for URSR evolution paths, as illustrated in Figure 2. This model encompassed a total of 17 theoretical hypotheses.

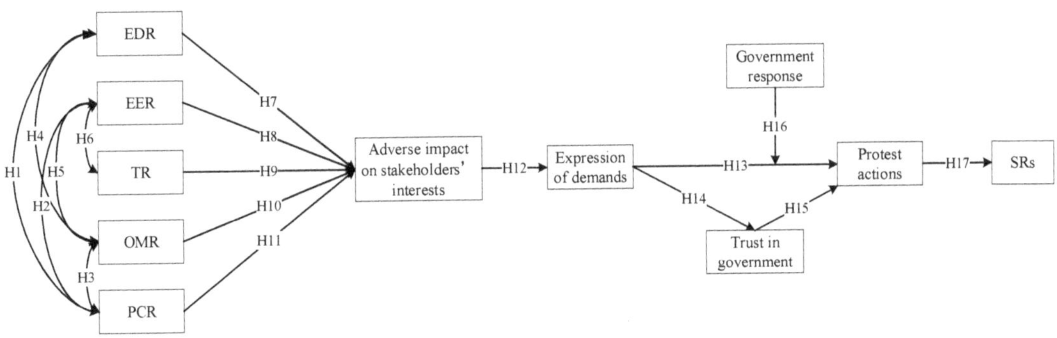

Figure 2. Theoretical model of URSR evolution paths.

4. Research Methodology

4.1. Research Framework

According to the literature review, this paper analyzes key stakeholders and SRs and proposes a theoretical framework for the evolution of URSRs. On this basis, hypotheses were made about the interrelationships among the five categories of SRs (i.e., EDR, EER, TR, OMR, and PCR) and the interrelationships between negatively impacted stakeholders' interests and stakeholders' protest actions, government actions, and SRs. Then, a theoretical model for URSR evolution paths was hypothesized and constructed. The stratified sampling method for the questionnaire survey was used to test the internal consistency of the survey and the rationality of the theoretical model, and then reliability and validity tests were conducted on this basis. The structural equation model (SEM) was employed to examine the above hypotheses and the theoretical model for URSR evolution paths using questionnaire survey data to obtain the evolution mechanisms of URSRs. Through modeling analysis and validation of SEM, the final model for the evolution mechanisms of URSRs was established. Then, the key SRs in the evolution mechanisms of URSRs were identified. Finally, suggestions on controlling URSRs were made. The research framework is organized as shown in Figure 3.

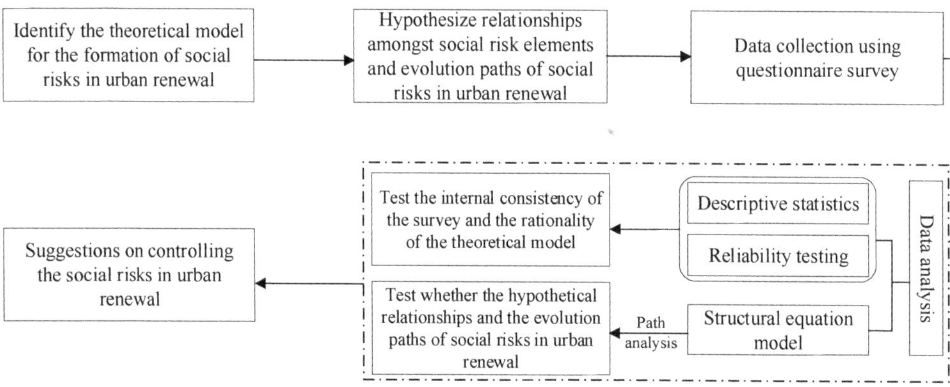

Figure 3. Research framework.

4.2. Questionnaire Design and Data Collection

The theoretical hypothesis model of URSR evolution paths depicted in Figure 2 was validated through a questionnaire survey. The questionnaire was divided into two parts. The initial section pertained to basic information from respondents, including their professional affiliations, years of work experience, and educational qualifications. The second part involved a survey on potential SR factors and stakeholder actions concerning UR projects. This section comprised a checklist of 36 items, with specific observation variables and measurement questions, as shown in Appendix A. Using a five-point Likert scale (1 for strongly disagree, 5 for strongly agree), respondents were required to evaluate the potential SR factors objectively, the levels of SR, the types of negative impacts on their interests, the forms through which their demands are expressed, the extent of government response, the level of trust that stakeholders have in the government, and the types of protest actions undertaken by stakeholders.

This study identified four categories of research subjects: academic scholars who have conducted research on topics such as UR and SRs, practitioners in the field of UR, government officials, and residents affected by UR activities. They were chosen as potential interviewees due to their stakeholder status in UR projects or their understanding of them. The questionnaire was distributed through various channels in cities such as Nanjing, Guangzhou, Beijing, and Hangzhou, China, from December 2022 to March 2023. By combining offline and online questionnaires, a total of 215 questionnaires were collected, out of which 205 were deemed valid, yielding a response rate of 95%. In accordance with the general guidelines of SEM, the dataset size should ideally exceed 100, preferably surpassing 200 [70,71]. Additionally, the sample size used for SEM should be 5 to 10 times the number of observed variables [72]. With 36 observed variables in this paper, the number of collected questionnaires met this criterion. Table 1 illustrates the distribution of the survey questionnaires. Among respondents, 89.27% held a college degree or above, and 88.29% had more than 3 years of work experience. Thus, the data quality was assured, given that the majority of respondents possessed a good understanding and sufficient experience.

Table 1. Basic information of the respondents.

Role	Scholars	Practitioners	Government Staff	Residents	N/A	Total
Number percentage	45 21.95	44 21.46	41 20	75 36.59	-- --	205 100.0
Working experience Number percentage	3 years or under 24 11.71	3–5 years 28 13.65	6–10 years 79 38.54	11–20 years 54 26.34	Over 20 years 20 9.76	Total 205 100.0
Degree of education Number Percentage	High school degree 21 10.24	Junior college degree 22 10.73	Bachelor's degree 42 20.49	Master's degree 86 41.95	Doctorate degree 34 16.59	Total 205 100.0

4.3. *Validation of the URSR Evolution Model through SEM*

Utilizing a combination of both offline and online questionnaires, a total of 215 responses were collected, with 205 of them being valid. Therefore, the questionnaire validity rate was 95%. Subsequently, this research used SPSS 25.0 software to conduct descriptive and reliability analysis of the data. Additionally, the SEM approach was employed to validate the hypothesized paths within the URSR evolution model.

4.3.1. Data Analysis

Statistical analysis was conducted on the sample data using SPSS software to calculate the mean value, standard deviation, skewness, and kurtosis of each measurement question for SRs, as well as stakeholders' risk perception and protest actions, as presented in Table 2.

Table 2. Results of descriptive statistics and reliability.

Latent Variable	Observed Variable	Mean Value	Standard Deviation	Skewness	Kurtosis	Cronbach's Alpha If Item Deleted	Cronbach's Alpha
EDR	X11	3.79	1.265	−0.909	−0.211	0.738	0.880
	X12	3.74	1.248	−0.848	−0.241	0.704	
	X13	3.72	1.224	−0.788	−0.238	0.685	
	X14	3.75	1.160	−0.797	−0.078	0.651	
EER	X21	3.73	1.289	−0.894	−0.285	0.765	0.920
	X22	3.75	1.253	−0.923	−0.136	0.748	
	X23	3.80	1.226	−0.852	−0.314	0.748	
	X24	3.86	1.219	−0.908	−0.162	0.707	
	X25	3.80	1.263	−0.890	−0.253	0.719	
TR	X31	3.88	1.293	−1.016	−0.078	0.708	0.843
	X32	3.74	1.251	−0.834	−0.360	0.747	

Table 2. *Cont.*

Latent Variable	Observed Variable	Mean Value	Standard Deviation	Skewness	Kurtosis	Cronbach's Alpha If Item Deleted	Cronbach's Alpha
OMR	X41	3.82	1.279	−0.916	−0.225	0.701	0.922
	X42	3.77	1.253	−0.766	−0.491	0.702	
	X43	3.66	1.294	−0.790	−0.390	0.720	
	X44	3.64	1.293	−0.716	−0.545	0.749	
	X45	3.70	1.202	−0.708	−0.427	0.691	
PCR	X51	3.77	1.245	−0.864	−0.218	0.754	0.851
	X52	3.76	1.301	−0.829	−0.434	0.760	
	X53	3.79	1.198	−0.841	−0.260	0.737	
SR levels	Y11	3.77	1.314	−0.930	−0.273	/	/
Adverse impacts on interests	S11	3.71	1.311	−0.842	−0.373	0.705	0.821
	S12	3.72	1.259	−0.877	−0.197	0.727	
Expression of demands	S21	3.78	1.220	−0.788	−0.370	0.781	0.916
	S22	3.72	1.289	−0.826	−0.388	0.699	
	S23	3.72	1.301	−0.901	−0.304	0.737	
	S24	3.81	1.215	−0.899	−0.139	0.738	
	S25	3.82	1.237	−0.922	−0.169	0.704	
Government response	S31	3.65	1.245	−0.778	−0.347	0.701	0.807
	S32	3.64	1.293	−0.810	−0.405	0.759	
Trust in the government	S41	3.82	1.305	−0.912	−0.310	0.739	0.860
	S42	3.80	1.255	−0.823	−0.378	0.697	
Stakeholders' protest actions	S51	3.76	1.313	−0.799	−0.476	0.708	0.910
	S52	3.74	1.275	−0.826	−0.345	0.703	
	S53	3.79	1.261	−0.816	−0.421	0.735	
	S54	3.77	1.253	−0.856	−0.264	0.736	
	S55	3.82	1.229	−0.849	−0.312	0.729	

It can be noted that the absolute skewness and kurtosis values of 36 measurement questions were both less than 1, indicating that the survey data followed a normal distribution [73]. The mean value of each measurement question set ranged from the lowest at 3.64 to the highest at 3.88, with no mean value score less than 1.5 ("not important"). Therefore, all of these 36 SR variables and the SR elements (latent variables) were important to construct the evolution model for URSRs.

Moreover, a reliability analysis was conducted to check the consistency of the survey results. Cronbach's alpha, a widely-used method for assessing the internal consistency of questionnaires, was utilized for this purpose [74]. As depicted in Table 2, all Cronbach's alpha coefficients (i.e., 0.880, 0.920, 0.843, 0.922, 0.851, 0.821, 0.916, 0.807, 0.860, and 0.910) were greater than 0.70, which was deemed acceptable, and none exceeded the maximum alpha value of 0.95 [75]. This indicated that the measurement questions of the scale met the reliability requirements, and the measurement scale of the URSR evolution model had good internal consistency [75].

Then, the Kaiser–Meyer–Olkin (KMO) test and Bartlett's sphericity test were adopted to check the construct validity of the measurement scale of the URSR evolution model. From Table 3, it can be seen that the KMO result was 0.901 (>0.8), and the significance result of Bartlett's Sphericity test was 0.000 (<0.01), both of which were considered suitable for SEM [76].

Table 3. Results of the KMO test and Bartlett's sphericity test.

Indicator	KMO Measure of Sampling Adequacy	Bartlett's Test of Sphericity		
		Approx. Chi-Square	Df	Sig.
Values	0.901	3587.430	561	0.000

4.3.2. SEM

Since Bentler advocated for the use of SEM in addressing latent variables within psychological science, it has evolved into a commonplace statistical analysis technique utilized in both theoretical investigations and empirical validations across various disciplines [77]. Because latent variables possess an abstract nature, they can be represented by multiple observed variables, which are comparatively simpler to measure [78]. SEM is capable of confirming relationships within the model structure, encompassing connections between observed variables and latent variables, as well as relationships among latent variables themselves. There is widespread recognition that SEM can yield reasonably accurate results in structural analysis [79]. Therefore, the adoption of SEM to construct the URSR evolution paths involved presenting the objective state of matters through causal hypotheses, followed by validation with quantitative data [80]. Based on the hypothesized relationships between latent variables, as well as between latent variables and observed variables, this study employed SEM to analyze the sample data obtained from the questionnaire survey, as shown in Figure 1. This analysis aimed to validate and refine the fit of the assumed URSR evolution model, ultimately establishing the most suitable evolution model. The SEM consists of two components: the measurement model and the structural model [23]. The measurement model, also known as confirmatory factor analysis, elucidates the connections between observed variables and latent variables. Meanwhile, the structural model, often referred to as the causal model, delineates the causal relationships between latent variables.

According to the hypothesized model shown in Figure 1, a total of 11 latent variables were identified, namely, EDR, EER, TR, OMR, PCR, SR, negative impacts on interests, expression of demands, government response, trust in government, and protest actions. Additionally, 36 observed variables were associated with these latent variables, as shown in Table 2. The interrelationships between the observed and latent variables were elucidated by SEM. A structural model, as depicted in Figure 4, was constructed using AMOS 26 software.

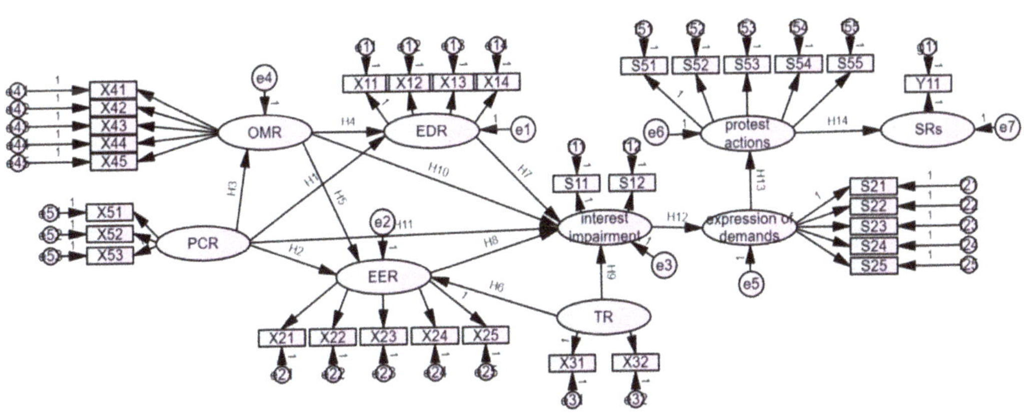

Figure 4. SEM-based URSR evolution model.

5. Result Analysis

To assess the adequacy of the model's fit, it is necessary to evaluate various fit indices, which address different aspects such as sample size effects, parsimony, and comparisons to null models [81]. In this study, eight different indices were chosen to assess the model's fit and its appropriateness. They included the Chi-Square to Degrees of Freedom ratio (CMIN/DF), the Goodness-of-Fit Index (GFI), the Adjusted Goodness-of-Fit Index (AGFI), the Root Mean Square Error of Approximation (RMSEA), the Normed Fit Index (NFI), the Incremental Fit Index (IFI), the Non-Normed Fit Index (NNFI), and the Comparative Fit Index (CFI). The model fitting results are presented in Table 4. It is evident that all fit indices of the model met the recommended standard values, indicating a good fit of the model [82].

Table 4. Model fitting results.

Fit Index	CMIN/DF	GFI	AGFI	RMSEA	NFI	IFI	NNFI	CFI
Recommended benchmarks	≤3	>0.9	>0.9	<0.05	>0.9	>0.9	>0.9	>0.9
Measured values	2.138	0.936	0.914	0.021	0.953	0.994	0.931	0.986
Judgment results	accept	accept	accept	accept	accept	accept	accept	accept

Due to the numerous factors and variables involved in this study, as well as the complexity of the evolution path, the 17 theoretical hypotheses depicted in Figure 1 were divided into three parts for analysis. These parts were the evolution paths of negatively impacted interests, the risk–action evolution paths, and the government regulation evolution paths.

5.1. SEM Results for Negatively Impacted Stakeholders' Interests Evolution Paths

The negatively impacted interests evolution paths consisted of 11 theoretical hypotheses. Table 5 provides the standardized path coefficients (SPCs) and significance test results for the relationships between variables.

Table 5. Path coefficients and significance test results for negatively impacted stakeholders' interests evolution paths.

Hypothesis	Model Path	SPC	S.E.	C.R.	*p*
H1	EDR←PCR	0.304	0.435	2.488	***
H2	EER←PCR	0.2545	0.356	2.53	***
H3	OMR←PCR	0.516	0.075	14.659	***
H4	EDR←OMR	0.187	0.409	2.912	***
H5	EER←OMR	0.24	0.334	2.439	***
H6	EER←TR	0.483	0.070	12.916	***
H7	adverse impacts on interests←EDR	0.8885	1.826	2.011	***
H8	adverse impacts on interests←EER	0.363	0.343	2.338	***
H9	adverse impacts on interests←TR	0.14	0.215	2.338	***
H10	adverse impacts on interests←OMR	0.479	2.906	2.342	***
H11	adverse impacts on interests←PCR	0.225	3.164	2.157	***

Note: *** indicates that $p < 0.001$.

The path coefficient analysis from Table 5 indicated that PCR significantly influenced EDR (SPC = 0.304), EER (SPC = 0.2545), and OMR (SPC = 0.516), all at the 0.001 significance level, validating hypotheses H1, H2, and H3. Similarly, OMR significantly influenced EDR (SPC = 0.187) and EER (SPC = 0.24), both at the 0.001 significance level, validating hypotheses H4 and H5. TR had a significant influence on EDR, with a coefficient of 0.483 at the 0.001 significance level, validating hypothesis H6.

Moreover, the analysis of the path coefficients for the five categories of SR factors on adversely impacted stakeholders' interests revealed that EDR, EER, TR, OMR, and PCR had SPCs of 0.8885, 0.363, 0.14, 0.479, and 0.225, respectively. All of these influence paths were significant at the 0.001 level, confirming the validity of hypotheses H7, H8, H9, H10, and H11. Hence, it could be concluded that all five categories of URSR factors impaired stakeholders' interests.

The impact of each category of factors could be categorized into three tiers, as depicted in Figure 5. The first tier comprised EDR (0.8885). Events such as forced demolition, inadequate compensation, and poor resettlement measures undermine the property rights of original owners, resulting in significant interest impairment [11]. The second tier consisted of EER (0.363) and OMR (0.479). Improper management by the leading authority in UR may affect the interests of vulnerable stakeholder groups [29,37]. For example, the deterioration of the urban environment during project construction can harm the interests of nearby residents and the general public [15]. The third tier encompassed PCR (0.225) and TR (0.14), with relatively smaller direct impacts on stakeholder groups' interests. It is essential to prioritize addressing factors that profoundly influence the interests of relevant stakeholder groups to safeguard stakeholders' interests.

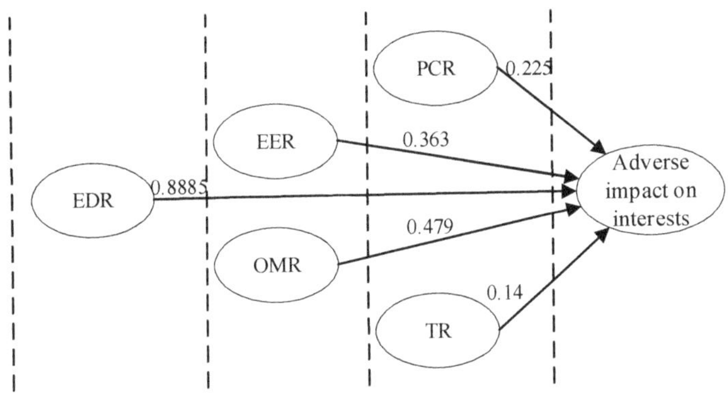

Figure 5. Risk factor–adverse impacts on interest evolution paths.

5.2. *SEM Results for Risk–Action Evolution Paths*

These paths encompassed five theoretical hypotheses. Table 6 provides the SPCs and significance test results for the relationships between variables.

Table 6. Path coefficients and significance test results for risk–action evolution paths.

Hypothesis	Model Path	SPCs	S.E.	C.R.	p
H12	expression of demands←adverse impacts on interests	0.5025	0.066	13.949	***
H13	protest actions←expression of demands	0.209	0.079	13.798	***
H14	trust in the government←expression of demands	0.4965	0.078	14.212	***
H15	protest actions←trust in government	−0.289	1.084	0.526	***
H17	SRs←protest actions	0.488	0.069	15.036	***

Note: *** indicates that $p < 0.001$.

According to Table 6, the SPC between adversely impacted stakeholders' interests and expression of demands was 0.5025, and the path was significant at the 0.001 level, validating hypothesis H12. The SPC between the expression of demands and protest actions was 0.209, and the path was significant at the 0.001 level, validating hypothesis H13. The SPC between expression of demands and trust in the government was 0.4965, and the path was significant at the 0.001 level, validating hypothesis H14. Trust in the government had an SPC of −0.289 on protest actions, and the path was significant at the 0.001 level, validating hypothesis H15. Finally, protest actions had an SPC of 0.488 on SRs, and the path was significant at the 0.001 level, validating hypothesis H17.

5.3. SEM Results for Government Regulation Evolution Paths

Government regulation corresponded to hypothesis H16: "government response plays a moderating role in the path from the expression of demands by stakeholders to their adoption of protest actions." In this context, the expression of demands and protest actions are the independent and the dependent variables, respectively, whereas government response serves as the moderating variable. The criteria to determine the presence of moderating effects are twofold: firstly, both the independent and moderating variables must have a significant direct impact on the dependent variable [83]; secondly, the independent variable and the interaction effect (independent effect × moderating effect) must have a significant indirect impact on the dependent variable [83].

From Table 7, it can be observed that the SPC between expression of demands and protest actions was 0.234, that the SPC between government response and protest actions was 0.384, and that the SPC between expression of demands × government response and protest actions was −0.081. All of these paths were significant at the 0.001 level. These results supported the moderating role of government response in the path from stakeholders' expression of demands to their adoption of protest actions, confirming the validity of hypothesis H16.

Table 7. Path coefficients and significance test results for government regulation.

Hypothesis	Model Path	SPC	S.E.	C.R.	p
H13	protest actions←expression of demands	0.234	1.667	13.3	***
H16	protest actions←government response	0.384	1.724	0.736	***
/	protest actions←expression of demands × government response	−0.081	0.275	−12.3	***

Note: *** indicates that $p < 0.001$.

From the comprehensive analysis of the fitting indices from the entire model and the path coefficients from the three evolution paths (negatively impacted stakeholders' interests, risk–action, and government regulation), a high degree of fit could be derived between the hypothesized URSR evolution path model and the actual data. All 17 hypotheses of the model were validated by the actual data. As a result, the ultimate URSR evolution path model (evolution mechanisms) could be established as follows: interactions between SR factors→negatively impacted stakeholders' interests→expression and protest by stakeholders→government intervention and adjustment→URSRs.

6. Discussion

Based on the validation results obtained from SEM, the ultimate URSR evolution path model was established. However, the path coefficients in the URSR evolution model not only reflected the direct effects between variables but also provided insight into the indirect effects of various SR factors [84]. Therefore, to measure the influence of each variable, this study considered the total effects, which encompassed both direct and indirect effects. The indirect effect refers to the influence of the independent variable on the dependent variable through the mediation of intermediary variables. The magnitude of an indirect effect is calculated as the product of the coefficients along the paths from the independent variable to the dependent variable [85]. In cases where multiple paths exist from the independent variable to the dependent variable, the indirect effects are computed for each path and then aggregated.

Therefore, this study identified key factors in the adversely impacted interests and risk–action evolution paths through an examination of the total effects of SR factors, with a view to proposing targeted governance strategies for addressing URSRs.

6.1. Identification of Key Risks in the Negatively Impacted Stakeholders' Interests Evolution Paths

The total, direct, and indirect effects of the negatively impacted stakeholders' interests evolution path system are detailed in Table 8. The various risks that contribute to the impairment of stakeholders' interests are analyzed.

Table 8. The total, direct, and indirect effect analysis of the adversely impacted stakeholders' interests evolution path system.

SR Element	Effect	PCR	OMR	TR	EDR	EER	Adverse Impacts on Interests
PR	Total effect	——	——	——	——	——	——
	Direct effect	——	——	——	——	——	——
	Indirect effect	——	——	——	——	——	——
OMR	Total effect	0.516	——	——	——	——	——
	Direct effect	0.516	——	——	——	——	——
	Indirect effect	0	——	——	——	——	——
EDR	Total effect	0.497	0.187	——	——	——	——
	Direct effect	0.304	0.187	——	——	——	——
	Indirect effect	0.193	0	——	——	——	——
EER	Total effect	0.502	0.24	0.483	——	——	——
	Direct effect	0.2545	0.24	0.483	——	——	——
	Indirect effect	0.2475	0	0	——	——	——
Adverse impacts on interests	Total effect	0.5	0.6735	0.4905	0.8885	0.363	——
	Direct effect	0.225	0.479	0.14	0.8885	0.363	——
	Indirect effect	0.2755	0.1945	0.3505	0	0	——

The most significant adverse impact on interests was attributed to EDR, with a coefficient of total effects of 0.8885 [21]. These effects were exclusively direct, underscoring the direct influence of EDR on stakeholder groups' interests. This highlighted that challenges such as compensation plans for expropriation and demolition, forced demolition, and resident resettlement, as well as conflicts arising from UR endeavors, notably the surge in rent due to extensive demolition and construction, can appreciably affect vulnerable urban populations, often leading to compromised interests [21,31]. To mitigate SRs arising from UR, it is crucial to establish robust preventive and control measures specifically targeting EDR [86].

OMR was closely followed as the second most impactful factor contributing to adversely impacted interest, registering a coefficient of total effects of 0.6735. These effects primarily materialized through direct means, accompanied by a coefficient of indirect effects of 0.1945. This indicated that the influence of OMR stems from the inadequate management practices of authoritative entities such as governmental bodies and developers [29,37]. Furthermore, OMR demonstrated a positive correlation with both EDR and EER, with coefficients of direct effects of 0.187 and 0.24, respectively, highlighting the essential role of effectively managing OMR in controlling URSRs. This aspect should be given special attention when establishing preventive and control measures [86]. To this end, government bodies should seek to enhance administrative efficiency while also considering organizational capacity as a criterion when selecting UR developers [24,34]. In tandem, vigilant regulatory measures should be implemented throughout the project's lifecycle to mitigate the emergence of OMR [87].

PCR and TR exerted a nearly equal influence on adversely impacted interests, with respective coefficients of total effects of 0.5 and 0.4905. In terms of the composition of the total effects, both PR and TR demonstrated notable indirect effects, reaching coefficients of 0.2755 and 0.3505, respectively. The coefficients of direct effects stood at 0.225 for PCR and 0.14 for TR, which signified that although both PCR and TR can directly lead to adverse impacts on stakeholders' interests, their primary influence was channeled through other risks that subsequently affect stakeholders' interests [29,37]. In regard to their impacts on other risks, PR significantly affected OMR, EDR, and EER, with coefficients of 0.516 (all directly), 0.497 (primarily directly), and 0.502 (primarily directly), respectively. TR, however, directly influenced EER, with a coefficient of 0.483. This underscored the relatively intricate paths through which PCR and TR impact stakeholder groups, with significant total effects. Consequently, measures should be taken to control their effects [86]. Maintaining a stable political environment and reinforcing oversight of construction techniques and safety is beneficial for promoting the successful implementation of UR initiatives [3,4].

The impact of EER on adversely impacted interests was comparatively modest, registering a coefficient of total effects of 0.363, with all effects being direct. As inferred from the earlier analysis, EER was subject to the combined influences of PCR, OMR, and TR. This underlined that EER control necessitates not only minimizing the negative effects of UR on both natural and cultural aspects but also entails addressing PCR, OMR, and TR [30,31].

6.2. Identification of Factors in the Risk–Action Evolution Paths

Table 9 illustrates the overall, direct, and indirect effects of the risk–action evolution system. The subsequent analysis provides insight into each stage within the process of transmitting SRs.

Table 9. The total, direct, and indirect effect analysis of risk–action evolution paths.

SR Element	Effect	Adverse Impacts on Interests	Expression of Demands	Trust in Government	Protest Actions	SRs
Adverse impacts on interests	Total effect	——	——	——	——	——
	Direct effect	——	——	——	——	——
	Indirect effect	——	——	——	——	——
Expression of demands	Total effect	0.5025	——	——	——	——
	Direct effect	0.5025	——	——	——	——
	Indirect effect	0	——	——	——	——
Trust in government	Total effect	0.499	0.4965	——	——	——
	Direct effect	0	0.4965	——	——	——
	Indirect effect	0.499	0	——	——	——
Protest actions	Total effect	0.4985	0.4965	−0.289	——	——
	Direct effect	0	0.209	−0.289	——	——
	Indirect effect	0.4985	0.2875	0	——	——
SRs	Total effect	0.4865	0.484	−0.282	0.488	——
	Direct effect	0	0	0	0.488	——
	Indirect effect	0.4865	0.484	−0.282	0	——

Negatively impacted interests, expression of demands, and protest actions all exerted significant influences on SRs, demonstrating comparable degrees of impact with coefficients of total effects measuring 0.4865, 0.484, and 0.488, respectively [21]. Regarding the composition of the coefficient of total effects, both interests were adversely impacted, and the expression of demands contributed to SRs indirectly, whereas protest actions had a direct effect on SRs. Specifically, negatively impacted interests fostered a positive influence on the expression of demands, trust in the government, and protest actions, with coefficients of 0.5025 (all directly), 0.499 (all indirectly), and 0.4985 (all indirectly), respectively. Expression of demands positively affected trust in the government and protest actions, with coefficients of 0.4965 (all directly) and 0.4965 (mainly indirectly), respectively. The foundation for managing URSRs lies in the sequential chain that begins with the expression of demands resulting from compromised interests, subsequently inciting protest actions and culminating in SRs [35,36]. Each step within this chain deserves focused attention.

Trust in the government exhibited a comparatively modest influence on SRs, resulting in a coefficient of total effects of −0.282, all through indirect means. Additionally, it directly impacted protest actions, with a coefficient of −0.289. This result highlighted the role of trust in the government in providing a buffer and facilitating influence within the process of protest action formation [51,67].

This study utilized stakeholder theory to identify potential URSR factors and employed SEM to explore the evolutionary mechanisms of URSRs. The research findings validated the interrelationships among the five categories of SRs (EDR, TR, OMR, PCR, and EER) and their impact on stakeholder interests [29]. This was consistent with the theoretical framework illustrated in Figure 1, which emphasized the interactions among these risk factors and their potential adverse effects on stakeholder interests, leading to stakeholder protests and government intervention, ultimately resulting in URSRs [28,51,56].

6.3. International Perspectives on URSRs and Stakeholder Involvement

Although this study focuses on the Chinese context, comparing the methods of managing URSRs and stakeholder participation with practices in other countries and regions is instructive. In many countries, such as the United Kingdom, European Union countries, and Turkey, there is a stronger emphasis on active public participation and comprehensive stakeholder involvement throughout the UR process [88–90]. The existence of comprehensive legal frameworks and established mechanisms ensures transparency of information, public consultation, and consideration of the interests of different stakeholders throughout the planning and implementation stages [91,92]. For example, when addressing the EDR, some countries have enacted robust legal protections and transparent procedures to uphold property rights and ensure fair compensation for displaced residents [92]. Public consultations and hearings are typically held to solicit feedback and address concerns before final demolition and relocation plans are determined [88,90]. Similarly, the EER of UR projects is rigorously assessed through environmental impact assessments and mitigation strategies [93]. Existing comprehensive frameworks address issues emphasized in this study, such as noise pollution, construction waste, and disruptions to local transportation networks (see Section 2.1). Moreover, ongoing stakeholder dialog and collaborative decision-making mechanisms are deeply ingrained in UR policies across the European Union [94]. This is particularly relevant for managing the OMR and the PCR, ensuring transparency of information, clear communication of responsibilities, and alignment with evolving urban development goals (see Sections 2.1 and 2.2). By contrast, as emphasized in this study, China's approach to UR has traditionally been more top–down and government-led, with relatively limited stakeholder and public participation [15,19]. However, in recognizing SRs and potential conflicts resulting from insufficient stakeholder participation, there has been a gradual shift toward more inclusive and participatory approaches in recent years.

6.4. Recommendations for Future UR Projects

Based on the identified key SR factors and the evolutionary mechanisms of SRs, the following recommendations can be made for future UR projects by drawing on international experiences and best practices. Firstly, establish a comprehensive and inclusive policy framework to address the concerns of all stakeholders, ensuring fair compensation, relocation assistance, and environmental protection measures [8,12,21,87]. Policies play critical roles in guiding, regulating, incentivizing, and balancing various stakeholders in UR [87]. A wise and effective policy framework helps ensure that UR is sustainable and socially beneficial [95]. Moreover, improve stakeholder participation and communication channels to facilitate dialog, address grievances promptly, and foster trust among government, developers, and affected residents [15,19,88–90]. Building an information-sharing platform is crucial during the implementation of UR projects. It not only facilitates information exchange among different participants but also provides decision makers with stakeholder-related information, thereby reducing SRs caused by information asymmetry and irrational decision-making [94]. Lastly, implement robust monitoring and mitigation strategies for SR factors, such as environmental impact assessments, cultural heritage preservation measures, and transparent organizational management practices [5,32,87,93]. By adopting these recommendations and building upon the research findings, future UR projects can better address complex social dynamics, mitigate potential risks, and ensure sustainable and socially responsible urban development.

7. Conclusions

This study commenced with an initial literature review to establish a clear theoretical model for understanding URSR evolution from the perspective of stakeholders. Drawing upon this theoretical model, 17 hypotheses pertinent to the evolution of URSRs were formulated. Relevant data on the evolution of SRs were gathered through the administration of survey questionnaires. Subsequently, a model corresponding to the formulated hypotheses was constructed using the SEM approach. The hypotheses were validated based on the structural equation model, leading to the identification of effective SR evolution paths. The URSR evolution data were utilized for model fitting. The results affirmed the reliability of all 17 research hypotheses. Consequently, the definitive model for URSR evolution was established: interactions among SR factors→adversely impacted stakeholders' interests→expression and protest by stakeholders→government intervention and adjustment→URSRs. Moreover, the utilization of SEM in the path validation process yielded the path coefficient corresponding to each URSR factor.

Furthermore, an in-depth examination was conducted to identify pivotal factors within the evolution paths of adversely impacted interests and risk–action. The analysis of the total effects of URSR factors unveiled a sequence of risks contributing to negative impacts on stakeholders' interests: EDR, OMR, PCR, TR, and EER. Within the risk–action evolution paths, a depiction of the stages in the transmission of SRs indicated that adversely impacted interests, expression of demands, and protest actions wield significant influence on SRs, whereas trust in the government exerts a comparably milder impact. Based on the identified key SR factors and the evolutionary mechanisms of SRs, this study proposes three recommendations for future UR projects: establishing a comprehensive and inclusive policy framework, improving stakeholder participation and communication channels, and implementing robust monitoring and mitigation strategies for SR factors.

Although this study has illuminated the paths leading to the evolution of URSRs, enhancing our comprehension of the processes unfolding behind URSRs, several limitations still warrant consideration. Further validation is required for the observed variables associated with the various risk factors along the SR evolution paths. Future research could explore the dynamics of SR in specific areas or types of UR projects by collecting and analyzing more refined data. Comparative studies of different urban environments could also provide valuable insight into the universality of the evolutionary model this study proposed. Additionally, as UR projects become increasingly complex, involving various stakeholders and overlapping interests, new methodological approaches such as agent-based modeling or social network analysis could be employed to capture the intricate interactions and feedback loops among stakeholders, thereby offering more dynamic approaches to understanding the propagation of SR.

Author Contributions: Conceptualization, H.X. and Q.L.; methodology, H.X.; software, H.X.; validation, H.X., J.Y. and L.Z.; formal analysis, P.C.; investigation, L.Z.; resources, P.C.; data curation, H.X.; writing—original draft preparation, H.X.; writing—review and editing, J.Y.; visualization, L.Z.; supervision, Q.L.; project administration, Q.L.; funding acquisition, J.Y., L.Z. and P.C. All authors have read and agreed to the published version of the manuscript.

Funding: This research was funded by the National Natural Science Foundation of China (72301131), the National Natural Science Foundation of China (72204113), the Qinglan Project of Jiangsu Province of China, and the Jiangsu Social Science (21GLC001).

Data Availability Statement: Data are contained within the article.

Acknowledgments: We sincerely appreciate all the respondents who participated in this research.

Conflicts of Interest: The authors declare no conflicts of interest.

Abbreviations

UR	Urban renewal
SRs	Social risks
URSRs	SRs in the UR process
SEM	Structural equation modeling
EDR	Expropriation and demolition
TR	Technical risk
OMR	Organizational management risk
PCR	Policy change risk
EER	External environmental risk
KMO	Kaiser–Meyer–Olkin
CMIN/DF	Chi-Square to Degrees of Freedom ratio
GFI	Goodness-of-Fit Index
AGFI	Adjusted Goodness-of-Fit Index
RMSEA	Root Mean Square Error of Approximation
NFI	Normed Fit Index
IFI	Incremental Fit Index
NNFI	Non-Normed Fit Index
CFI	Comparative Fit Index
SPCs	Standardized path coefficients

Appendix A

Latent Variable		Observed Variable	Measurement Question
EDR	X11	Unreasonable demolition compensation plan	The demolition plan for this project is unreasonable.
	X12	Forced demolition	The project involves forced demolition, land occupation, and unfair land acquisition compensation.
	X13	Resident resettlement not implemented	After the project's relocation, the promised resettlement measures were not carried out, resulting in the original property owner's original lifestyle being disrupted, the residents' social network being disrupted, and psychological problems such as a sense of deprivation.
	X14	Rent increase	This project's implementation has caused difficulties for low-income groups in renting due to the imbalance in the rental market.
EER	X21	Construction environmental pollution	This project will cause noise pollution, water pollution, dust pollution, construction waste, radiation hazards, etc., due to engineering construction.
	X22	Traffic congestion	During the construction process of this project, construction vehicles, materials, or machinery may occupy roads, causing traffic congestion.
	X23	Destruction of cultural landscapes	The implementation of this project will result in the removal or damage of the cultural landscape within the area.
	X24	Ethnic cultural conflicts	The implementation of this project involves ethnic minorities, who may have conflicting beliefs and cultures due to different religious beliefs, especially in demolition negotiations and cultural integration in resettlement areas.
	X25	Destroy geomancy	The demolition and reconstruction of this project have damaged the geomancy of houses, ancestral graves, towns, and other areas.
TR	X31	Construction safety risks	The project is located in a densely populated urban area, as insufficient safety protection measures can lead to safety problems such as falls from heights, traffic accidents, etc.
	X32	Inadequate technical specifications	Some new technologies may lack proper specifications, or the construction may fail to meet technical standards.

Latent Variable		Observed Variable	Measurement Question
OMR	X41	Obstruction in public expression channels	This project may trigger public opinions, which may not be expressed smoothly or cannot be responded to and dealt with in a timely manner.
	X42	The workflow is not standardized	The urban renewal work program of this project has defects, or the steps in practice may be chaotic, which may lead to hidden dangers.
	X43	Information opacity	Local governments' failure to disclose information regarding urban renewal violates the public's right to know and can raise doubts about the rationality of the project.
	X44	Unreasonable project planning	Due to an insufficient understanding of factors such as the surrounding environment and project positioning by developers, project construction planning and housing price evaluation are unreasonable.
	X45	Improper construction management	Poor management by developers of costs, schedules, and relevant partners in the construction process has resulted in a loss of control over the project.
PCR	X51	Unclear subject of responsibility	Due to ambiguous responsibilities among the government, developers, and other entities, the project implementation process has encountered a "three no matter" zone.
	X52	Insufficient safeguards	The exit mechanism, mandatory implementation conditions, and mandatory implementation methods during the implementation process of the project are not complete in terms of guarantee measures.
	X53	Unstable policy environment	Due to limited guidance from local urban renewal policies on practice or changes in government leadership, project policies may be discontinuous.
SR levels	Y11	Social risk magnitude	Overall, what do you think the SR levels that this project may lead to are: ____.
Adverse impact on interests	S11	Damage to economic interests	I think my economic interests have been harmed.
	S12	Emotional interest damage	I think my emotional interests have been compromised.
Expression of demands	S21	Government hotline	I expressed my demands through the government hotline.
	S22	Leader's mailbox	I expressed my demands through the leader's Email.
	S23	Discipline inspection report	I expressed my demands through disciplinary reporting.
	S24	Complaint letter and visit	I expressed my demands through complaint letters and visits.
	S25	Media exposure	I expressed my demands through media exposure.
Government response	S31	Process response	I made a request and received a response.
	S32	Result response	My request has been fulfilled.
Trust in government	S41	Problem-solving ability	I believe in the local government's ability to resolve problems and conflicts.
	S42	Ability to arrange work	I believe that the local government's work arrangements are reasonable.
Protest actions	S51	Public Sitting and Walking	I will take public sitting and walking to resist.
	S52	Public demonstrations	I will take the form of a demonstration to protest.
	S53	Collective petitions or individual petitions	I will fight through collective petitions or individual petitions.
	S54	Monkeywrenching	I will disrupt and obstruct the actions of the government and developers.
	S55	Violent conflict	I will fight through violent conflicts with the government and developers.

References

1. Hyra, D.S. Conceptualizing the New Urban Renewal: Comparing the Past to the Present. *Urban Aff. Rev.* **2012**, *48*, 498–527. [CrossRef]
2. Nel· lo, O. The Challenges of Urban Renewal. Ten Lessons from the Catalan Experience. *Análise Soc.* **2010**, *45*, 685–715.
3. Zhao, Y.; An, N.; Chen, H.; Tao, W. Politics of Urban Renewal: An Anatomy of the Conflicting Discourses on the Renovation of China's Urban Village. *Cities* **2021**, *111*, 103075. [CrossRef]

4. Liu, G.; Yi, Z.; Zhang, X.; Shrestha, A.; Wei, L. An Evaluation of Urban Renewal Policies of Shenzhen, China. *Sustainability* **2017**, *9*, 1001. [CrossRef]
5. Yildiz, S.; Kivrak, S.; Arslan, G. Factors Affecting Environmental Sustainability of Urban Renewal Projects. *Civ. Eng. Environ. Syst.* **2017**, *34*, 264–277. [CrossRef]
6. Xiao, Y.; Zhong, J.-L.; Zhang, Q.-F.; Xiang, X.; Huang, H. Exploring the Coupling Coordination and Key Factors between Urbanization and Land Use Efficiency in Ecologically Sensitive Areas: A Case Study of the Loess Plateau, China. *Sustain. Cities Soc.* **2022**, *86*, 104148. [CrossRef]
7. Zeng, P.; Wei, X.; Duan, Z. Coupling and Coordination Analysis in Urban Agglomerations of China: Urbanization and Ecological Security Perspectives. *J. Clean. Prod.* **2022**, *365*, 132730. [CrossRef]
8. Zhuang, T.; Qian, Q.K.; Visscher, H.J.; Elsinga, M.G.; Wu, W. The Role of Stakeholders and Their Participation Network in Decision-Making of Urban Renewal in China: The Case of Chongqing. *Cities* **2019**, *92*, 47–58. [CrossRef]
9. Yazar, M.; Hestad, D.; Mangalagiu, D.; Saysel, A.K.; Ma, Y.; Thornton, T.F. From Urban Sustainability Transformations to Green Gentrification: Urban Renewal in Gaziosmanpaşa, Istanbul. *Clim. Chang.* **2020**, *160*, 637–653. [CrossRef]
10. Nachmany, H.; Hananel, R. The Fourth Generation: Urban Renewal Policies in the Service of Private Developers. *Habitat Int.* **2022**, *125*, 102580. [CrossRef]
11. Mai, Y.; Wu, J.; Zhang, Q.; Liang, Q.; Ma, Y.; Liu, Z. Confront or Comply? Managing Social Risks in China's Urban Renewal Projects. *Sustainability* **2022**, *14*, 12553. [CrossRef]
12. Liu, H.; Long, H.; Li, X. Identification of Critical Factors in Construction and Demolition Waste Recycling by the Grey-DEMATEL Approach: A Chinese Perspective. *Environ. Sci. Pollut. Res.* **2020**, *27*, 8507–8525. [CrossRef] [PubMed]
13. Shi, Q.; Liu, Y.; Zuo, J.; Pan, N.; Ma, G. On the Management of Social Risks of Hydraulic Infrastructure Projects in China: A Case Study. *Int. J. Proj. Manag.* **2015**, *33*, 483–496. [CrossRef]
14. Yu, T.; Shen, G.Q.; Shi, Q.; Zheng, H.W.; Wang, G.; Xu, K. Evaluating Social Sustainability of Urban Housing Demolition in Shanghai, China. *J. Clean. Prod.* **2017**, *153*, 26–40. [CrossRef]
15. Bai, Y.; Wu, S.; Zhang, Y. Exploring the Key Factors Influencing Sustainable Urban Renewal from the Perspective of Multiple Stakeholders. *Sustainability* **2023**, *15*, 10596. [CrossRef]
16. Bai, X.; Shi, P.; Liu, Y. Society: Realizing China's Urban Dream. *Nature* **2014**, *509*, 158–160. [CrossRef] [PubMed]
17. Wan, C.; Su, S. China's Social Deprivation: Measurement, Spatiotemporal Pattern and Urban Applications. *Habitat Int.* **2017**, *62*, 22–42. [CrossRef]
18. Zhuang, T.; Qian, Q.K.; Visscher, H.J.; Elsinga, M.G. An Analysis of Urban Renewal Decision-Making in China from the Perspective of Transaction Costs Theory: The Case of Chongqing. *J. Hous. Built Environ.* **2020**, *35*, 1177–1199. [CrossRef]
19. Yu, Y.; Wu, Y.; Yu, N.; Wan, J. Fuzzy Comprehensive Approach Based on AHP and Entropy Combination Weight for Pipeline Leak Detection System Performance Evaluation. In Proceedings of the 2012 IEEE International Systems Conference SysCon 2012, Vancouver, BC, Canada, 19–22 March 2012; pp. 1–6.
20. Liu, W.; Yang, J.; Gong, Y.; Cheng, Q. An Evaluation of Urban Renewal Based on Inclusive Development Theory: The Case of Wuhan, China. *ISPRS Int. J. Geo-Inf.* **2022**, *11*, 563. [CrossRef]
21. Yu, T.; Shen, G.Q.; Shi, Q.; Lai, X.; Li, C.Z.; Xu, K. Managing Social Risks at the Housing Demolition Stage of Urban Redevelopment Projects: A Stakeholder-Oriented Study Using Social Network Analysis. *Int. J. Proj. Manag.* **2017**, *35*, 925–941. [CrossRef]
22. Maculan, L.S.; Dal Moro, L. Strategies for Inclusive Urban Renewal. In *Sustainable Cities and Communities*; Filho, W.L., Azul, A.M., Brand, L., Ozuyar, P.G., Wall, T., Eds.; Springer Nature: Cham, Switzerland, 2020; pp. 662–672.
23. Lu, Y.; Liu, J.; Yu, W. Social Risk Analysis for Mega Construction Projects Based on Structural Equation Model and Bayesian Network: A Risk Evolution Perspective. *Eng. Constr. Archit. Manag.* **2023**. [CrossRef]
24. Liu, Z.; Zhu, Z.; Wang, H.; Huang, J. Handling Social Risks in Government-Driven Mega Project: An Empirical Case Study from West China. *Int. J. Proj. Manag.* **2016**, *34*, 202–218. [CrossRef]
25. Mehdipanah, R.; Marra, G.; Melis, G.; Gelormino, E. Urban Renewal, Gentrification and Health Equity: A Realist Perspective. *Eur. J. Public Health* **2018**, *28*, 243–248. [CrossRef]
26. Kasperson, J.X.; Kasperson, R.E.; Pidgeon, N.; Slovic, P. The Social Amplification of Risk: Assessing Fifteen Years of Research and Theory. In *The Feeling of Risk*; Routledge: Abingdon, UK, 2013; pp. 317–344.
27. Li, C.; Xi, Z. Social Stability Risk Assessment of Land Expropriation: Lessons from the Chinese Case. *Int. J. Environ. Res. Public Health* **2019**, *16*, 3952. [CrossRef] [PubMed]
28. Zhou, L.; Zhu, D.; Shen, W. Social Stability Risk Assessment of Disaster-Preventive Migration in Ethnic Minority Areas of Southwest China. *Int. J. Environ. Res. Public Health* **2022**, *19*, 6192. [CrossRef]
29. He, Z.; Huang, D.; Zhang, C.; Fang, J. Toward a Stakeholder Perspective on Social Stability Risk of Large Hydraulic Engineering Projects in China: A Social Network Analysis. *Sustainability* **2018**, *10*, 1223. [CrossRef]
30. Yin, H.; Zhao, S.; Wu, Y. Low-Income Groups' Housing Issues Research. In Proceedings of the ICCREM 2013, Karlsruhe, Germany, 10–11 October 2013; pp. 658–669.
31. Xu, K.; Shen, G.Q.; Liu, G.; Martek, I. Demolition of Existing Buildings in Urban Renewal Projects: A Decision Support System in the China Context. *Sustainability* **2019**, *11*, 491. [CrossRef]
32. Gbadegesin, J.T.; Aluko, B.T. The Programme of Urban Renewal for Sustainable Urban Development in Nigeria: Issues and Challenges. *Pakistan J. Soc. Sci.* **2010**, *7*, 244–253. [CrossRef]

33. Rafindadi, A.D.; Napiah, M.; Othman, I.; Mikić, M.; Haruna, A.; Alarifi, H.; Al-Ashmori, Y.Y. Analysis of the Causes and Preventive Measures of Fatal Fall-Related Accidents in the Construction Industry. *Ain Shams Eng. J.* **2022**, *13*, 101712. [CrossRef]
34. Migchelbrink, K.; Van de Walle, S. A Systematic Review of the Literature on Determinants of Public Managers' Attitudes toward Public Participation. *Local Gov. Stud.* **2022**, *48*, 1–22. [CrossRef]
35. Yu, J.; Leung, M.; Jiang, X. Impact of Critical Factors within Decision Making Process of Public Engagement and Public Consultation for Construction Projects–Case Studies. *Int. J. Constr. Manag.* **2022**, *22*, 2290–2299. [CrossRef]
36. Ye, L.; Peng, X.; Aniche, L.Q.; Scholten, P.H.T.; Ensenado, E.M. Urban Renewal as Policy Innovation in China: From Growth Stimulation to Sustainable Development. *Public Adm. Dev.* **2021**, *41*, 23–33. [CrossRef]
37. Wang, W.; Guo, X.; Cao, Q.; Tang, A. A Stakeholder Perspective on Social Stability Risk of Public–Private Partnerships Project for Water Environmental Governance in China: A Social Network Analysis. *Front. Ecol. Evol.* **2023**, *10*, 1022383. [CrossRef]
38. Wang, S.; Palazzo, E. Sponge City and Social Equity: Impact Assessment of Urban Stormwater Management in Baicheng City, China. *Urban Clim.* **2021**, *37*, 100829. [CrossRef]
39. Liu, J.; Love, P.E.D.; Sing, M.C.P.; Smith, J.; Matthews, J. PPP Social Infrastructure Procurement: Examining the Feasibility of a Lifecycle Performance Measurement Framework. *J. Infrastruct. Syst.* **2017**, *23*, 4016041. [CrossRef]
40. Jones, T.M.; Harrison, J.S.; Felps, W. How Applying Instrumental Stakeholder Theory Can Provide Sustainable Competitive Advantage. *Acad. Manag. Rev.* **2018**, *43*, 371–391. [CrossRef]
41. Wu, F.; Zhang, F.; Webster, C. Informality and the Development and Demolition of Urban Villages in the Chinese Peri-Urban Area. *Urban Stud.* **2013**, *50*, 1919–1934. [CrossRef]
42. Guo, P.; Li, Q.; Guo, H.; Li, H. Quantifying the Core Driving Force for the Sustainable Redevelopment of Industrial Heritage: Implications for Urban Renewal. *Environ. Sci. Pollut. Res.* **2021**, *28*, 48097–48111. [CrossRef]
43. Di Maddaloni, F.; Davis, K. Project Manager's Perception of the Local Communities' Stakeholder in Megaprojects. An Empirical Investigation in the UK. *Int. J. Proj. Manag.* **2018**, *36*, 542–565. [CrossRef]
44. Worley, L.C.; Underwood, K.L.; Diehl, R.M.; Matt, J.E.; Lawson, K.S.; Seigel, R.M.; Rizzo, D.M. Balancing Multiple Stakeholder Objectives for Floodplain Reconnection and Wetland Restoration. *J. Environ. Manag.* **2023**, *326*, 116648. [CrossRef]
45. Ma, L.; Christensen, T. Government Trust, Social Trust, and Citizens' Risk Concerns: Evidence from Crisis Management in China. *Public Perform. Manag. Rev.* **2019**, *42*, 383–404. [CrossRef]
46. Holzmann, R.; Jørgensen, S. Social Risk Management: A New Conceptual Framework for Social Protection, and Beyond. *Int. Tax Public Financ.* **2001**, *8*, 529–556. [CrossRef]
47. Paek, H.; Hove, T. Mediating and Moderating Roles of Trust in Government in Effective Risk Rumor Management: A Test Case of Radiation-contaminated Seafood in South Korea. *Risk Anal.* **2019**, *39*, 2653–2667. [CrossRef] [PubMed]
48. Zhu, Z.; Liu, Y.; Kapucu, N.; Peng, Z. Online Media and Trust in Government during Crisis: The Moderating Role of Sense of Security. *Int. J. Disaster Risk Reduct.* **2020**, *50*, 101717. [CrossRef]
49. Al-Omoush, K.S.; Garrido, R.; Cañero, J. The Impact of Government Use of Social Media and Social Media Contradictions on Trust in Government and Citizens' Attitudes in Times of Crisis. *J. Bus. Res.* **2023**, *159*, 113748. [CrossRef]
50. Hassan, M.S.; Al Halbusi, H.; Najem, A.; Razali, A.; Fattah, F.A.M.A.; Williams, K.A. Risk Perception, Self-Efficacy, Trust in Government, and the Moderating Role of Perceived Social Media Content during the COVID-19 Pandemic. *Chang. Soc. Personal.* **2021**, *5*, 9–35. [CrossRef]
51. Hassan, M.S.; Al Halbusi, H.; Najem, A.; Razali, A.; Williams, K.A. Impact of Risk Perception on Trust in Government and Self-Efficiency during Covid-19 Pandemic: Does Social Media Content Help Users Adopt Preventative Measures? *2022*. Available online: https://www.researchsquare.com/article/rs-43836/v3 (accessed on 8 June 2023).
52. Li, T.H.Y.; Ng, S.T.; Skitmore, M. Conflict or Consensus: An Investigation of Stakeholder Concerns during the Participation Process of Major Infrastructure and Construction Projects in Hong Kong. *Habitat Int.* **2012**, *36*, 333–342. [CrossRef]
53. Freeman, E.R. *Strategic Management: A Stakeholder Approach*; Cambridge University Press: Cambridge, UK, 2010.
54. Mainardes, E.W.; Alves, H.; Raposo, M. Stakeholder Theory: Issues to Resolve. *Manag. Decis.* **2011**, *49*, 226–252. [CrossRef]
55. Ye, X.; Shi, S.; Chong, H.-Y.; Fu, X.; Liu, L.; He, Q. Empirical Analysis of Firms' Willingness to Participate in Infrastructure PPP Projects. *J. Constr. Eng. Manag.* **2018**, *144*, 4017092. [CrossRef]
56. Zhang, Z.; Wang, X.; Su, C.; Sun, L. Evolutionary Game Analysis of Shared Manufacturing Quality Synergy under Dynamic Reward and Punishment Mechanism. *Appl. Sci.* **2022**, *12*, 6792. [CrossRef]
57. Zhang, X. Social Risks for International Players in the Construction Market: A China Study. *Habitat Int.* **2011**, *35*, 514–519. [CrossRef]
58. AlKheder, S.; Alzarari, A.; AlSaleh, H. Urban Construction-Based Social Risks Assessment in Hot Arid Countries with Social Network Analysis. *Habitat Int.* **2023**, *131*, 102730. [CrossRef]
59. Zheng, S.; Xu, K.; He, Q.; Fang, S.; Zhang, L. Investigating the Sustainability Performance of PPP-Type Infrastructure Projects: A Case of China. *Sustainability* **2018**, *10*, 4162. [CrossRef]
60. Xu, T. Media, Trust in Government, and Risk Perception of COVID-19 in the Early Stage of Epidemic: An Analysis Based on Moderating Effect. *Healthcare* **2021**, *9*, 1597. [CrossRef]
61. Pirson, M.; Malhotra, D. Foundations of Organizational Trust: What Matters to Different Stakeholders? *Organ. Sci.* **2011**, *22*, 1087–1104. [CrossRef]

62. Mirzakhani, A.; Turró, M.; Jalilisadrabad, S. Key Stakeholders and Operation Processes in the Regeneration of Historical Urban Fabrics in Iran. *Cities* **2021**, *118*, 103362. [CrossRef]
63. Hu, Y.; Hooimeijer, P.; Bolt, G.; Sun, D. Uneven Compensation and Relocation for Displaced Residents: The Case of Nanjing. *Habitat Int.* **2015**, *47*, 83–92. [CrossRef]
64. Chen, R.; Fan, R.; Wang, D.; Yao, Q. Exploring the Coevolution of Residents and Recyclers in Household Solid Waste Recycling: Evolutionary Dynamics on a Two-Layer Heterogeneous Social Network. *Waste Manag.* **2023**, *157*, 279–289. [CrossRef]
65. Han, Z.; Lu, X.; Hörhager, E.I.; Yan, J. The Effects of Trust in Government on Earthquake Survivors' Risk Perception and Preparedness in China. *Nat. Hazards* **2017**, *86*, 437–452. [CrossRef]
66. Hassan, A.; Abdelghany, K.; Semple, J. Dynamic Road Pricing for Revenue Maximization. *Transp. Res. Rec. J. Transp. Res. Board* **2013**, *2345*, 100–108. [CrossRef]
67. Nam-Speers, J.; Berry, F.S.; Choi, D. Examining the Role of Perceived Risk and Benefit, Shared Concern for Nuclear Stigmatization, and Trust in Governments in Shaping Citizen Risk Acceptability of a Nuclear Power Plant. *Soc. Sci. J.* **2023**, *60*, 695–714. [CrossRef]
68. Fei, L.; Mingzhu, H. An Analysis of the Evolution Mechanism of Social Stability Risk in Major Railway Projects--Based on Social Combustion Theory. *J. Chongqing Jiaotong Univ. Soc. Sci. Ed.* **2022**, *22*, 25.
69. Ma, Q.; Wu, W.; Liu, Y.; Liang, Z.; Kou, L. Impact of the Synergy between Technology Management and Technological Capability on New Product Development: A System Dynamics Approach. *J. Syst. Eng. Electron.* **2022**, *33*, 105–119. [CrossRef]
70. Rigdon, E. *Structural Equation Modeling: Nontraditional Alternatives*; Wiley: Hoboken, NJ, USA, 2014; Volume 4, ISBN 9780470013199.
71. Tabachnik, B. *Using Multivariate Statistics*; Pearson: Boston, MA, USA, 2007.
72. Wolf, E.J.; Harrington, K.M.; Clark, S.L.; Miller, M.W. Sample Size Requirements for Structural Equation Models: An Evaluation of Power, Bias, and Solution Propriety. *Educ. Psychol. Meas.* **2013**, *73*, 913–934. [CrossRef] [PubMed]
73. Muthen, B.; Kaplan, D. A Comparison of Some Methodologies for the Factor Analysis of Non-normal Likert Variables: A Note on the Size of the Model. *Br. J. Math. Stat. Psychol.* **1992**, *45*, 19–30. [CrossRef]
74. Brown, J.D. The Cronbach Alpha Reliability Estimate. *JALT Test. Eval. SIG Newsl.* **2002**, *6*, 17–18.
75. Bujang, M.A.; Omar, E.D.; Baharum, N.A. A Review on Sample Size Determination for Cronbach's Alpha Test: A Simple Guide for Researchers. *Malaysian J. Med. Sci. MJMS* **2018**, *25*, 85. [CrossRef]
76. Williams, B.; Onsman, A.; Brown, T. Exploratory Factor Analysis: A Five-Step Guide for Novices. *J. Emerg. Prim. Health Care* **1996**, *19*, 42–50. [CrossRef]
77. Bentler, P.M. Multivariate Analysis with Latent Variables: Causal Modeling. *Annu. Rev. Psychol.* **1980**, *31*, 419–456. [CrossRef]
78. Black, W.C.; Babin, B.J.; Anderson, R.E. *Multivariate Data Analysis: A Global Perspective*; Pearson: Boston, MA, USA, 2010; ISBN 0135153093.
79. Bagozzi, R.P.; Yi, Y. Specification, Evaluation, and Interpretation of Structural Equation Models. *J. Acad. Mark. Sci.* **2012**, *40*, 8–34. [CrossRef]
80. Hoyle, R.H. *Structural Equation Modeling: Concepts, Issues, and Applications*; Sage: Thousand Oaks, CA, USA, 1995; ISBN 0803953186.
81. Sivo, S.A.; Xitao, F.A.N.; Witta, E.L.; Willse, J.T. The Search for "Optimal" Cutoff Properties: Fit Index Criteria in Structural Equation Modeling. *J. Exp. Educ.* **2006**, *74*, 267–288. [CrossRef]
82. David, G. *Multicollinearity & Singularity*; Statistical Associates Publishers: Asheboro, NC, USA, 2012; pp. 18–19.
83. Namazi, M.; Namazi, N.-R. Conceptual Analysis of Moderator and Mediator Variables in Business Research. *Procedia Econ. Financ.* **2016**, *36*, 540–554. [CrossRef]
84. Kline, R.B. *Principles and Practice of Structural Equation Modeling*; Guilford Publications: New York, NY, USA, 2023; ISBN 1462551912.
85. Jin, C.; Li, B.; Ye, Z.; Xiang, P. Identifying the Non-Traditional Safety Risk Paths of Employees from Chinese International Construction Companies in Africa. *Int. J. Environ. Res. Public Health* **2021**, *18*, 1990. [CrossRef] [PubMed]
86. Xie, H.; Zheng, S.; Zhai, Y.; Yuan, J.; Li, Q. Unveiling Urban Regeneration Risks in China: A Social Perspective. *Sustainability* **2024**, *16*, 1671. [CrossRef]
87. Pérez, M.G.R.; Laprise, M.; Rey, E. Fostering Sustainable Urban Renewal at the Neighborhood Scale with a Spatial Decision Support System. *Sustain. Cities Soc.* **2018**, *38*, 440–451. [CrossRef]
88. Bügl, R.; Stauffacher, M.; Kriese, U.; Pollheimer, D.L.; Scholz, R.W. Identifying Stakeholders' Views on Sustainable Urban Transition: Desirability, Utility and Probability Assessments of Scenarios. *Eur. Plan. Stud.* **2012**, *20*, 1667–1687. [CrossRef]
89. Šiugždinienė, J.; Gaulė, E.; Rauleckas, R. In search of smart public governance: The case of Lithuania. *Int. Rev. Adm. Sci.* **2019**, *85*, 587–606. [CrossRef]
90. Mavrodieva, A.V.; Daramita, R.I.F.; Arsono, A.Y.; Yawen, L.; Shaw, R. Role of civil society in sustainable urban renewal (Machizukuri) after the Kobe Earthquake. *Sustainability* **2019**, *11*, 335. [CrossRef]
91. Cihangir Çamur, K. Transformation of Urban Regeneration Legislation, Practice, and Planning Principles Conflict in a Construction-Led Economy: Ankara Case in Court Decisions. *Planlama* **2021**, *31*, 95–107. [CrossRef]
92. Kuyucu, T. The Great Failure: The Roles of Institutional Conflict and Social Movements in the Failure of Regeneration Initiatives in Istanbul. *Urban Aff. Rev.* **2022**, *58*, 129–163. [CrossRef]
93. Tam, V.W.Y.; Hao, J.J.L. Adaptive reuse in sustainable development. *Int. J. Constr. Manag.* **2019**, *19*, 509–521. [CrossRef]

94. Zhuang, T.; Qian, Q.K.; Visscher, H.J.; Elsinga, M.G. Stakeholders' expectations in urban renewal projects in China: A key step towards sustainability. *Sustainability* **2017**, *9*, 1640. [CrossRef]
95. Zheng, W.; Shen, G.Q.; Wang, H.; Hong, J.; Li, Z. Decision support for sustainable urban renewal: A multi-scale model. *Land Use Policy* **2017**, *69*, 361–371. [CrossRef]

buildings

MDPI

Article

A Bayesian Network Model of Megaproject Social Responsibility Behavior and Project Performance: From the Perspective of Resource-Based Theory

Yuhua Wu [1,*], Zhao Zhou [2], Linlin Xie [2], Bo Xia [3] and Mian Huang [2]

1 Guangdong Yongsheng Construction Engineering Co., Ltd., Maoming 525000, China
2 School of Civil Engineering & Transportation, South China University of Technology, Guangzhou 510641, China; llxie@scut.edu.cn (L.X.)
3 School of Civil Engineering and Built Environment, Queensland University of Technology, Brisbane, QLD 4001, Australia; paul.xia@qut.edu.au
* Correspondence: 13929796066@163.com

Abstract: Megaproject Social Responsibility (MSR) is widely acknowledged as contributing to project performance. However, the effect of Megaproject Social Responsibility Behavior (MSRB) implemented by organizations participating in construction on project performance remains a subject of considerable debate, and the intrinsic mechanism of MSRB's effect on the performance of megaprojects has not been elucidated. Therefore, this study employs resource-based theory to investigate the mechanism underlying MSRB's effect on project performance, taking into account both internal and external social capital as well as resource integration capacity as pivotal influences. Drawing on sample data from 206 experienced project managers across the various parties involved, this study develops a Bayesian network model to elucidate the MSRB effect mechanism. Through inference and sensitivity analysis, this study discovers variations in the enhancement effects across the four dimensions of MSRB on project performance. Notably, a combination strategy yields superior enhancement effects. Furthermore, when project performance is suboptimal, resource integration capacity emerges as a significant mediator between MSRB and project performance. Conversely, at high levels of project performance, MSRB directly contributes to enhancing project outcomes. The findings of this study offer valuable insights for the governance of MSR and the enhancement of project performance in megaprojects.

Keywords: megaproject social responsibility behaviors; project performance; resource-based theory; bayesian network

1. Introduction

Megaproject Social Responsibility (MSR) is frequently lauded as a reliable method for integrating economic, social, environmental, and legal aspects into project activities with the aim of enhancing project performance [1–3]. In 2015, Zeng et al. conducted the first systematic analysis of the social responsibility of major infrastructures, defining MSR as "the policies and practices of stakeholders throughout the entire megaproject life cycle, which reflect responsibilities for the well-being of the wider society" [4]. Recent years have seen a significant emergence of MSR practices within the engineering community. Driven by a concern for societal welfare, organizations participating in megaprojects strive to undertake practical actions to address relevant MSR issues, actions that are termed Megaproject Social Responsibility Behavior (MSRB) [5,6].

Recent studies suggest that deficiencies, lapses, or misconceptions in the social responsibility behaviors of stakeholders involved in megaprojects might undermine the effectiveness of project investments, potentially disrupting the project's social image and diminishing its performance. For instance, there have been construction accidents and casualties resulting from the neglect of production safety awareness by the construction party [7,8]. A lack of awareness of community responsibility within management circles has led to protests by non-governmental organizations, staff strikes, social conflicts, and, in extreme cases, the project's termination [9]. MSRB encompasses a myriad of stakeholders, each with differing motives, varied expressive forms, and significant resource investments [5,6]. MSRB exhibits considerable multivariate heterogeneity, complex situational dependence, and dynamic evolution. Consequently, the specific effects of implementing MSRB are still largely undetermined [10], and its impact on project performance continues to be a "black box", warranting further investigation. To address this issue, the object of this study is to reveal the mechanism of the impact of MSRB on megaproject performance, which ultimately assists administrators in identifying key performance improvement pathways and expands strategies to overcome megaproject performance challenges.

In recent years of studying MSRB practice, scholars have identified positive effects of MSRB. Such behaviors promote project performance [2], enhance organizational financial and social performance [10], and contribute to industry sustainability [11]. Wang et al. investigated behavioral drivers to explore the dynamic effects of megaproject citizenship behaviors (MCB) as part of MSRB on project performance, discovering that the actual potential growth rate of promotions positively influences both MCB and project performance [12]. However, there exist mechanisms that exert local negative and indirect effects on the enhancement process of performance. Guo et al. demonstrated that most dimensions of project citizenship behaviors are positively correlated with sustainable project performance. Conversely, proactive responsibility behaviors exhibited a negative effect [13]. He et al. identified a positive and significant influence of MSRB and innovations on project performance, with innovative capacity serving as a mediator [2]. Wang et al. reported that environmental responsibility behaviors aimed at the broader public and local communities did not yield the anticipated project environmental performance [14].

However, some scholars have pointed out that the results produced by MSRBs of project organizations should be measured more rationally and objectively [15]. Environmental responsibility behaviors could potentially pursue project environmental performance while ignoring or forgoing project economic benefits [16]. Unilateral social responsibility behaviors may exist to camouflage and defraud project MSRs, resulting in a waste of public resources in society [6,16]. To maintain a balance between performance and MSR that includes economic, social, and environmental goals, project organizations may need to consume a large amount of resources to maintain internal and external relationships [17]. It can be seen that the research on MSRB's effect is rather fragmented, and project performance has not yet been considered comprehensively. In order to further clarify the issue of the effect of social responsibility behavior on project performance, it is necessary to consider both multidimensional MSRB and a more comprehensive assessment of megaproject performance, so as to effectively reveal the mechanism of the effect of MSRB on the performance of megaprojects.

To achieve the above research objective, this study seeks to explore the mechanisms by which MSRB affects project performance from the perspective of resource-based theory. The implementation of MSRB by organizations participating in a project is intrinsically linked to the resources and capabilities available to the project, and the realization of MSR requires both intangible and tangible resources [18]. Moreover, participating organizations can integrate resources and capabilities from internal and external stakeholders to achieve MSR [19]. Intangible resources such as knowledge, reputation, and trust represent the key resources created by MSRB for project participating organizations, which typically stem from the internal and external social networks of the project embedded in the project organization, and are also referred to as a project's social capital [20]. These intangible resources are valuable, scarce, and difficult to replicate, thereby forming the basis for enabling positive project performance [21]. Moreover, good project performance also depends on effective resource integration [22]. Firstly, MSRB promotes participation in organizational interactions, thereby increasing access and space for resource acquisition [23]. Secondly, MSRB can promote staff quality training and thus improve the resource integration ability of the project team itself, while simultaneously prompting the megaproject team to absorb and exert the resource allocation and utilization ability of participating organizations [24,25]. A megaproject requires more comprehensive and complex resources in comparison to a general project; thus, the adaptive resource integration capacity is critical to realizing the value of the project [26]. Through the appropriate combination of resources and resource integration capacity, a project can create the value shared by the economy and society [23]. MSRB influences multiple aspects of a project's social capital and resource integration capacity, which will ultimately have an impact on the project's performance. This study seeks to uncover the mechanism of that impact.

Therefore, this study endeavors to investigate the mechanism of the impact of MSRB on project performance based on the resource-based theory, integrating the resource perspective and the capability perspective. The structure of this paper is as follows: The Section 2 outlines the theoretical foundation. The Section 3 introduces the research methodology of this study, including variable measurements, data collection, and model construction. The Section 4 presents the results of Bayesian network inference analysis. The Section 5 offers a discussion of the results, and the Section 6 summarizes the conclusions drawn from this study.

2. Literature Review

2.1. MSRB and Project Performance

According to recent academic discourse, social responsibility behaviors in megaprojects manifest in varied forms, including contractual social responsibility behaviors [27], megaproject citizenship behaviors [28], environmental responsibility behaviors [14,29,30], and greenwashing behaviors [31–33]. Contemporary studies of these behaviors fall into two primary categories: research focusing on the motivation behind MSR behaviors [3,5,29,30,34,35] and research examining the effects of MSR behaviors [2,10–12,14], which encompass outcomes related to project sustainability, organizational performance, and industry performance. Specifically, within the framework of global sustainable development, expectations regarding the value of megaprojects have surged, with MSR increasingly serving as a barometer for a project's "value" [36]. Consequently, organizational behaviors that could potentially enhance project value have attracted significant scholarly attention. Therefore, the impact of MSRB on project performance has garnered particular focus among researchers.

Project performance measurement generally refers to the process of determining the success of a project organization or individual in achieving its desired objectives after taking a series of management actions. The construction of megaprojects is large in scale, long in cycle, complex in environment, and widespread in social impact, and the evaluation of its project success focuses on the success of the systematic and comprehensive process. To measure megaproject performance, studies have proposed a performance framework including cost, quality, safety, social response, and contractual relationships [37]. Compared with traditional project performance indicators, due to the high consumption of natural resources in megaprojects [14], high technical management difficulty [38], and numerous participating parties [39], elements such as effective utilization of resources, environmental impacts, innovation outcomes, collaborative cooperation effects, and stakeholders' satisfaction have also been added to the project performance indicator system by numerous scholars [40–43]. It has been shown that there is a positive relationship between the implementation of social responsibility and project performance in megaprojects [44]. Specifically, in order to realize MSR, megaprojects contractors will stipulate the MSRBs to be performed by subcontractors in the contract, and at the same time set up clauses to encourage the implementation of MSRBs which will effectively enhance the project performance standard and promote the achievement of performance through the contract [27]. In addition, participants in megaprojects invest resources and efforts beyond the scope of the contract, including responding positively to the owner's initiatives, advising the owner proactively, taking measures to protect the environment, employing local labor, and striving to deliver the project successfully and to achieve performance that exceeds expectations [28,35,45,46].

Further studies have shown that the mechanism of MSRB's influence on project performance is nonlinear and comprehensive, involving the intervention of numerous internal and external project factors. Wang et al. identified internal and external factors, such as laws and regulations, public satisfaction, and project culture, as elements in the feedback loop of MCB's influence on megaproject performance. They found that the opportunity for potential promotion is a crucial factor in enhancing organizational citizenship behaviors, thereby facilitating project performance [12]. Guo et al.'s study demonstrated that most dimensions of project citizenship behaviors were positively associated with sustainable project performance, whereas proactive responsibility behaviors had a negative impact [13]. This finding suggests that the influence of MSRB on project performance may not be entirely positive. Wang et al. differentiated between environmental responsibility behaviors directed at the subject matter and highlighted that such behaviors failed to yield desirable project environmental performance [14], or even masked poor environmental performance [31]. Furthermore, studies have demonstrated that the influence mechanism of MSRB is not a single direct relationship, as social capital, dynamic capability, and innovation capacity have been proven to play mediating roles [2,23]. Thus, in this paper, we suggest that MSRB has both a positive and a negative impact on project performance.

Most of the above studies on the influence relationship utilize structural equation modeling or multiple regression analysis models, often for a single form of MSRB. However, the effects triggered by different MSRBs vary, and there may be a superimposed or constrained relationship between MSRBs. Research on the overall intrinsic influence of MSRBs on the performance of megaprojects is lacking. Above all, this paper summarizes MSRBs into four dimensions: economic and qualitative, political and communal, legal and regulatory, and environmental and ethical behaviors. We define megaproject performance as the degree of cost, quality, schedule, safety, green, innovation, and stakeholder satisfaction in the construction process of a megaproject. In the following section, this paper will introduce the resource-based theory to reveal the influence mechanism of different dimensions of MSRBs on megaproject performance.

2.2. Resource-Based Theory

Formerly known as the resource-based view, the resource-based theory was initially proposed by Wernerfelt in 1984 within the field of corporate strategic management, emphasizing the fundamental role of the resource perspective on corporate strategic options [47]. The resource-based theory posits that resources are the foundation for organizational survival and growth, emphasizing that organizations must possess heterogeneous resources and effectively utilize and assimilate these critical resources through appropriate capabilities in pursuit of sustainable competitive advantages.

Taking an organization's possession of heterogeneous resources as the fundamental premise [47], the resource-based theory focuses on analyzing how to enhance competitive advantage and organizational performance through the utilization of those heterogeneous resources. Social responsibility behaviors accumulate social capital [48,49], a form of heterogeneous and valuable intangible resource that is highly specific, complex, and difficult to imitate or substitute [50]. Social capital plays a crucial role in achieving organizational performance [21,51]. Equally crucial as heterogeneous resources is an organization's resource integration capacity. No organization possesses all the required resources at all times, and resources themselves are not productive. Thus, an organization's capacity to acquire and utilize resources is critical for gaining a competitive advantage. Megaprojects, with their complex and heterogeneous stakeholder networks and high degree of asset specificity, require even greater attention to resource allocation and effective utilization. The implementation of social responsibility in megaprojects can help consolidate the stakeholder networks of participating organizations, enabling them to acquire additional resources [4]. In summary, in analyzing the mechanism of the impact of MSRB on project performance, this paper takes social capital and resource integration capacity as the influencing factors of the role of MSRB on project performance.

For megaprojects with one-time and temporary characteristics, resource-based theory has been applied in a significant number of research projects, such as Dzeng et al.'s research on the strategic management of contractors in megaprojects [52], Govan et al.'s investigation of project risk management [53], and Shi et al.'s examination of the impact of competence on subcontracting in construction projects [54]. In the realm of construction project management, RBT serves as a prevalent viewpoint for formulating a conceptual framework for performance research [55]. Emphasizing that any organization harbors distinctive resources and capabilities [47], a megaproject typically draws upon resources and capabilities from multiple participating organizations [52], thus rendering the megaproject an amalgamation of resources and capabilities contributed by various organizations. The management of large-scale construction projects involving multiple participants entails a strategic decision-making process designed to guarantee that the entire construction project team possesses sufficient resources and capabilities [52], serving as the foundation for enhancing project performance [50,56]. Additionally, Mansour et al.'s investigation examined 20 construction project management studies that applied RBT and delineated a conceptual structural model illustrating how resources and capacity influence project performance [55]. This is coupled with the assertion that RBT effectively elucidates the intermediaries of social responsibility behaviors impacting organizational performance [23,49,57]. Therefore, it is appropriate to introduce resource-based theory in this study.

2.2.1. Social Capital of Megaproject

The concept of "social capital" emerged from community studies to underscore that the network of interpersonal relationships within a community is a relational resource that fosters the development of individuals within the community [58]. Viewing megaprojects as social actors integrated into social networks, the essence of social capital lies in being a relational network that enables social actors to mobilize and access resources. From a resource perspective, megaproject social responsibility behavior can assist actors in building and sustaining a wider and more stable social relationship network [59]. These relationship networks represent high-quality intangible resources, which serve as a crucial assurance of project success.

When focusing on the project aspect, the social capital of a megaproject is the sum of actual or potential resources of the project team, which are related to social networks and stakeholder relationships [23]. Previous research has noted that an increase in a project's social capital has a positive effect on project management performance improvement [60]. Organizations can build a network of social relationships through the implementation of social responsibility, which in turn leads to the formation of social capital [48]. First, organizations implementing social responsibility behaviors in megaprojects can establish and maintain good relationships with units in industry, the government and related departments at the project site, the community, and the public. Good relationships are conducive to strengthening the cooperative relationship between the participating organizations and the upstream and downstream enterprises, facilitating exchanges between the participating organizations and governmental departments, and reducing the asymmetric nature of the information between the participating organizations and other stakeholders, which in turn reduces the uncertainty and transaction costs [10]. Second, the social responsibility behavior of megaprojects establishes or consolidates the organization's stakeholder network, so that the participating organizations can expand more channels through this network, from which they can obtain knowledge, human resources, financial resources, and other resources, so as to more effectively achieve the project performance objectives. It can be seen that the social responsibility behavior of megaprojects helps to improve stakeholder relations and promote the accumulation of social capital inside and outside the project, which in turn reduces transaction costs and improves the efficiency of project construction.

Thus, in this paper, we hypothesize that MSRB has a positive impact on project social capital, but also that project social capital has a positive impact on project performance.

In early studies, scholars usually regarded social capital as a social network [61] or trust [62], but the measurement of a single dimension tends to lead to a failure to truly reflect the level of organizational social capital. Nahapiet and Ghoshal proposed a three-dimensional model of social capital that includes structure, relationship, and cognition [63], which is effective in conceptualizing social capital and has received widespread scholars' support. In the three-dimensional classification, the structural dimension refers to the connections between individuals or organizations, which in the case of projects are expressed as network connections between project participants [64]. The relational dimension refers to the resources and information in the relational network, such as trust, norms, and reciprocity, of which trust is the central element. The cognitive dimension refers to the common knowledge, common understanding, and common vision of some thing or goal among the members of the social network. Bian et al. proposed categorizing the social capital of enterprises into three types, including vertical relations, horizontal relations, and social relations, from the perspective of the connection between the enterprise and the external economy and society. The phrase "vertical relationship" refers to the relationship between enterprises and relevant government departments and their subordinate departments. A "horizontal relationship" is a variety of legitimate connections between enterprises and other enterprises, industry associations, research institutes, and so on, such as collaborative relationships, holding relationships, etc. The phrase "social connections" refers to the social interactions and connections of enterprise operators and managers in the non-economic field [65]. According to the degree of the connection's closeness, some scholars also divide social capital into categories based on the perspective of individuals inside and outside the organization or inside and outside the team. Adler and Kwon believe that social capital can be divided into two categories: bonding and bridging [66]. This perspective was applied by Huang et al. in their analysis of the relationship between a construction project team's social capital and project performance [67], while Kim and Cannella stated that social capital can be divided into internal social capital involving interpersonal connections in the central organization, and external social capital involving other connections outside the organization [68].

Megaprojects often have prominent economic and political significance, and the involvement of non-project participants such as the government, industry associations, and the public in the project tends to be more profound than in general projects. The influence of external stakeholders such as the government, industry associations, and the public on project construction management in the social network of megaprojects cannot be ignored. However, the interactions among the participants of megaprojects contain links including both formal and informal relationships, while the links between project participants and external parties are relatively homogenous. Therefore, with reference to the previous studies, this study divides the social capital of megaprojects into project internal social capital (ISC) and project external social capital (ESC). Project internal social capital refers to the internal links of the project organization, consisting of each participant and the actual or potential resources generated therefrom, with reference to Nahapiet and Ghoshal's model [63]. Project external social capital refers to the links between project participants, the world outside the project, and the actual or potential resources generated therefrom, and includes three dimensions, namely, vertical, horizontal, and social relationships [65].

2.2.2. Resource Integration Capacity

Resource integration refers to an intricate and complex dynamic process in which an organization identifies, selects, absorbs, configures, activates, and organically integrates resources from different sources and of different kinds, making them more flexible, rational, systematic, and valuable, so as to form a unique core resource system that is more conducive to its own development [69]. Resource integration capacity is the ability of an organization to effectively plan and configure various resources within or between organizations to achieve comprehensive benefits [70]. In the field of construction engineering, a construction project is a temporary organization formed by multiple units, and its resource integration is the process of configuring relevant resources within the participants, between the participants, and outside the project organization according to the project objectives, then obtaining the overall optimal integration of resources through the organization's institutional arrangements and management coordination [71]. According to previous studies, the resource integration capacity of a project organization is formed in the resource integration process of resource acquisition, allocation, and utilization [72,73]. Resource acquisition is the process of identifying and acquiring the required resources from a network of multiple internal and external megaproject stakeholders [74]. Resource allocation refers to the process of adjusting and combining the resources to make them more valuable so as to maximize their effectiveness [75]. Resource utilization refers to the process of applying the allocated resources to create maximum value for the organization [76].

Actively performing social responsibility is conducive to improving the resource integration capacity of the organization [23]. In the process of performing social responsibility, project participants can broaden the channels of resource integration through strong relationships with external stakeholders, attracting and acquiring more resources conducive to project construction. These resources include tangible resources, such as government subsidies, environment-friendly materials, and information on innovative technologies, as well as intangible resources, such as project reputation and learning opportunities. Furthermore, when the organization takes social responsibility initiatives, the level of internal employee competence may also be improved. Organizational social responsibility behaviors include creating a good environment and training opportunities for employees, and therefore can play a role in gathering innovative resources for the organization [24], as well as being an important means of improving organizational learning and knowledge creation [77]. In megaprojects, Wang et al. found that engineering environmental responsibility practices toward internal stakeholders is positively associated with their organizational citizenship behaviors [14]. The positive image that MSRB creates for the project has the potential to enhance the project identity of project participants, who in turn are more motivated to complete their work. The positive image created by MSRB can also promote internal knowledge sharing. The project gains excellent organizational and employee resources, and its resource integration capacity is also improved. Thus, we suggest that MSRB has a positive impact on the project's resource integration capacity.

High resource integration capacity can enhance the performance of megaprojects. The resource composition and management of megaprojects are extremely complex. Project resources are of great significance to megaprojects, but just having resources does not guarantee the successful application of those resources [78]. In particular, megaprojects face the complex internal and external environments of dynamic changes, and it is difficult to adapt static resources to meet the needs of the dynamic development of the organization. Only through the integration of resources, so that resources are constantly updated, combined, and matched to form a new resource system that adapts to the needs of megaprojects, can resources be fully and effectively utilized. Putra et al. found that the effective integration of resources in engineering projects can not only solve the problems encountered in the project, the ability of the organization to utilize its resources can also improve the project's application and cost control capabilities, which in turn promote the project's profitability [79]. The sustainable growth and success of the project requires the project organization to integrate internal and external resources. The project organization can use its own organizational management mechanism, combined with external value networks, to achieve the acquisition and absorption of internal and external resources. Integrating resources via the transfer of material and information resources and the sharing of knowledge resources increases the value of cooperation [80]. Thus, we suggest that project resource integration capacity has a positive impact on project performance.

3. Research Method

3.1. Variables and Measurements

Through our literature study, we identified 15 factors (see Table 1), and due to the abstract nature of the above variable factors in the practice of megaprojects, this study used the method of a questionnaire survey, which is commonly applied in related studies, to collect the variable data. The scale for this study was adapted from scales used in related studies. A Likert five-point scale was applied to investigate the data needed for this study.

Table 1. Research Scale.

	Factor	Code	Item	Reference
Megaproject social responsibility behaviors (MSRB)	Economic and qualitative responsibility behaviors	EQRB1	The organization conducts full and effective economic and technical feasibility studies	Lin et al. [34]; Xie et al. [3]
		EQRB2	The organization develops and implements a reasonable and complete cost-control plan	
		EQRB3	The organization develops and implements a reasonable and complete plan	
		EQRB4	The organization develops and implements a reasonable and complete schedule management plan	
		EQRB5	There are many innovations and technological advances	
	Political and communal responsibility behaviors	PCRB1	The organization hires local staff to drive local employment	Lin et al. [34]; Xie et al. [3]
		PCRB2	The organization organizes public welfare activities and provides public welfare services in the relevant communities where the works are carried out	
		PCRB3	The organization deals with related accidents and public incidents in a timely manner	
		PCRB4	The organization takes a variety of measures to promote public participation	
		PCRB5	The organization takes an active role in anti-corruption and anti-corruption activities	
	Legal and regulatory responsibilityvbehaviors	LRRB1	The organization discloses project information to the public in a timely manner	Lin et al. [34]; Xie et al. [3]
		LRRB2	All project participants strictly comply with laws and regulations.	
		LRRB3	All project participants follow industry standards	
		LRRB4	All project participants adhere to the principle of fair competition within the industry	
	Environmental and ethical responsibility Behaviors	EERB1	The organization provides a safe and healthy working environment for its employees	Lin et al. [34]; Xie et al. [3]
		EERB2	The organization pays attention to the protection of the ecological environment in the area where the project is to be built (e.g., prevention of river pollution and destruction of vegetation, etc.).	
		EERB3	The organization pays attention to the protection of the environment in the local community (e.g., prevention of pollution from noise, dust, etc.).	Lin et al. [34]; Xie et al. [3]
		EERB4	The organization uses resources rationally and avoids wasting them during the construction of the project.	

Table 1. *Cont.*

	Factor	Code	Item	Reference
Internal social capital (ISC)	Structural dimension	SISC1	During project implementation, there are frequent formal exchanges between the various parties involved in the construction, such as holding project seminars, organizing exchange and learning activities, etc.	Leana et al. [81]; Wang et al. [82]
		SISC2	During project implementation, there are frequent informal exchanges among the participants, such as exchanging information on the construction of the project in informal settings, such as at the site, lounge, gathering places, etc.	
		SISC3	During project implementation, members of the parties often use web-based media tools (telephone, e-mail, WeChat, etc.) to communicate about the project construction work	
	Relational dimension	RISC1	During the project implementation, the participants will trust each other	Leana et al. [81]; Wang et al. [82]
		RISC2	The project participants keep their promises to each other.	
		RISC3	When one participant has a difficulty or dispute, the other participants are willing to help	
	Cognitive dimension	CISC1	The participants have similar views on the development prospects of the project	Leana et al. [81]; Wang et al. [82]
		CISC2	Participants have a shared understanding of the key points of the project (e.g., core technologies)	
		CISC3	Participants reach consensus, especially on key project decisions	
External social capital (ESC)	Vertical relationships	VESC1	During the construction of the project, the project participants maintain good relations with government departments	Leana et al. [81]; Change et al. [83]
		VESC2	During the construction of the project, the relevant government departments provide resources and support for the construction of the project	
		VESC3	During the construction of the project, the project participants maintain close communication with government departments	
	Horizontal relationships	HESC1	Project participants have a wide range of contacts with other relevant organizations in the industry	Leana et al. [81]; Change et al. [83]
		HESC2	Project participants exchange project construction experience with other related organizations in the industry	
		HESC3	Project participants and other relevant organizations in the industry organize study tours and learning activities for each other.	
	Social relationships	SESC1	During the construction of the project, the project participants establish good relations with the communities where the work is being carried out	Leana et al. [81]; Change et al. [83]
		SESC2	During project construction, project participants establish good relations with non-governmental organizations	
		SESC3	During project construction, the project participants establish good relations with the media	
		SESC4	During project construction, project participants establish good relations with the public	

Table 1. *Cont.*

Factor		Code	Item	Reference
Megaproject resource Integration capacity (MRI)		MRI1	The organization is able to clearly identify resource needs and allocate them appropriately.	GE et al. [73]
		MRI2	The organization has access to the resources the organization wants	
		MRI3	The organization is able to divest useless resources in a timely manner	
		MRI4	The organization believes that useful resources can help the project to be completed better	
		MRI5	The organization is able to leverage consolidated resources to gain access to other useful resources	
		MRI6	The organization is able to leverage consolidated resources to help create greater value for the project	
Megaproject performance (MP)		MP1	The project is on schedule or will be completed on schedule	He et al. [2]
		MP2	The project expenditure is within budget	
		MP3	The project's quality is in accordance with the relevant national standards and contractually agreed standards	
		MP4	The project's safety management is well executed	
		MP5	The project has a high level of resource utilization efficiency and environmental protection	
		MP6	The project has an innovative contribution to theory and practice in the field of technology and management	
		MP7	The project participants, government departments and the public are satisfied with the project construction process and deliverables	
		MP8	Participants consider continued cooperation in the future	
Internal social capital (ISC)	Structural dimension	SISC1	During project implementation, there are frequent formal exchanges between the various parties involved in the construction, such as holding project seminars, organizing exchange and learning activities, etc.	Leana et al. [81]; Wang et al. [82]
		SISC2	During project implementation, there are frequent informal exchanges among the participants, such as exchanging information on the construction of the project in informal settings, such as at the site, lounge, gathering places, etc.	
		SISC3	During project implementation, members of the parties often use web-based media tools (telephone, e-mail, WeChat, etc.) to communicate about the project construction work	
	Relational dimension	RISC1	During the project implementation, the participants will trust each other	Leana et al. [81]; Wang et al. [82]
		RISC2	The project participants keep their promises to each other.	
		RISC3	When one participant has a difficulty or dispute, the other participants are willing to help	
	Cognitive dimension	CISC1	The participants have similar views on the development prospects of the project	Leana et al. [81]; Wang et al. [82]
		CISC2	Participants have a shared understanding of the key points of the project (e.g., core technologies)	
		CISC3	Participants reach consensus, especially on key project decisions	

Table 1. *Cont.*

	Factor	Code	Item	Reference
External social capital (ESC)	Vertical relationships	VESC1	During the construction of the project, the project participants maintain good relations with government departments	Leana et al. [81]; Change et al. [83]
		VESC2	During the construction of the project, the relevant government departments provide resources and support for the construction of the project	
		VESC3	During the construction of the project, the project participants maintain close communication with government departments	
	Horizontal relationships	HESC1	Project participants have a wide range of contacts with other relevant organizations in the industry	Leana et al. [81]; Change et al. [83]
		HESC2	Project participants exchange project construction experience with other related organizations in the industry	
		HESC3	Project participants and other relevant organizations in the industry organize study tours and learning activities for each other.	
	Social relationships	SESC1	During the construction of the project, the project participants establish good relations with the communities where the work is being carried out	Leana et al. [81]; Change et al. [83]
		SESC2	During project construction, project participants establish good relations with non-governmental organizations	
		SESC3	During project construction, the project participants establish good relations with the media	
		SESC4	During project construction, project participants establish good relations with the public	
Megaproject resource integration capacity (MRI)		MRI1	The organization is able to clearly identify resource needs and allocate them appropriately.	GE et al. [73]
		MRI2	The organization has access to the resources the organization wants	
		MRI3	The organization is able to divest useless resources in a timely manner	
		MRI4	The organization believes that useful resources can help the project to be completed better	
		MRI5	The organization is able to leverage consolidated resources to gain access to other useful resources	
		MRI6	The organization is able to leverage consolidated resources to help create greater value for the project	
Megaproject performance (MP)		MP1	The project is on schedule or will be completed on schedule	He et al. [2]
		MP2	The project expenditure is within budget	
		MP3	The project's quality is in accordance with the relevant national standards and contractually agreed standards	
		MP4	The project's safety management is well executed	
		MP5	The project has a high level of resource utilization efficiency and environmental protection	
		MP6	The project has an innovative contribution to theory and practice in the field of technology and management	
		MP7	The project participants, government departments and the public are satisfied with the project construction process and deliverables	
		MP8	Participants consider continued cooperation in the future	

To ensure that the scale collected reliable variable data, three owners, contractors, and supervisory experts who have been working in the construction industry for more than 15 years were invited to test our scale before distributing the questionnaire to ensure that the questionnaire was aligned with practical scenarios. The initial question items of the questionnaire were determined through scale development and design, and the first draft of the questionnaire was formed. In addition, before the official questionnaire was distributed, we invited five academic and industry experts to evaluate the content and academic expression of our scale to ensure that the questionnaire was aligned with practical scenarios. After making modifications and improvements based on the experts' comments, we conducted the distribution and collection of the questionnaire.

3.2. Sampling and Data Collection

The questionnaire was distributed to managers of megaprojects. The managers were involved in many types of megaprojects, including urban complex projects, transportation infrastructure, industrial energy, water conservancy, and hydropower. The projects are distributed in South China, East China, North China, Central China, Northeast China, Northwest China, and Southwest China. The investment amount of most individual projects is over 4 billion dollars. In terms of investment attributes, government investment projects account for more than half of the projects, which is in line with the law of the investment attributes of megaprojects. It can be seen that the samples researched in this study are in line with our research study's needs and have strong representativeness. Finally, a total of 345 questionnaires were recovered, including 247 paper questionnaires and 98 electronic questionnaires. A total of 206 valid questionnaires were obtained after screening, with an effective rate of 59.7%. Table 2 shows the descriptive statistics of the respondents. The diversity of respondents in terms of their positions, years of experience, roles in their organizations, as well as the types of megaprojects they were involved in and the attributes of their project investments reflect the views of respondents from different backgrounds, which better ensures the generalizability and reliability of the research results. Specifically, in terms of the individual characteristics of the respondents, the ratio of male respondents (181) to female respondents (25) among the 206 respondents who participated in this research is about 7:1. In terms of the number of years of working experience in the engineering field, the research subjects in the valid samples are more experienced in engineering. In terms of the level of education, 71.4% of the respondents have a bachelor's degree and 23.3% have a master's degree or above, indicating a relatively high level of education.

Table 2. Statistics of Interviewed Experts.

Characteristic		Item	Frequency	Proportion (%)
Individual characteristic (N = 206)	Gender	Male	181	87.86
		Female	25	12.14
	Education attainment	College and below	11	5.3
		Undergraduate	147	71.4
		Bachelor's degree	44	21.4
		Ph. D.	4	1.9
	Working experience	<5 years	82	39.8
		5–10 years	62	30.1
		11–15 years	40	19.4
		16–20 years	12	5.8
		>20 years	10	4.9
	Management level	Top manager	20	9.7
		Middle manager	54	26.2
		Grassroots Manager	126	61.2
		Other	6	2.9

Table 2. Cont.

Characteristic		Item	Frequency	Proportion (%)
Individual characteristic (N = 206)	Project Roles	Government	7	3.4
		Owner	90	43.7
		Contractor	66	32.0
		Supervisor	11	5.3
		Designer	16	7.8
		Consultant	6	2.9

3.3. Data Pre-Processing

This study constructs a Bayesian network based on questionnaire data. The variables in the questionnaire are observed and assessed through a five-level Likert scale, with 1–5 indicating complete nonconformity, nonconformity, general, conformity, and full conformity, respectively. In this paper, the Cronbach α coefficient is used as a criterion to test the reliability of the questionnaire, and to test the internal consistency of the measurement data of each variable of the questionnaire, including EQRB, PCRB, LRRB, EERB, MSRB, ISC, ESC, MRI, and MP. The value of the Cronbach α coefficient is related to the degree of correlation between the questions of the questionnaire, and the larger the coefficient is, the better the internal consistency between the questions and the better the reliability. In existing studies, it is widely recognized that when the Cronbach α coefficient of the questionnaire measurement items exceeds 0.7, the reliability of these items is considered relatively good. The analysis of the valid data of the recovered questionnaire shows that the Cronbach α coefficient of the overall data of MSRB is 0.818, the Cronbach α coefficient of ISC is 0.762, the Cronbach α coefficient of ESC is 0.808, and the Cronbach α coefficient of MRI is 0.861. The Cronbach α coefficient of MP is 0.868. All of these data have a good level of reliability. Further testing the internal consistency of the dimensions of MSRB, the Cronbach α coefficients of EQRB, PCRB, LRRB, and EERB are 0.848, 0.742, 0.828, and 0.841, respectively, with a good level of reliability. The reliability test showed that the five state values of the latent variables measured by directly observing the variables could be used in the Bayesian network model. We obtained the representative value of each variable by calculating the mean score of the relevant questions of the questionnaire and rounding it up, where the score of MSRB is obtained by rounding up the mean score of all the questions of the four dimensions of social responsibility behavior. In this study, the node mean scores are normalized to the max-min value and divided into five equally spaced score intervals rated as Higher, High, Medium, Low, and Lower. From this, discretized sample data can be obtained.

3.4. Bayesian Network

The Bayesian network (BN), first introduced by the American scholar Pears in 1988, is a probabilistic graphical model that integrates the theories and techniques of probability theory and graph theory, with a robust modeling capability and a comprehensive inference mechanism. A Bayesian network integrates prior knowledge and sample data to infer and represent the complex correlations between multiple variables in a large framework, which is suitable for making probabilistic inferences regarding complex uncertain events.

A Bayesian network is a pair consisting of a directed acyclic graph G and a probability distribution P, i.e., $BN = (G, P)$: (1) directed acyclic graph $G = (V, E)$, where V denotes the set of nodes and the nodes represent random variables; E denotes the set of directed arcs, and the directed arcs connecting the nodes are used to describe the probabilistic dependencies between the nodes. If there is no directed arc from node A to node B, it indicates that the two random variables represented by node A and node B do not have a dependency relationship and are either independent of each other or conditionally independent given the conditions of the other random variables in the graph. If there exists a directed arc from node A to node B pointing from A to B, it indicates that there is a dependency relationship between node A and node B and that node A has a direct causal effect on node B. In Bayesian networks, nodes with incoming arrows are considered child nodes and nodes with outgoing arrows are considered parent nodes. The node that has no parent node is called the root node and is the underlying node in the network, which has an associated prior probability distribution. The rest of the nodes are called non-root nodes. (2) Probability distribution *P*, which is used to describe the degree of association between nodes, has a probability distribution value for each node. The conditional probability is denoted as $P(A \mid B)$, which reflects the existence of a conditional probability value between the pair of nodes A and B.

A Bayesian network is a decomposition of the joint probability distribution. A necessary condition for deciding whether a directed acyclic graph $G = (V, E)$ and a conditional probability P can be combined into a pair $BN = (G, P)$ is whether P conforms to the joint probability distribution decomposition rule specified by $G = (V, E)$:

$$PX1, X2, \ldots, Xn = \prod_{i=1}^{n} P(Xi|\ X1,\ X2,\ \ldots, Xi-1) = \prod_{i=1}^{n} P(Xi|\ Pa(Xi)) \quad (1)$$

In Equation (1), $P(X1, X2, \ldots, Xn)$ is the joint probability distribution of the variables, $Pa(Xi)$ denotes the set of parent nodes of node Xi, and $P(Xi \mid Pa(Xi))$ is the conditional probability distribution of node Xi given its parent nodes. The theoretical basis of Bayesian network inference is to satisfy the conditional independence assumption, upon which the joint probability of each node in the network model can be expressed as the product of its conditional probabilities. If $G = (V, E)$ and P satisfy Equation (1), then $G = (V, E)$ and P can be combined into a $B = (G, P)$. Bayesian networks represent correlations between variables using conditional probabilities and visualize the values in graphical form.

Bayesian networks have a strong advantage in risk assessment [84,85], effective strategy reasoning [86,87], and in complexity studies in the engineering and construction field. In addition, studies have shown that Bayesian networks can be beneficially applied to studies for stakeholder influence relationships [88] and project performance improvement [89,90]. Therefore, it is feasible for this study to explore the role path of megaproject social responsibility behaviors on project performance enhancement using a Bayesian network through its forward prediction, reverse diagnosis, and other functions.

3.4.1. BN Model Structure

In the above study, we identified 15 variables. Although numerous existing studies have shown that these variables are interconnected, there is no recognized explanatory model. Given this, we aim to establish a causal model of these variables using literature research methods and structural learning methods to obtain a directed acyclic graph (DAG). There are two main approaches to BN structural learning: the scoring search-based approach and the constraint-based methods. The scoring search algorithm assigns a score to each BN structure and then selects the model with the highest score. Constraint-based search algorithms perform a set of conditional independence analyses on the data. The combination of these two approaches provides a hybrid learning framework. The search space is first reduced through dependency analysis, and then the highest-rated network structure is identified using a scoring search [91]. The MMHC (Max-Min Hill-Climbing) algorithm used in this study is a hybrid structure learning method. It integrates the MMPC (Max-Min Parent-Child) algorithm (for reducing the search space) and the HC (Hill-Climbing) algorithm (for identifying the optimal network structure within a restricted space) [92].

To align the BN model with the objectives of this study, we utilized the R language bnlearn toolkit to customize the structure learning algorithm. Based on the potential connections between MSRB and each factor identified in the literature review, we established an undirected correlation and determined the causal relationships between the factors through algorithmic learning. We also constrained the learning conditions such that MP is a parent node and MSRB is a child node, in line with the study's objectives. Finally, 660 structural tests were performed using the MMHC method for structure learning, resulting in a DAG with 15 nodes and 20 edges (see Figure 1). As inferred from the DAG, MSRB generates causal relationships for MRI, ESC, and ISC. Interestingly, the structure learning results indicate that CISC has a strong probabilistic causal relationship with MP.

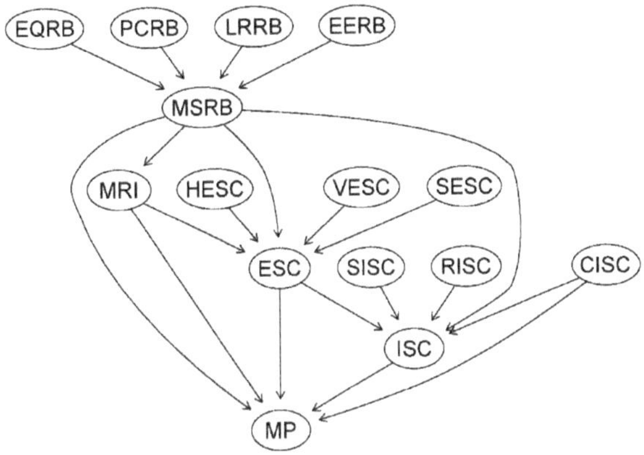

Figure 1. Directed Acyclic Graphical Model.

3.4.2. Parametric Learning

Conditional probability distributions, also referred to as belief relations, are typically denoted by CPT. In the conditional probability distribution of a Bayesian network, the sum of the probabilities of all states of a given node, under all possible combinations of the node's parent nodes, equals 1. Using the MSRB node as an example, the value of its CPT can be computed based on the variable data gathered from the questionnaire research. The parent nodes of the MSRB node encompass the four dimensions of the social responsibility behaviors of the megaprojects. Each parent node state comprises five levels: Higher, High, Medium, Low, and Lower, and the total number of possible combination states of all parent nodes is 625. Table 3 lists some of the conditional probability cases of megaproject social responsibility behaviors in this study, with the sum of probabilities in each row equating to 1.

We apply GENIE to learn the parameters of the dataset based on the DAG graph obtained by structural learning. We then calculate the conditional probability table of each node, enabling us to obtain the chartbar graph of this BN model (see Figure 2). The sum of the probabilities of the different states of each node, as calculated from the questionnaire research data, is 100%. Based on the Bayesian network model constructed from the questionnaire research data, we can infer that the evaluation of the overall social responsibility behavior of the project by the managers of megaprojects is predominantly positive, with a small number of negative evaluations. Specifically, 28.50% of the managers perceive the project's social responsibility behavior performance as being of a higher level, 36.47% view it as high level, and 21.10% consider the level of project social responsibility behavior as low or very low. This shows that in the internal management of megaprojects for the implementation of social responsibility is still more important, in the four dimensions of social responsibility behavior, the belief state of LRRB is the highest, followed by EQRB which are currently the most basic social responsibility behavior of megaprojects, with a high degree of completion and evaluation, which is in line with Ma's research [18]. The state probability distribution of PCRB is scattered, which shows that the implementation of PCRB of different megaprojects is uneven, and the managers of different projects have different praise and criticism for it. The probability of a high level is highest for MRI, ISC and ESC, and the average level of MRI, ESC, and ISC is high. Due to the wide range of questionnaire data sources, there is diversity in the evaluation of these three nodes, and the distribution of probability is also wide. In addition, on the whole, managers evaluate the performance of megaprojects well, and the probability that MP will fall into each of the five states of Higher level, High level, Medium level, Low level, and Lower level is 22.42%, 24.56%, 19.35%, 17.02%, and 16.65%, respectively. On the whole, the performance of the majority of megaprojects is evaluated to be good.

Table 3. MSRB's Conditional probability table (partial).

MSRB State					Condition			
Higher	**High**	**Medium**	**Low**	**Lower**	**EQRB**	**EERB**	**PCRB**	**LRRB**
97.14%	0.71%	0.71%	0.71%	0.71%	Higher	Higher	Higher	Higher
77.50%	15.00%	2.50%	2.50%	2.50%	High	Higher	Higher	Higher
20.00%	20.00%	20.00%	20.00%	20.00%	Medium	Higher	Higher	Higher
20.00%	20.00%	20.00%	20.00%	20.00%	Low	Higher	Higher	Higher
20.00%	20.00%	20.00%	20.00%	20.00%	Lower	Higher	Higher	Higher
60.00%	10.00%	10.00%	10.00%	10.00%	Higher	Higher	Higher	High
30.00%	55.00%	5.00%	5.00%	5.00%	High	Higher	Higher	High
20.00%	20.00%	20.00%	20.00%	20.00%	Medium	Higher	Higher	High
20.00%	20.00%	20.00%	20.00%	20.00%	Low	Higher	Higher	High
20.00%	20.00%	20.00%	20.00%	20.00%	Lower	Higher	Higher	High
20.00%	20.00%	20.00%	20.00%	20.00%	Higher	Higher	Higher	Medium
20.00%	20.00%	20.00%	20.00%	20.00%	High	Higher	Higher	Medium
20.00%	20.00%	20.00%	20.00%	20.00%	Medium	Higher	Higher	Medium
20.00%	20.00%	20.00%	20.00%	20.00%	Low	Higher	Higher	Medium
20.00%	20.00%	20.00%	20.00%	20.00%	Lower	Higher	Higher	Medium
20.00%	20.00%	20.00%	20.00%	20.00%	Higher	Higher	Higher	Low
20.00%	20.00%	20.00%	20.00%	20.00%	High	Higher	Higher	Low
20.00%	20.00%	20.00%	20.00%	20.00%	Medium	Higher	Higher	Low
20.00%	20.00%	20.00%	20.00%	20.00%	Low	Higher	Higher	Low
20.00%	20.00%	20.00%	20.00%	20.00%	Lower	Higher	Higher	Low
20.00%	20.00%	20.00%	20.00%	20.00%	Higher	Higher	Higher	Lower
20.00%	20.00%	20.00%	20.00%	20.00%	High	Higher	Higher	Lower
20.00%	20.00%	20.00%	20.00%	20.00%	Medium	Higher	Higher	Lower
20.00%	20.00%	20.00%	20.00%	20.00%	Low	Higher	Higher	Lower
20.00%	20.00%	20.00%	20.00%	20.00%	Lower	Higher	Higher	Lower

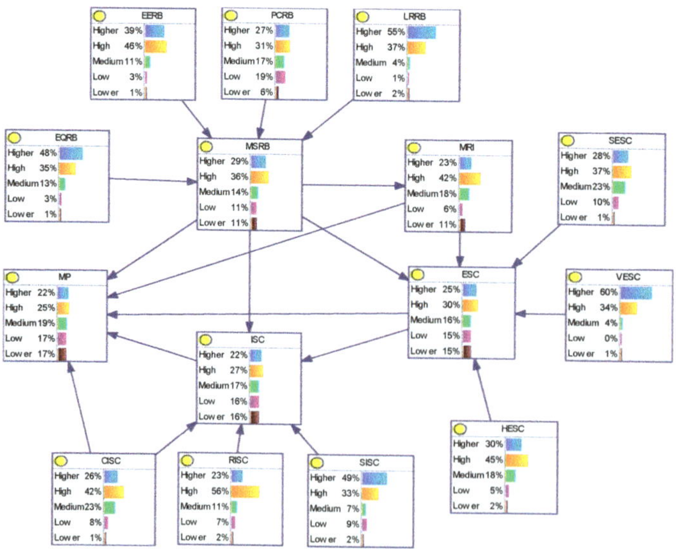

Figure 2. Causal reasoning model.

4. Results

4.1. Model Validation

This study validates 206 samples of data. The greater the validity of the model, the greater the consistency between the model simulation results and the real level of MP under the given variable conditions and the closer the model's prediction results are to the real level. The data of variable nodes are fed into the constructed Bayesian network model, and the probability distribution of variable nodes "MP" is computed. See Figure 2.

The performance of the trained classifier can be assessed by calculating the prediction accuracy and kappa coefficient from the confusion matrix [93]. Tables 4 and 5 display the confusion matrix and accuracy for MSRB and MP. With accuracies of 0.951 and 0.825 for MSRB and MP, 0.888 for both nodes, and kappa coefficients of 0.928 and 0.744 (kappa coefficient > 0.6 is considered to be a reliable measure of accuracy [94,95]), it is evident from Tables 4 and 5 that this BN network exhibits superior classification performance, which can be utilized for subsequent inference analysis.

Table 4. MSRB's confusion matrix.

MSRB		Predicted				
		Higher	High	Medium	Low	Lower
Actual	Higher	77	3	0	0	0
	High	5	67	1	0	0
	Medium	0	1	45	0	0
	Low	0	0	0	5	0
	Lower	0	0	0	0	2
Accuracy		0.963 (77/80)	0.918 (67/73)	0.978 (45/46)	1 (5/5)	1 (2/2)
		0.951 (196/206)				
Kappa		0.928				

Table 5. MP's confusion matrix.

MP		Predicted				
		Higher	High	Medium	Low	Lower
Actual	Higher	60	7	0	0	0
	High	8	68	1	1	0
	Medium	3	11	33	0	0
	Low	1	1	2	5	0
	Lower	0	1	0	0	4
Accuracy		0.896 (60/67)	0.872 (68/78)	0.702 (33/47)	0.556 (5/9)	0.8 (4/5)
		0.825 (170/206)				
Kappa		0.744				

4.2. Bayesian Network Inference

4.2.1. Predictive Inference

Predictive inference can be based on the directed arc direction of connected nodes to reason about the outcome, given the latest cause information. The purpose of predictive analysis is to forecast future outcomes given different antecedent information. This approach can effectively analyze the influence of various states of evidence nodes on the state of target nodes. In the Bayesian network model, we set different levels of social responsibility behaviors of various dimensions as evidence nodes. We can then calculate the probability distribution of the MP level after updating the network.

(1) Simple strategy for single precondition changes

To estimate the impact of changes in the states of EQRB, PCRB, LRRB, and EERB on MP, the probability of the "higher" state of each dimension of social responsibility behavior is set to 100% to simulate the scenario of adopting a single strategy. Figure 3 displays the posterior probability of MP corresponding to adjusting the probability of the higher-level state of LRRB to 100%. It can be noted that when the LRRB is set to a 100% higher level, the probability of MP being at a higher level increases from 22.42% to 23.60%.

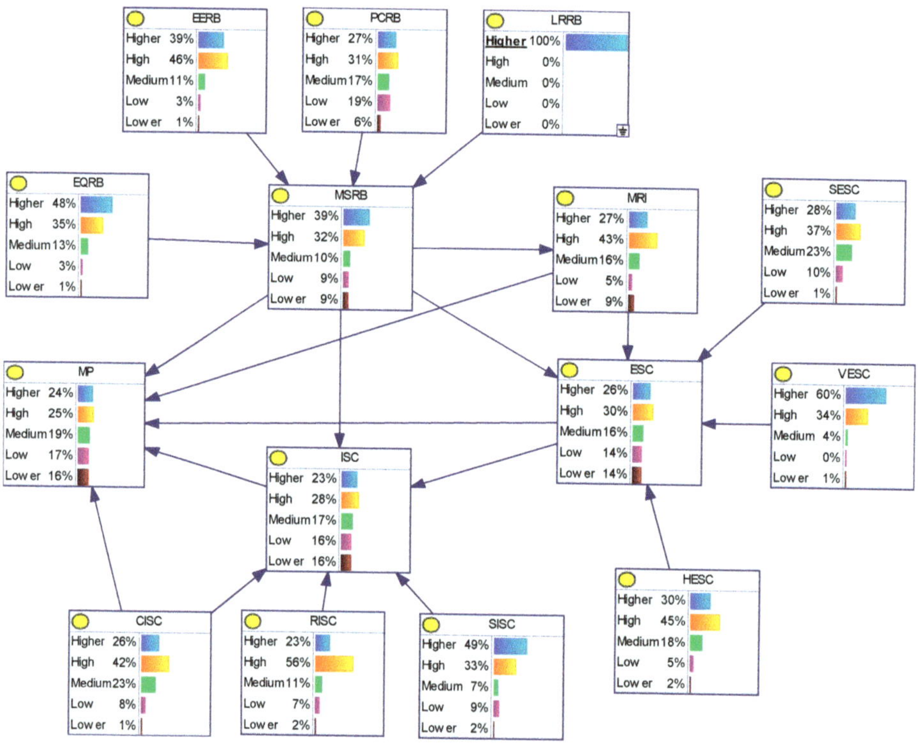

Figure 3. Project performance changes induced by changing single-dimensional social responsibility behaviors.

We separately adjust the level of social responsibility behavior for each dimension to a higher level and calculate the conditional probability distributions for the other nodes. The inference results are shown in Table 6. Each social responsibility behavior node influences the other nodes. PCRB, LRRB, EERB, and EQRB at higher levels contribute in decreasing order to the overall level of social responsibility behavior, while the latter two help avoid a low level of overall MSRB. The predictive analysis of these four nodes indicates that all dimensions of social responsibility behaviors influence MRI, ESC, ISC, and MP, with a particularly significant effect on MRI. However, the enhancement of social capital and resource integration capacity did not result in a comparable enhancement of MP, suggesting that its effect on MP is not linear. Megaprojects that adopt partial social responsibility measures can achieve a slight improvement in project performance due to the measures' combined effect on the ISC, ESC, and MRI.

Table 6. Predictive results of changing one-factor beliefs.

State	Belief changes				
	Belief change: "EQRB = Higher"				
	MSRB	MRI	ESC	ISC	MP
Higher	9.83%	3.73%	0.99%	0.55%	1.06%
High	−3.10%	0.65%	0.08%	0.19%	0.01%
Medium	−3.86%	−1.90%	−0.41%	−0.33%	−0.55%
Low	−1.44%	−0.90%	−0.33%	−0.21%	−0.28%
Lower	−1.43%	−1.57%	−0.33%	−0.21%	−0.25%
	Belief change: "EERB = Higher"				
	MSRB	MRI	ESC	ISC	MP
Higher	11.01%	3.74%	0.89%	0.51%	1.20%
High	−7.11%	−0.65%	−0.37%	−0.12%	−0.29%
Medium	−3.56%	−2.31%	−0.25%	−0.26%	−0.59%
Low	−0.18%	−0.40%	−0.13%	−0.07%	−0.19%
Lower	−0.17%	−0.38%	−0.13%	−0.06%	−0.12%
	Belief change: "PCRB = Higher"				
	MSRB	MRI	ESC	ISC	MP
Higher	14.60%	5.13%	1.24%	0.71%	1.57%
High	−8.16%	−0.22%	−0.27%	0.00%	−0.27%
Medium	−4.41%	−2.86%	−0.41%	−0.39%	−0.76%
Low	−1.02%	−0.83%	−0.28%	−0.16%	−0.31%
Lower	−1.01%	−1.22%	−0.28%	−0.16%	−0.23%
	Belief change: "LRRB = Higher"				
	MSRB	MRI	ESC	ISC	MP
Higher	10.87%	4.03%	1.05%	0.59%	1.17%
High	−4.16%	0.44%	0.00%	0.15%	−0.05%
Medium	−4.11%	−2.13%	−0.41%	−0.34%	−0.60%
Low	−1.31%	−0.87%	−0.32%	−0.20%	−0.28%
Lower	−1.30%	−1.47%	−0.32%	−0.19%	−0.24%

(2) Combined strategies for multiple precondition changes

The multifactor strategy involves changing the level of social responsibility behavior performance in several dimensions simultaneously to improve the level of project performance. The single-factor strategy demonstrated that the probability of project performance being at a higher level reached a maximum 24.00% when controlling the level of social responsibility behaviors in one dimension alone. When a strategy of controlling several variables simultaneously was implemented, project performance might be further improved. After making several predictions by integrating control variables, it is observed that the probability of project performance being at a higher level reaches a maximum of 29.70% when controlling the four dimensions of social responsibility behaviors to reach a higher level, at which time the probability distribution of the MSRB is 97.14% for a higher level, and the probability of all other statuses is 0.71%, so it may be difficult to significantly improve the MP by improving the overall MSRB.

EQRB, EERB, PCRB, and LRRB are in the higher-level state. The probability of different combinations of strategies corresponding to higher levels of project performance is displayed in Table 7. It can be observed that the combination where the levels of PCRB, EQRB, and EERB are all at higher levels is the most effective strategy. Among the combinations that simultaneously control three dimensions of social responsibility behavioral performance, the combination with EQRB, EERB, and LRRB at higher levels simultaneously has the lowest effect, but it has a better inhibitory effect on low-level performance. The increase of the probability of project performance at a higher level when controlling for the strategy of two dimensions of social responsibility behaviors performance at a higher level simultaneously ranged from 2.64% to 3.37%, with little difference in effect.

Table 7. Posterior probability of MP under combinatorial condition.

Combinatorial Condition	State of MP				
	Higher	High	Medium	Low	Lower
"EQRB = Higher" + "PCRB = Higher" + "EERB = Higher"	5.22%	−1.09%	−2.35%	−1.05%	−0.73%
"PCRB = Higher" + "LRRB = Higher" + "EERB = Higher"	5.17%	−0.83%	−2.36%	−1.13%	−0.85%
"EQRB = Higher" + "PCRB = Higher" + "LRRB = Higher"	5.11%	−0.45%	−2.38%	−1.25%	−1.03%
"EQRB = Higher" + "EERB = Higher" + "LRRB = Higher"	4.99%	−0.45%	−2.33%	−1.21%	−1.00%
"EQRB = Higher" + "EERB = Higher"	2.64%	−0.49%	−1.23%	−0.53%	−0.39%
"EQRB = Higher" + "PCRB = Higher"	2.88%	−0.42%	−1.35%	−0.63%	−0.48%
"EQRB = Higher" + "LRRB = Higher"	2.92%	−0.21%	−1.40%	−0.71%	−0.60%
"PCRB = Higher" + "EERB = Higher"	3.37%	−0.71%	−1.54%	−0.66%	−0.46%
"LRRB = Higher" + "EERB = Higher"	2.72%	−0.28%	−1.30%	−0.63%	−0.51%
"PCRB = Higher" + "LRRB = Higher"	3.20%	−0.35%	−1.51%	−0.74%	−0.60%

4.2.2. Diagnostic Inference

In Bayesian networks, the consequence variable is designated as the evidence node and the cause variable as the query variable. This process of reasoning from consequence to cause is referred to as diagnostic analysis or backward reasoning. We set the probability that the MP of the target node is at a higher level to 100%, implying that the MP level is good and the project has a high likelihood of success. We then observe the performance of other nodes in the Bayesian network model of this study. The inference results are shown in Figure 4. By comparing Figure 4 with the belief data in Figure 2, we plot Table 8 to observe the results of diagnostic inference.

Table 8. Beliefs Changes in Diagnostic Inference.

Nodes	Belief Changes				
	Higher	High	Medium	Low	Lower
EQRB	2.27%	−1.30%	−0.77%	−0.15%	−0.04%
EERB	2.10%	−1.26%	−0.68%	−0.12%	−0.04%
PCRB	1.87%	0.22%	−0.65%	−1.08%	−0.36%
LRRB	2.90%	−2.59%	−0.17%	−0.06%	−0.07%
MSRB	9.63%	−4.86%	−2.44%	−1.18%	−1.15%
MRI	6.44%	−1.86%	−2.64%	−0.72%	−1.22%
ESC	7.68%	−2.83%	−1.55%	−1.71%	−1.59%
ISC	4.81%	1.41%	−2.56%	−1.89%	−1.78%

As observed from Figure 4 and Table 8, when the probability of MP at a higher level is 100% and is used as an evidence node for backward reasoning, the incremental increase in the probability of a higher level of MSRB, ESC, ISC, and MRI decreases in order. This indicates that when MP is at a higher level, it is evident that the MSRB is playing a role in the higher level of MP. Among the four types of social responsibility behaviors, LRRB has the highest probability of being at a higher level, at 58.24%, followed by EQRB at 50.33%, then EERB at 41.42%, and finally PCRB at 28.57%. Therefore, the probability of the occurrence of higher levels of MP in these four dimensions of social responsibility behaviors is likely to be associated with all four dimensions of social responsibility behaviors, and the above order reflects the different requirements of higher levels of project performance for higher levels of social responsibility behaviors in each dimension.

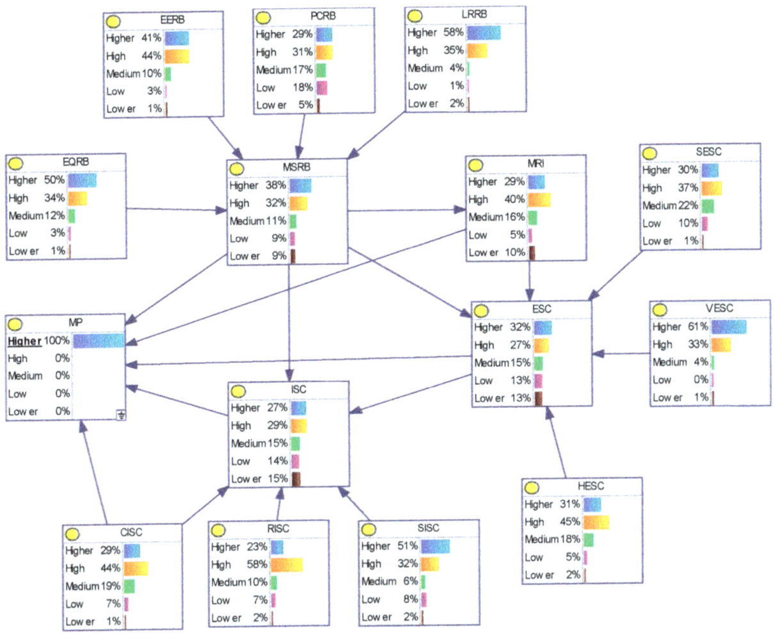

Figure 4. Results of Diagnostic Reasoning for Changing MP Node Beliefs.

4.2.3. Sensitivity Analysis

Sensitivity analysis refers to the degree of impact on other evidence nodes when certain nodes undergo slight changes. In the Bayesian network model, sensitivity analysis constitutes an important component of the model analysis calculation. The sensitivity analysis calculation of the model can determine the degree of influence and the influence pattern on other nodes when any node changes. To explore which nodes in the Bayesian network model of the relationship between MSRB and MP are more sensitive to the impact of the MP node, we set the MP node as the target node for sensitivity analysis. Figure 5, which is obtained as a result, indicates that the darker the node, the higher its sensitivity to MP.

As observed from Figure 5, for node MP, the sensitivity of the dimensions of social responsibility behavior, MRI, and ISC is greater. Among the four dimensions of social responsibility behavior nodes, the sensitivity of LRRB is the greatest. Furthermore, a tornado diagram can be generated by the sensitivity analysis function in GENIE, which can visualize the sensitivity of MP to each parent node at each level, with red bars representing a negative correlation and green bars representing a positive correlation in the diagram. In Figure 6a, MP is most sensitive to LRRB when MP is at a higher level, and the enhancement of the higher-level probability of LRRB may produce the greatest probability of a higher level of MP. Secondly, higher levels of MP require that the CISC be controlled at a high level, and a moderate increase in the level can also have a negative effect. EQRB and EERB can help MP achieve a higher level.

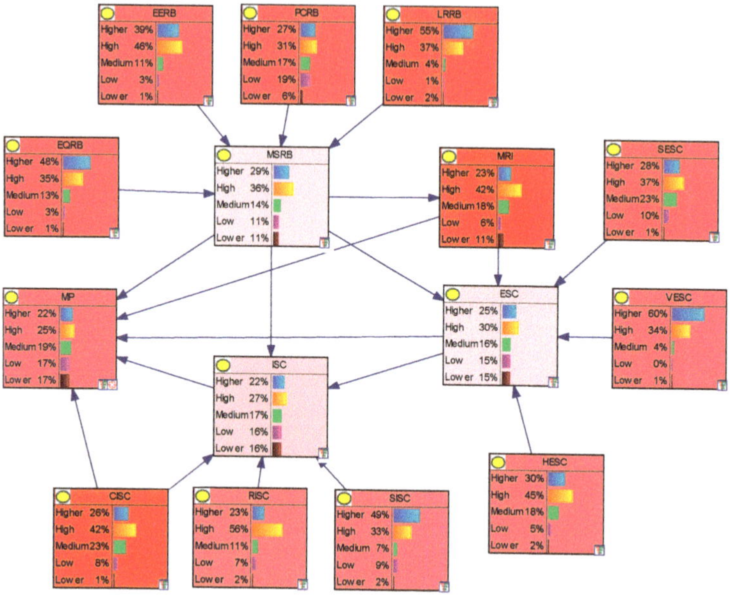

Figure 5. Sensitivity analysis result of project performance as a target node.

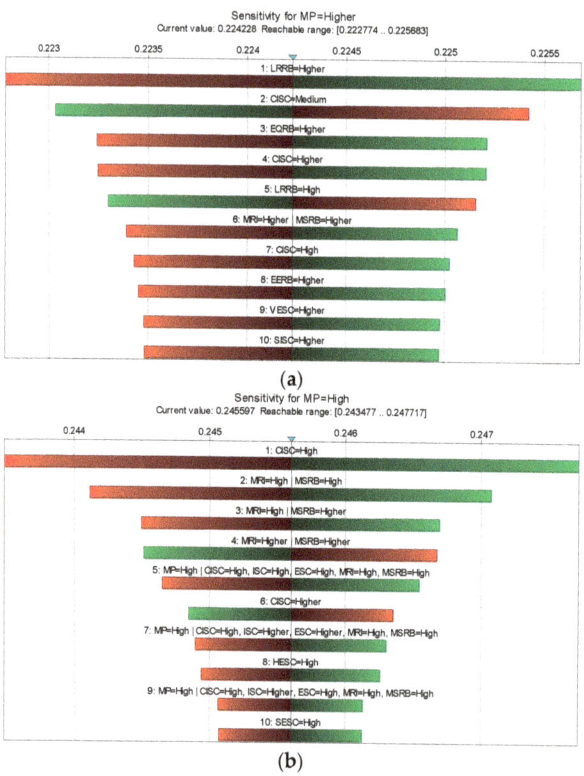

Figure 6. *Cont.*

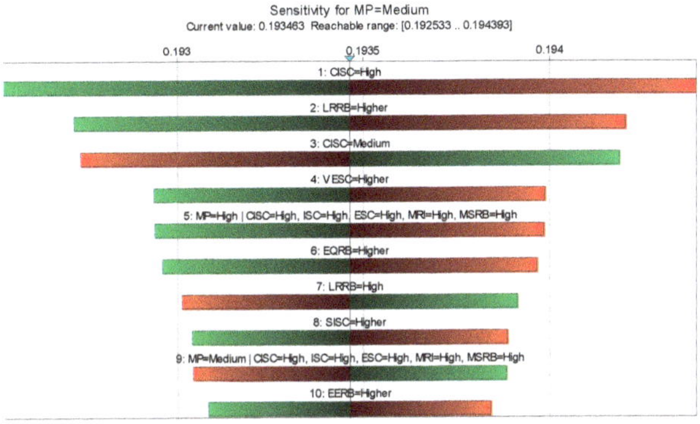

(c)

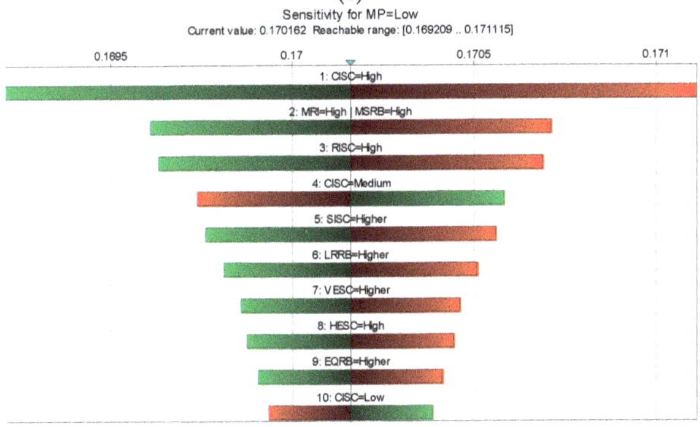

(d)

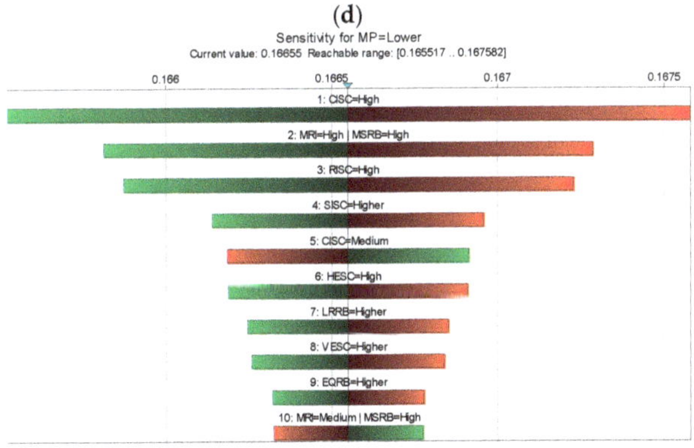

(e)

Figure 6. Tornado diagram of MP: (**a**) Sensitivity for "MP = Higher"; (**b**) Sensitivity for "MP = High"; (**c**) Sensitivity for "MP = Medium"; (**d**) Sensitivity for "MP = Low"; (**e**) Sensitivity for "MP = Lower".

In Figure 6b, when project performance is at a high level, an increase in CISC can contribute the most to MP improvement, indicating that the consistency of project organization cognition is crucial in megaprojects. MRI begins to play a mediating role, and a high level of MRI enhances a high level of MP, when MSRB is at a high or higher level. In Figure 6c–e, it can be observed that each dimension of ISC has a significant impact on medium, low, and lower levels of MP, while the impact of MSRB on MP diminishes with decreasing levels of MP. This impact also shifts from a direct impact to an indirect impact through the mediating role of MRI.

The results of the sensitivity analysis indicate that when the MP is at a high level, the fulfillment of MSRB can exert a greater positive effect on MP. Among these, LRRB has the most significant influence, followed by EQRB, EERB, and PCRB. As the MP level decreases, the ISC gradually becomes the dominant factor affecting project performance. This suggests that the ISC is a fundamental factor for megaprojects seeking to achieve a high MP level. Concurrently, the influence of MSRB weakens and transforms into an indirect influence mediated by MRI. It can be inferred that the positive impact of MSRB on MP may be predicated on superior basic project performance.

5. Discussion and Implications

5.1. Discussion

This study establishes a Bayesian network to analyze the impact of MSRB on MP. After structure learning and parameter learning, the Bayesian network model demonstrates good predictive performance. As referenced in [94], the model performs well and is capable of effective predictive analysis. This study collected 206 valid questionnaires from managers of megaprojects across various regions of China. The data from these questionnaires revealed the current fulfillment of social responsibility behaviors of megaprojects, as well as the diversity of resources and capabilities of these projects. The a priori probability of parent nodes of the MSRB corroborates the hierarchical relationship of the MSRB of each dimension as outlined in the Ma's study [18]. PCRB remains the project's desired responsibility behavior, which is currently performed in a more general way. However, the data from this study indicate that the megaprojects in the sample perform better in LRRB than EQRB, suggesting that the current LRRB of megaprojects forms the basis of MSRB. In addition, the questionnaire data also reveal that although most of the megaprojects possess a high level of social capital and resource integration capacity, there is considerable uncertainty, particularly regarding internal social capital.

Inferential analysis can quickly evaluate the effectiveness of adopting certain measures through belief updating. Inferential analysis can lead to the following findings: First, MSRB enhancement does enhance project performance, a finding that supports He's research [2]. This study also attempts to explore the mediating role played by social capital and resource integration capacity. The results show that MSRB can significantly affect resource integration capacity, and it can also weakly affect the internal and external social capital of the project. However, these indirect effects bring about an enhancement of project performance, but not significantly. Secondly, the reasoning analysis of the four dimensions of MSRB revealed that PCRB is a bonus, where the performance of PCRB acts as icing on the cake and is expected by the national government when the other dimensions of behavior are performed well, while EERB plays a similar role to the former. In contrast, while the enhancement of EQRB and LRRB do not significantly contribute to superior performance, they enable the project to reduce the likelihood of low performance.

The sensitivity analysis yielded the following findings: firstly, the sensitivity of project performance to LRRB and EQRB is high, corroborating the findings of the inference analysis. LRRB and EQRB are social responsibility behaviors that participating organizations of megaprojects should adhere to [18]. Secondly, it is interesting to note that MP exhibits the highest sensitivity to CISC. The situation of project performance is closely tied to the cognitive consistency of participating organizations. Thirdly, when the level of MP is not high, the impact of MSRB on MP performance diminishes, transitioning from a direct impact to an indirect impact mediated by resource integration capacity.

5.2. Theoretical Implications

The findings of this study make some theoretical contributions to the field of megaproject social responsibility research. For the first time, this study demonstrates the positive impact of MSRB on project performance from a resource-based theory perspective. This paper introduces resource and capability factors, thereby enriching the theoretical model of MSRB's effect on project performance. In response to Ma et al.'s call for a multi-theoretical approach to understanding megaproject social responsibility [18], this paper advances the application of resource-based theory in megaproject management studies.

This paper further extends the study of MSRB's effects. Different from He et al.'s research, which found the mediating influence of innovation capability on the impact of MSR practices on project performance [2], our findings reveal the complexity of mediating factors. This study shows that both the internal and external social capital of a megaproject, as well as its resource integration capacity, can mediate the impact of MSRB in certain contexts. Additionally, we applied the Bayesian network analysis method in this study to provide an adjustable analysis function for studying MSRB's effects, enhancing the practical relevance of our findings. Therefore, this paper has made some theoretical explorations in the study of megaproject social responsibility behavior.

5.3. Practical Implications

The results of the study have some implications for megaproject management.

Through the predictive inference result, it can be observed that project performance is jointly dominated by internal and external social capital, and the participating organizations of megaprojects are able to use their strong social capital to create competitive advantages. However, in the context of a large number of contractual and economic relationships between the participating organizations of megaprojects, megaprojects are limited in their attempts to reduce the transaction costs and achieve better performance through MSR [10].

For MSR strategy building, EQRB and LRRB can be considered basic strategies, while PCRB and EERB can be considered further strategies. Strategy implementation for different behaviors needs to be considered in the context of project status and timing. At the initial stage of a megaproject, the general level of project performance is often dominated by the project's resources and capabilities. While MSRB can play a role, it is subject to the intermediary influence of resources and capabilities. Once the MP reaches a certain level, MSRB will exert a direct promotional effect. Therefore, the mediating role of the internal and external social capital and resource integration capacity of the project is likely to be concentrated in the early stage of the project. When the project achieves a certain level of performance, the direct facilitating effect of MSRB on MP will gradually become more prominent. A combined strategy of multidimensional social responsibility behaviors provides a greater enhancement of project performance than a single-factor strategy, and managers need to attend to the integration of each dimension of social responsibility implementation to promote the implementation of overall MSR [18].

Moreover, the study's findings remind megaproject managers of the importance of organizational cognition. For megaproject organizations, in addition to the need for a scientific structure and the establishment of good internal relations, it is crucial for participating organizations to strengthen the common vision, collective values, and cultural identity to achieve project performance [96].

6. Conclusions

In the context of sustainable development, megaproject social responsibility has garnered widespread attention in the field of engineering management. Managers have gradually begun to implement social responsibility behaviors in projects, which have significantly impacted the management of megaprojects. Project performance is the focal point of managers' attention. The type of changes produced by the influence of social responsibility behavior is a question that many scholars are currently attempting to answer. However, current research often overlooks considerations of overall social responsibility behavior. At the same time, it stops at identifying the existence of the effect and so far lacks a deep exploration of the mechanism of the influence. Therefore, the aim of this study is to uncover the mechanism of the impact of social responsibility behaviors of megaprojects on project performance. This study introduces the resource-based theory, incorporates the factors of internal and external social capital and resource integration capacity of the project, and employs the Bayesian network research method.

Based on the questionnaire data, this study conducted structural learning and parametric learning, established a Bayesian network model with robust predictive performance of the impact of social responsibility behaviors on project performance, and derived the following findings through inference analysis and sensitivity analysis: (1) MSRB can exert a direct and positive impact on project performance, and the impacts of different dimensions of social responsibility behaviors vary, exhibiting a hierarchical relationship. (2) Resource integration capacity plays a significant positive mediating role, while the mediating role of social capital is minimal. (3) The impact of MSRB on project performance fluctuates with the level of project performance, with a smaller impact at low levels of project performance and a larger impact at higher levels of project performance.

The findings of this study provide a solid reference for research on the mechanism of the effect of social responsibility behavior in megaprojects, and the conclusions of the study offer valuable insights for managers of megaprojects in making sustainable development decisions.

7. Limitations and Future Research

The sample data utilized in this study are exclusively cross-sectional, which precludes a dynamic examination of the mechanisms through which megaproject social responsibility behavior impacts project performance. The examination of the dynamic evolution of megaproject social responsibility behaviors represents a promising avenue for future research. Furthermore, while this study is primarily centered on the context of megaprojects within China, it is important to note that the practices of megaproject management may exhibit variability across different countries or regions, influenced by distinct institutional and cultural environments. Therefore, it is recommended that future research broaden the scope of sample collection, integrating case data from international contexts for comparative analysis. This would enable the derivation of more reliable and generalizable findings.

Author Contributions: Conceptualization, Y.W. and L.X.; methodology, Z.Z. and M.H.; formal analysis, Y.W. and Z.Z.; investigation, Y.W. and M.H.; resources, L.X.; writing—original draft preparation, Y.W. and Z.Z.; writing—review and editing, L.X. and B.X.; supervision, B.X.; project administration, L.X.; funding acquisition, L.X. All authors have read and agreed to the published version of the manuscript.

Funding: This research was funded by National Natural Science Foundation of China (grant number 72271097) and technical sevice project of Guangdong Yongsheng Construction Engineering Corporation Limited (grant number 20231624).

Data Availability Statement: The data presented in this study are available on request from the corresponding author. The data are not publicly available due to privacy.

Conflicts of Interest: Author Yuhua Wu was employed by the company Guangdong Yongsheng Construction Engineering Co., Ltd. The remaining authors declare that the research was conducted in the absence of any commercial or financial relationships that could be construed as a potential conflict of interest.

References

1. Liu, Z.; Wang, L.; Sheng, Z.; Gao, X. Social Responsibility in Infrastructure Mega-Projects: A Case Study of Ecological Compensation for *Sousa Chinensis* during the Construction of the Hong Kong-Zhuhai-Macao Bridge. *Front. Eng. Manag.* **2018**, *5*, 98–108. [CrossRef]
2. He, Q.; Chen, X.; Wang, G.; Zhu, J.; Yang, D.; Liu, X.; Li, Y. Managing Social Responsibility for Sustainability in Megaprojects: An Innovation Transitions Perspective on Success. *J. Clean Prod.* **2019**, *241*, 118395. [CrossRef]
3. Xie, L.; Ju, T.; Xia, B. Institutional Pressures and Megaproject Social Responsibility Behavior: A Conditional Process Model. *Buildings* **2021**, *11*, 140. [CrossRef]
4. Zeng, S.X.; Ma, H.Y.; Lin, H.; Zeng, R.C.; Tam, V.W.Y. Social Responsibility of Major Infrastructure Projects in China. *Int. J. Proj. Manag.* **2015**, *33*, 537–548. [CrossRef]
5. Xie, L.; Ju, T.; Han, T.; Hou, L. A Meta-Network-Based Management Framework for Megaproject Social Responsibility Behaviour in China. *Eng. Constr. Archit. Manag.* **2023**, *30*, 4415–4434. [CrossRef]
6. Xie, L.; Han, T.; Chu, H.; Xia, B. Behavior Selection of Stakeholders toward Megaproject Social Responsibility: Perspective from Social Action Theory. *Adv. Civ. Eng.* **2019**, *2019*, e4956067. [CrossRef]
7. Xue, X.; Zhang, R.; Zhang, X.; Yang, R.J.; Li, H. Environmental and Social Challenges for Urban Subway Construction: An Empirical Study in China. *Int. J. Proj. Manag.* **2015**, *33*, 576–588. [CrossRef]

8. Ma, H.; Zeng, S.; Lin, H.; Chen, H.; Shi, J.J. The Societal Governance of Megaproject Social Responsibility. *Int. J. Proj. Manag.* **2017**, *35*, 1365–1377. [CrossRef]
9. Lee, C.; Won, J.W.; Jang, W.; Jung, W.; Han, S.H.; Kwak, Y.H. Social Conflict Management Framework for Project Viability: Case Studies from Korean Megaprojects. *Int. J. Proj. Manag.* **2017**, *35*, 1683–1696. [CrossRef]
10. Ma, H.; Sun, D.; Zeng, S.; Lin, H.; Shi, J.J. The Effects of Megaproject Social Responsibility on Participating Organizations. *Proj. Manag. J.* **2021**, *52*, 418–433. [CrossRef]
11. Ma, H.; Liu, Z.; Zeng, S.; Lin, H.; Tam, V.W.Y. Does Megaproject Social Responsibility Improve the Sustainability of the Construction Industry? *Eng. Constr. Archit. Manag.* **2020**, *27*, 975–996. [CrossRef]
12. Wang, T.; He, Q.; Lu, Y.; Yang, D. How Does Organizational Citizenship Behavior (OCB) Affect the Performance of Megaprojects? Insights from a System Dynamic Simulation. *Sustainability* **2018**, *10*, 1708. [CrossRef]
13. Guo, S.; Wang, X.; Fu, L.; Liu, Y. How Individual's Proactive Behavior Helps Construction Sustainability: Exploring the Effects of Project Citizenship Behavior on Project Performance. *Sustainability* **2019**, *11*, 6922. [CrossRef]
14. Wang, G.; He, Q.; Meng, X.; Locatelli, G.; Yu, T.; Yan, X. Exploring the Impact of Megaproject Environmental Responsibility on Organizational Citizenship Behaviors for the Environment: A Social Identity Perspective. *Int. J. Proj. Manag.* **2017**, *35*, 1402–1414. [CrossRef]
15. Javeed, S.A.; Teh, B.H.; Ong, T.S.; Chong, L.L.; Abd Rahim, M.F.B.; Latief, R. How Does Green Innovation Strategy Influence Corporate Financing? Corporate Social Responsibility and Gender Diversity Play a Moderating Role. *Int. J. Environ. Res. Public Health* **2022**, *19*, 8724. [CrossRef] [PubMed]
16. Chen, Y.; Jin, S. Corporate Social Responsibility and Green Technology Innovation: The Moderating Role of Stakeholders. *Sustainability* **2023**, *15*, 8164. [CrossRef]
17. Halkos, G.; Skouloudis, A. Corporate Social Responsibility and Innovative Capacity: Intersection in a Macro-Level Perspective. *J. Clean. Prod.* **2018**, *182*, 291–300. [CrossRef]
18. Ma, H.; Lv, K.; Zeng, S.; Lin, H.; Shi, J.J. Climbing the Pyramid of Megaproject Social Responsibility: Impacts of External Stakeholders and Project Complexity. *J. Constr. Eng. Manag.* **2022**, *148*, 04022116. [CrossRef]
19. Miller, R.; Hobbs, B. Governance Regimes for Large Complex Projects. *Proj. Manag. J.* **2005**, *36*, 42–50. [CrossRef]
20. Bourdieu, P. The Forms of Capital. In *Handbook of Theory and Research for the Sociology of Education*; Richardson, J.G., Ed.; Bloomsbury Academic: London, UK, 1986; pp. 241–258.
21. Mathur, G.; Jugdev, K.; Shing Fung, T. Project Management Assets and Project Management Performance Outcomes: Exploratory Factor Analysis. *Manag. Res. Rev.* **2013**, *36*, 112–135. [CrossRef]
22. Sirmon, D.G.; Hitt, M.A. Managing Resources: Linking Unique Resources, Management, and Wealth Creation in Family Firms. *Entrep. Theory Pract.* **2003**, *27*, 339–358. [CrossRef]
23. Zhao, Z.; Meng, F.; He, Y.; Gu, Z. The Influence of Corporate Social Responsibility on Competitive Advantage with Multiple Mediations from Social Capital and Dynamic Capabilities. *Sustainability* **2019**, *11*, 218. [CrossRef]
24. Yuan, B.; Cao, X. Do Corporate Social Responsibility Practices Contribute to Green Innovation? The Mediating Role of Green Dynamic Capability. *Technol. Soc.* **2022**, *68*, 101868. [CrossRef]
25. Jiang, W.; Chai, H.; Shao, J.; Feng, T. Green Entrepreneurial Orientation for Enhancing Firm Performance: A Dynamic Capability Perspective. *J. Clean. Prod.* **2018**, *198*, 1311–1323. [CrossRef]
26. Pathak, A.; Zhang, L.; Ganapati, N.E. Understanding Multisector Stakeholder Value Dynamics in Hurricane Michael: Toward Collaborative Decision-Making in Disaster Contexts. *Nat. Hazards Rev.* **2020**, *21*, 04020032. [CrossRef]
27. Xue, F.; Chen, G.; Huang, S.; Xie, H. Design of Social Responsibility Incentive Contracts for Stakeholders of Megaprojects under Information Asymmetry. *Sustainability* **2022**, *14*, 1465. [CrossRef]
28. Ma, L.; Fu, H. A Governance Framework for the Sustainable Delivery of Megaprojects: The Interaction of Megaproject Citizenship Behavior and Contracts. *J. Constr. Eng. Manag.* **2022**, *148*, 04022004. [CrossRef]
29. Xie, L.; Huang, M.; Xia, B.; Skitmore, M. Megaproject Environmentally Responsible Behavior in China: A Test of the Theory of Planned Behavior. *Int. J. Environ. Res. Public Health* **2022**, *19*, 6581. [CrossRef]
30. Xie, L.; Xu, T.; Ju, T.; Xia, B. Explaining the Alienation of Megaproject Environmental Responsibility Behavior: A Fuzzy Set Qualitative Comparative Analysis Study in China. *Eng. Constr. Archit. Manag.* **2022**, *30*, 2794–2813. [CrossRef]
31. He, Q.; Wang, Z.; Wang, G.; Xie, J.; Chen, Z. The Dark Side of Environmental Sustainability in Projects: Unraveling Greenwashing Behaviors. *Proj. Manag. J.* **2022**, *53*, 349–366. [CrossRef]
32. He, Q.; Wang, Z.; Wang, G.; Zuo, J.; Wu, G.; Liu, B. To Be Green or Not to Be: How Environmental Regulations Shape Contractor Greenwashing Behaviors in Construction Projects. *Sustain. Cities Soc.* **2020**, *63*, 102462. [CrossRef]
33. Wang, Z.; He, Q.; Locatelli, G.; Wang, G.; Li, Y. Exploring Environmental Collaboration and Greenwashing in Construction Projects: Integrative Governance Framework. *J. Constr. Eng. Manag.* **2023**, *149*, 04023109. [CrossRef]
34. Lin, H.; Sui, Y.; Ma, H.; Wang, L.; Zeng, S. CEO Narcissism, Public Concern, and Megaproject Social Responsibility: Moderated Mediating Examination. *J. Manag. Eng.* **2018**, *34*, 04018018. [CrossRef]
35. Yang, D.; He, Q.; Cui, Q.; Hsu, S.-C. Non-Economic Motivations for Organizational Citizenship Behavior in Construction Megaprojects. *Int. J. Proj. Manag.* **2020**, *38*, 64–74. [CrossRef]
36. Turner, R.; Zolin, R. Forecasting Success on Large Projects: Developing Reliable Scales to Predict Multiple Perspectives by Multiple Stakeholders over Multiple Time Frames. *Proj. Manag. J.* **2012**, *43*, 87–99. [CrossRef]

37. Christiansen, J.C. Obtaining Mega Performance on a Megaproject. *Trans. AACE Int.* **1994**, *1994*, TR3.1.
38. Davies, A.; Manning, S.; Söderlund, J. When Neighboring Disciplines Fail to Learn from Each Other: The Case of Innovation and Project Management Research. *Res. Policy* **2018**, *47*, 965–979. [CrossRef]
39. Li, T.H.Y.; Ng, S.T.; Skitmore, M. Evaluating Stakeholder Satisfaction during Public Participation in Major Infrastructure and Construction Projects: A Fuzzy Approach. *Autom. Constr.* **2013**, *29*, 123–135. [CrossRef]
40. Yun, L.; Wan, J.; Wang, G.; Bai, J.; Zhang, B. Exploring the Missing Link between Top Management Team Characteristics and Megaproject Performance. *Eng. Constr. Archit. Manag.* **2020**, *27*, 1039–1064. [CrossRef]
41. Pesämaa, O.; Larsson, J.; Eriksson, P.E. Role of Performance Feedback on Process Performance in Construction Projects: Client and Contractor Perspectives. *J. Manag. Eng.* **2018**, *34*, 04018023. [CrossRef]
42. Rezvani, A.; Khosravi, P.; Ashkanasy, N.M. Examining the Interdependencies among Emotional Intelligence, Trust, and Performance in Infrastructure Projects: A Multilevel Study. *Int. J. Proj. Manag.* **2018**, *36*, 1034–1046. [CrossRef]
43. Wu, G.; Liu, C.; Zhao, X.; Zuo, J. Investigating the Relationship between Communication-Conflict Interaction and Project Success among Construction Project Teams. *Int. J. Proj. Manag.* **2017**, *35*, 1466–1482. [CrossRef]
44. Flyvbjerg, B. What You Should Know about Megaprojects and Why: An Overview. *Proj. Manag. J.* **2014**, *45*, 6–19. [CrossRef]
45. Heere, B.; Xing, X. BOCOG's Road to Success: Predictors of Commitment to Organizational Success among Beijing Olympic Employees. *Eur. Sport Manag. Q.* **2012**, *12*, 161–181. [CrossRef]
46. Anvuur, A.M.; Kumaraswamy, M.M. Effects of Teamwork Climate on Cooperation in Crossfunctional Temporary Multi-Organization Workgroups. *J. Constr. Eng. Manag.* **2016**, *142*, 04015054. [CrossRef]
47. Wernerfelt, B. A Resource-Based View of the Firm. *Strateg. Manag. J.* **1984**, *5*, 171–180. [CrossRef]
48. Liu, Y.; Chen, Y.; Ren, Y.; Jin, B. Impact Mechanism of Corporate Social Responsibility on Sustainable Technological Innovation Performance from the Perspective of Corporate Social Capital. *J. Clean. Prod.* **2021**, *308*, 127345. [CrossRef]
49. Mohtsham Saeed, M.; Arshad, F. Corporate Social Responsibility as a Source of Competitive Advantage: The Mediating Role of Social Capital and Reputational Capital. *J. Database Mark. Cust. Strateg. Manag.* **2012**, *19*, 219–232. [CrossRef]
50. Barney, J.B. How a Firm's Capabilities Affect Boundary Decisions. *Sloan Manag. Rev.* **1999**, *40*, 137–145.
51. Branco, M.C.; Rodrigues, L.L. Corporate Social Responsibility and Resource-Based Perspectives. *J. Bus. Ethics* **2006**, *69*, 111–132. [CrossRef]
52. Dzeng, R.-J.; Wen, K.-S. Evaluating Project Teaming Strategies for Construction of Taipei 101 Using Resource-Based Theory. *Int. J. Proj. Manag.* **2005**, *23*, 483–491. [CrossRef]
53. Govan, P.; Damnjanovic, I. The Resource-Based View on Project Risk Management. *J. Constr. Eng. Manag.* **2016**, *142*, 04016034. [CrossRef]
54. Shi, C.; Chen, Y.; Hua, Y.; Tang, Y. Understanding Subcontracting Organizational Arrangements for Construction Projects in China: Integrating Capabilities and Uncertainty. *Eng. Constr. Archit. Manag.* **2022**, *30*, 2381–2399. [CrossRef]
55. Mansour, H.; Aminudin, E.; Mansour, T.; Abidin, N.I.A.B.; Lou, E. Resource-Based View in Construction Project Management Research: A Meta-Analysis. *IOP Conf. Ser. Earth Environ. Sci.* **2022**, *1067*, 012057. [CrossRef]
56. Afzal, F.; Lim, B. Organizational Factors Influencing the Sustainability Performance of Construction Organizations. *Sustainability* **2022**, *14*, 10449. [CrossRef]
57. McWilliams, A.; Siegel, D.S.; Wright, P.M. Corporate Social Responsibility: Strategic Implications. *J. Manag. Stud.* **2006**, *43*, 1–18. [CrossRef]
58. Jacobs, J. *The Death and Life of Great American Cities*; Vintage; Pimlico: London, UK, 2016; ISBN 0-525-43285-X.
59. Mok, K.Y.; Shen, G.Q.; Yang, J. Stakeholder Management Studies in Mega Construction Projects: A Review and Future Directions. *Int. J. Proj. Manag.* **2015**, *33*, 446–457. [CrossRef]
60. Di Vincenzo, F.; Mascia, D. Social Capital in Project-Based Organizations: Its Role, Structure, and Impact on Project Performance. *Int. J. Proj. Manag.* **2012**, *30*, 5–14. [CrossRef]
61. Burt, R.S. The Network Structure of Social Capital. *Res. Organ. Behav.* **2000**, *22*, 345–423. [CrossRef]
62. Fukuyama, F. Social Capital, Civil Society and Development. *Third World Q.* **2001**, *22*, 7–20. [CrossRef]
63. Nahapiet, J.; Ghoshal, S. Social Capital, Intellectual Capital, and the Organizational Advantage. *Acad. Manag. Rev.* **1998**, *23*, 242–266. [CrossRef]
64. Koh, T.Y.; Rowlinson, S. Relational Approach in Managing Construction Project Safety: A Social Capital Perspective. *Accid. Anal. Prev.* **2012**, *48*, 134–144. [CrossRef]
65. Bian, Y.; Xie, J.; Yang, Y.; Hao, M. Local Embeddedness, Corporate Social Capital and Chinese Enterprises: The Case of Shaanxi FDI Firms. *Chin. Manag. Stud.* **2019**, *13*, 860–876. [CrossRef]
66. Adler, P.S.; Kwon, S.-W. Social Capital: Prospects for a New Concept. *Acad. Manag. Rev.* **2002**, *27*, 17–40. [CrossRef]
67. Huang, Y.; Shi, Q.; Pena-Mora, F.; Lu, Y.; Shen, C. Exploring the Impact of Information and Communication Technology on Team Social Capital and Construction Project Performance. *J. Manag. Eng.* **2020**, *36*, 04020056. [CrossRef]
68. Kim, Y.; Cannella, A.A., Jr. Toward a Social Capital Theory of Director Selection. *Corp. Gov. Int. Rev.* **2008**, *16*, 282–293. [CrossRef]
69. Hitt, M.; Ireland, R.; Camp, M.; Sexton, D. Strategic Entrepreneurship: Entrepreneurial Strategies for Wealth Creation. *Strateg. Manag. J.* **2001**, *22*, 479–491. [CrossRef]
70. Collis, D.; Montgomery, C.A. Competing on Resources: Strategy in the 1990s. *Harv. Bus. Rev.* **1995**, *73*, 118–129.

71. Jia, J.; Li, Y.; Gan, X.; Ma, G. From Local to the Whole: Understanding Construction Projects as Multiteam Systems. *Int. J. Manag. Proj. Bus.* **2022**, *16*, 355–373. [CrossRef]
72. Sirmon, D.G.; Hitt, M.A.; Ireland, R.D. Managing Firm Resources in Dynamic Environments to Create Value: Looking Inside the Black Box. *Acad. Manag. Rev.* **2007**, *32*, 273–292. [CrossRef]
73. Ge, B.; Dong, B. Resource Integration Process and Venture Performance: Based on the Contingency Model of Resource Integration Capability. In Proceedings of the 2008 International Conference on Management Science and Engineering 15th Annual Conference Proceedings, Long Beach, CA, USA, 10–12 September 2008; pp. 291–297.
74. Wang, Z.; Zhang, Y.; Zheng, K.; Zeng, R.; Yuan, H.; Liu, J. A Review of Mega-Project Management Research from an Organization Science Perspective: Current Status and Future Directions. *Dev. Built Environ.* **2023**, *16*, 100254. [CrossRef]
75. Newbert, S.L. Empirical Research on the Resource-based View of the Firm: An Assessment and Suggestions for Future Research. *Strateg. Manag. J.* **2007**, *28*, 121–146. [CrossRef]
76. Brush, C.G.; Greene, P.G.; Hart, M.M. From Initial Idea to Unique Advantage: The Entrepreneurial Challenge of Constructing a Resource Base. *Acad. Manag. Perspect.* **2001**, *15*, 64–78. [CrossRef]
77. Lv, H.; Li, D. Impacts of Heterogeneous Green Consumers on Green Innovation in Electric Vehicle and Charging Pile Firms. *Sustain. Prod. Consum.* **2021**, *28*, 1216–1231. [CrossRef]
78. Chang, C.-H. Enhancing New Product Development Performance from Adaptive Ability and Relationship Learning: The Mediation Role of Resource Integration. *Total Qual. Manag. Bus. Excell.* **2017**, *28*, 62–75. [CrossRef]
79. Putra, D.; Rahayu, R.; Putri, A. The Influence of Enterprise Resource Planning (ERP) Implementation System on Company Performance Mediated by Organizational Capabilities. *J. Account. Invest.* **2021**, *22*, 221–241. [CrossRef]
80. Jun-yuan, F.; Hui-ping, D. Construction Enterprise Resource Integration Based on the Improvement of the Value Creation Capabilities. In Proceedings of the 2007 International Conference on Management Science and Engineering, Harbin, China, 20–22 August 2007; pp. 2299–2305.
81. Leana, C.R.; Pil, F.K. Social Capital and Organizational Performance: Evidence from Urban Public Schools. *Organ. Sci.* **2006**, *17*, 353–366. [CrossRef]
82. Wang, G.; Locatelli, G.; Wan, J.; Li, Y.; Le, Y. Governing Behavioral Integration of Top Management Team in Megaprojects: A Social Capital Perspective. *Int. J. Proj. Manag.* **2021**, *39*, 365–376. [CrossRef]
83. Chang, K.-C.; Wong, J.-H.; Li, Y.; Lin, Y.-C.; Chen, H.-G. External Social Capital and Information Systems Development Team Flexibility. *Inf. Softw. Technol.* **2011**, *53*, 592–600. [CrossRef]
84. Wang, Q.; Zhang, J.; Zhu, K.; Guo, P.; Shen, C.; Xiong, Z. The Safety Risk Assessment of Mine Metro Tunnel Construction Based on Fuzzy Bayesian Network. *Buildings* **2023**, *13*, 1605. [CrossRef]
85. Chemweno, P.; Pintelon, L.; Van Horenbeek, A.; Muchiri, P. Development of a Risk Assessment Selection Methodology for Asset Maintenance Decision Making: An Analytic Network Process (ANP) Approach. *Int. J. Prod. Econ.* **2015**, *170*, 663–676. [CrossRef]
86. Nepal, B.; Yadav, O.P. Bayesian Belief Network-Based Framework for Sourcing Risk Analysis during Supplier Selection. *Int. J. Prod. Res.* **2015**, *53*, 6114–6135. [CrossRef]
87. Rezakhani, P. Hybrid Fuzzy-Bayesian Decision Support Tool for Dynamic Project Scheduling and Control under Uncertainty. *Int. J. Constr. Manag.* **2022**, *22*, 2864–2876. [CrossRef]
88. Yu, T.; Man, Q.; Wang, Y.; Shen, G.Q.; Hong, J.; Zhang, J.; Zhong, J. Evaluating Different Stakeholder Impacts on the Occurrence of Quality Defects in Offsite Construction Projects: A Bayesian-Network-Based Model. *J. Clean. Prod.* **2019**, *241*, 118390. [CrossRef]
89. Ghorbany, S.; Noorzai, E.; Yousefi, S. BIM-Based Solution to Enhance the Performance of Public-Private Partnership Construction Projects Using Copula Bayesian Network. *Expert Syst. Appl.* **2023**, *216*, 119501. [CrossRef]
90. de Oliveira, M.A.; Possamai, O.; Dalla Valentina, L.V.O.; Flesch, C.A. Applying Bayesian Networks to Performance Forecast of Innovation Projects: A Case Study of Transformational Leadership Influence in Organizations Oriented by Projects. *Expert Syst. Appl.* **2012**, *39*, 5061–5070. [CrossRef]
91. Fuster-Parra, P.; García-Mas, A.; Ponseti, F.J.; Leo, F.M. Team Performance and Collective Efficacy in the Dynamic Psychology of Competitive Team: A Bayesian Network Analysis. *Hum. Mov. Sci.* **2015**, *40*, 98–118. [CrossRef]
92. Claeskens, G.; Hjort, N.L. *Model Selection and Model Averaging*; Cambridge Books; Cambridge University Press: Cambridge, UK, 2008.
93. Marcot, B.G. Metrics for Evaluating Performance and Uncertainty of Bayesian Network Models. *Ecol. Model.* **2012**, *230*, 50–62. [CrossRef]
94. Goyal, P.; Chanda, U. A Bayesian Network Model on the Association between CSR, Perceived Service Quality and Customer Loyalty in Indian Banking Industry. *Sustain. Prod. Consum.* **2017**, *10*, 50–65. [CrossRef]
95. Landis, J.R.; Koch, G.G. The Measurement of Observer Agreement for Categorical Data. *Biometrics* **1977**, *33*, 159. [CrossRef]
96. Wan, J.; Le, Y.; Li, J.; Wang, G. Linking Top Management Team's Internal Social Capital and Project Performance in a Mega Construction Project: The Mediating Role of Behavioral Integration. In Proceedings of the ICCREM 2019: Innovative Construction Project Management and Construction Industrialization, Banff, AB, Canada, 21–24 May 2019; pp. 314–324.

buildings

MDPI

Article

How to Foster Project Organization Resilience in the Construction Industry: The Role of Data Governance Capabilities

Yuanyuan Hua [1,*], Manlin Kang [1], Hongjiang Yao [2] and Yafan Fu [3]

1 School of Urban Economics and Management, Beijing University of Civil Engineering and Architecture, Beijing 102612, China; 2112010023003@stu.bucea.edu.cn
2 School of Management, Shandong University, Jinan 250100, China; yaohj_sdu@sdu.edu.cn
3 School of Economics and Management, North China Electric Power University, Beijing 102206, China; yafanfu@ncepu.edu.cn
* Correspondence: huayuanyuan@bucea.edu.cn

Abstract: With the ongoing changes in global economic and social environments, project organization resilience has emerged as a core competency in addressing environmental uncertainty, thereby becoming a prominent research area within construction management. This resilience is intricately linked to data resources; however, construction enterprises face systemic challenges in governance—fragmented data standards, siloed storage, and underutilized analytics—which limit their ability to translate data into crisis-responsive actions. Consequently, it is of paramount importance to investigate how data governance capabilities influence project organization resilience. This study initially identifies five dimensions of data governance capabilities within construction enterprises, spanning the three stages of planning, flow, and application: top-level design, data standard management, data collection, data storage, and data application. These dimensions are derived through a combination of literature review and expert interviews. Subsequently, this study establishes a theoretical model titled "Data Governance Capability—Project Organization Resilience", exploring the relationships among these data governance capability dimensions and their correlations with project organization resilience. Data were collected through 142 valid questionnaires from practitioners in the Chinese construction industry and analyzed by the Partial Least Squares Structural Equation Modeling (PLS-SEM) method. The results indicate a significant relationship between data governance capabilities and project organization resilience, contributing to the research on the antecedents of the latter. This study offers a scale for construction enterprises to systematically assess data governance capabilities and provides guidance on enhancing organization resilience by progressively nurturing these capabilities.

Keywords: data governance capabilities; project organization resilience; construction management; project management

1. Introduction

In today's VUCA (Volatility, Uncertainty, Complexity, Ambiguity) world, global supply chain restructuring, frequent public health crises, and disruptive digital technology innovations have intensified the dynamism and vulnerability of construction projects [1]. Existing studies indicate that adverse factors or crises, such as policy changes, project funding issues, market fluctuations, and major unforeseen events, can lead to difficulties

https://doi.org/10.3390/buildings15081219

in project initiation, supply chain disruptions, schedule delays, and cost overruns [2]. Resilience has gradually transcended the traditional scope of risk management, becoming a core strategic objective for project organizations to achieve counter-cyclical growth and sustainable development in times of crisis [3,4]. Therefore, enhancing the ability to respond to crises and disruptions is crucial in construction projects.

Existing studies have examined the antecedents of project organization resilience from various perspectives. First, the resource perspective highlights the role of strategic resource adequacy and allocation—such as financial assets, digital infrastructure, and human resources—in buffering against crises [5,6]. Second, psychological and cognitive perspectives stress leadership effectiveness and collective mindfulness in enhancing crisis response capacity, thereby mitigating the impact of adverse factors [7,8]. Additionally, the institutional perspective underscores the importance of social capital and relational networks in fostering rapid recovery after experiencing disruptions [9]. Finally, capability perspectives emphasize strong situational awareness and digital technology innovation as key enablers of resilience, by identifying and responding to risks at an early stage and leveraging new capabilities and adaptive strategies to absorb and capitalize on environmental changes [10,11].

While these studies provide valuable insights, two critical gaps persist. First, existing research predominantly focuses on resilience mechanisms at the enterprise or supply chain level [12–14]. Project organizations, as core units in the construction industry, lack a systematic examination of how to build their own resilience. This gap limits both theoretical and practical efforts to effectively enhance the adaptive capacity of project organizations. Second, although scholars widely acknowledge the enabling effect of digital technologies (such as BIM and the IoT) on resilience, the systematic impact of data governance capabilities remains understudied, especially since data governance has emerged as a hot topic now in the wave of digital transformation. For instance, while Hilkenmeier et al. (2021) found that data-driven decision-making frameworks enhance project organization agility [15], and Gölzer et al. (2017) linked optimized data processes to organizational stability [16], no study has holistically examined how structured data governance frameworks contribute to project organization resilience across the data lifecycle (planning–flow–application) from a capability-building perspective. Existing frameworks primarily focus on individual technological tools (e.g., data collection devices) or collaborative models (e.g., multi-party data-sharing platforms). To address those issues, this study aims to explore how construction enterprises enhance project organization resilience through a holistic data governance capability approach.

However, blindly implementing data governance may introduce challenges such as increased market volatility pressures, organizational rigidity, and insufficient strategic planning, which could negatively impact technological innovation and sustainable development. Existing data governance frameworks are often generic and lack specificity for construction enterprises. The sector-specific characteristics of data governance in construction (e.g., multi-entity collaboration and dynamic task environments) have not been fully incorporated into current frameworks, leading to misalignment between governance capabilities and resilience enhancement. Therefore, it is essential to identify the key dimensions of data governance tailored to the construction industry and develop a corresponding data governance capability scale.

The remainder of this study was organized as follows. Firstly, we defined and conceptualized data governance capabilities, and developed an initial measurement scale by conducting a systematic literature review. Secondly, we constructed a cross-level theoretical model linking data governance capabilities and project organization resilience, proposing specific hypotheses to elucidate the relationships between the dimensions of data governance capabilities and their impact on project organization resilience. Then, we refined the initial dimensions and indicators through expert interviews and validated the measurement framework using pre-test data and factor analysis, ultimately deriving the final data governance capability scale for construction enterprises. Finally, the data were analyzed by the PLS-SEM method to examine the proposed hypotheses. The Discussion and Conclusion sections show how this research enriches the existing literature and provide theoretical and practical guidance. The research design process diagram is shown in Figure 1.

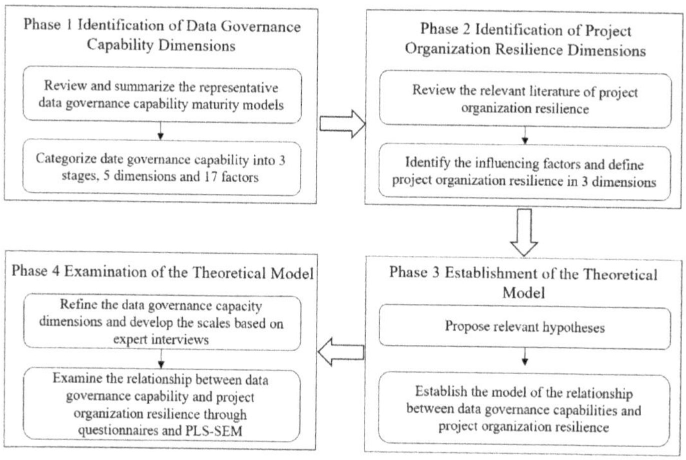

Figure 1. Research design process.

2. Data Governance

2.1. Definition and Connotations of Data Governance

Data governance refers to an organization's core capability to plan, control, and extract value from data assets through a systematic framework [17]. Its primary goal is to maximize the value of data, minimize risks and costs, and support the organization in decision-making and process management [18,19]. While there is no universally accepted definition in academia, theoretical advancements have progressively deepened its connotations. Early research took a narrow perspective, defining data governance as the control activities of data and its application processes, emphasizing technical measures to ensure data quality and security [20,21]. As digital transformation has advanced, the definition has evolved into a broader perspective, encompassing (1) authority and accountability over data assets [22], organizational structures and the distribution of decision rights in data processes [23], and (3) the alignment between data strategy and business goals [24]. The theoretical evolution of data governance can be categorized into three progressive stages, as summarized in Table 1. This evolution reflects a paradigm shift from "managing data" to "governing through data", where data reconfigure organizational decision-making.

Table 1. Evolutionary stages of data governance.

Stage	Core Focus	Source
Technical Governance	Human–machine interaction and data quality	[20,25]
Organizational Design	Institutionalized structures and processes	[19,26,27]
Strategic Governance	Cross-functional collaboration and value creation	[24]

Distinct from traditional data management, data governance emphasizes governance across the entire data lifecycle. It is an integrated governance framework that includes policies, processes, standards, and organizational structures [19]. Data governance extends beyond mere data collection and storage; it involves establishing decision-making authority and responsibilities, formulating standards and guidelines, and ensuring data quality, security, and compliance [19]. Existing research primarily focuses on four key perspectives, namely, the governance process, governance standards, constituent elements, and governance outcomes, which are summarized in Table 2.

Table 2. Perspectives on data governance.

Perspective	Key Definitions	Source
Process	Cross-functional coordination of data lifecycle management that covers rule-making, division of responsibilities, organization and implementation, etc.	[28,29]
Standards	Their core objective is to improve data quality, which requires a governance framework integrating data infrastructure, standards, policies, etc.	[30]
Components	Data strategy, standardized processes, and governance mechanisms including structural, procedural, and relational mechanisms.	[21]
Outcomes	Operational optimization, cost reduction, strategic decision level, and competitiveness enhancement.	[26,31]

In the construction context, data governance needs to adapt to two key characteristics: project-driven nature and multi-stakeholder complexity. Since construction enterprises operate based on engineering projects, data originate from fragmented projects, necessitating governance capabilities that ensure quality, security, and the fluidity of data among various stakeholders across the full lifecycle (planning → flow → application). In summary, this study proposes that data governance in construction enterprises should encompass two main aspects: first, establishing digital platforms and standards for data transmission, storage, and utilization; second, optimizing workflows, redefining accountability, and aligning strategies to mitigate risks and unlock value in dynamic environments.

2.2. Identification of Data Governance Capability Dimensions

In construction enterprises, data governance capability is reflected not only in the effective management of the data lifecycle but also in the ability to extract value from data, enhance data-driven decision-making, and improve the efficiency of all stakeholders. Achieving effective data governance requires a high level of data governance capability maturity. To measure data governance activities, scholars have introduced the capability maturity model (CMM) into related research, leading to the development of the data governance capability maturity model. These models serve as essential tools for assessing an enterprise's data governance capabilities by providing structured criteria for evaluating the development and management of software platforms. The following are the current representative models and their data governance capability dimensions (Figure 2):

Currently, there is no universally accepted classification of data governance capability dimensions in academia, so this study draws upon the most widely accepted data governance capability maturity models to establish a framework for data governance capability dimensions within the construction field. Given that data governance in construction management is still in its early stages, an overly detailed classification may hinder practical implementation. Therefore, based on value chain theory and data lifecycle theory, this study categorizes data governance into three stages: planning stage, flow stage, and application stage. By synthesizing the common dimensions across various data governance maturity models (Figure 2), we identify five key capability dimensions for data governance in the construction sector:

- Planning Stage: top-level design capability and data standard management capability.

- Flow Stage: data collection capability and data storage capability.
- Application Stage: data application capability.

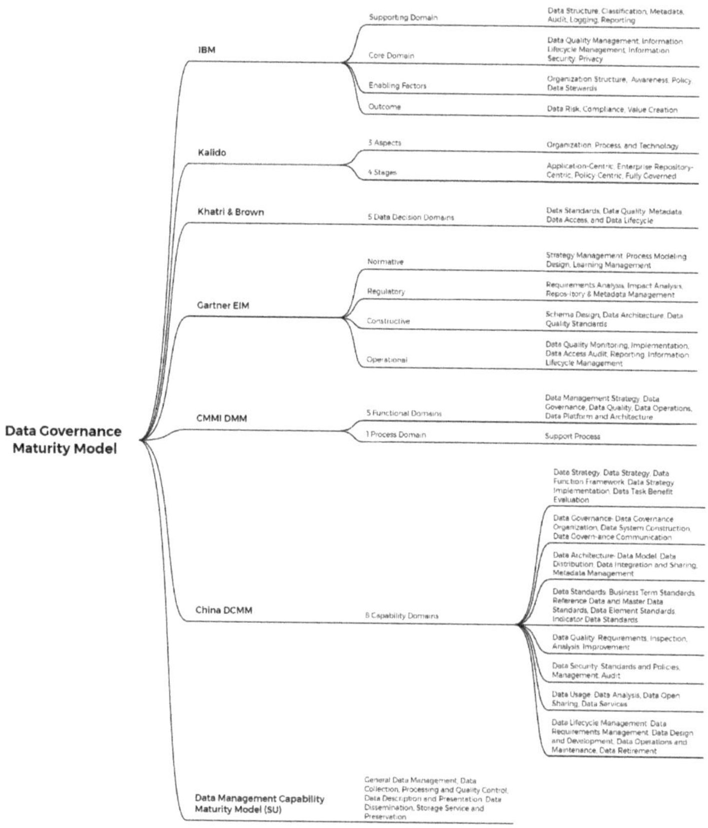

Figure 2. Summary of data governance capability maturity models.

Top-level design capability is the foundational guarantee for enterprises to implement data governance. It refers to an organization's ability to formulate data governance strategies, establish governance teams, and develop governance frameworks based on digital transformation trends and internal resources [32]. Data standard management capability is critical for ensuring data consistency and accuracy. This capability involves developing standardized data definitions, classifications, and formatting guidelines. Data collection capability ensures that data are authentic, reliable, easily accessible, and timely. Effective data collection is crucial for high-quality data analysis and decision-making, as incomplete or inaccurate data can lead to inefficient management and operational risks. Data storage capability involves establishing standardized data storage protocols and equipping enterprises with specialized storage infrastructure to ensure data security, quality, and traceability. Enhancing this capability enables organizations to better protect and utilize their data assets, reducing data loss risks while improving data retrieval efficiency. Data application capability refers to an enterprise's ability to analyze and leverage data for decision-making and business optimization.

After reviewing the content of the seven data governance capability maturity models mentioned earlier, it is evident that different countries and industries emphasize different aspects of data governance and adopt varying classification standards. There are significant differences among these maturity models in terms of capability dimensions and capability items, primarily reflected in the categories and number of capability dimensions and items they define. Additionally, data governance encompasses a wide range of aspects, with complex and extensive content, which further contributes to the variations among models. Moreover, differences in evaluation objectives and areas of focus across models further exacerbate these discrepancies. Despite their distinct approaches to capability domain classification, all these models commonly address key areas such as data strategy, data organization, data quality, data standards, data application, and data infrastructure.

Based on the findings above, data governance capabilities were consolidated and categorized according to the characteristics of the construction industry, resulting in five dimensions and 17 factors. The classification of data governance capability dimensions for construction enterprises is shown in Table 3.

Table 3. Data governance capability dimension division.

No.	Stage	Capability Dimension	Factor	Source
1	Planning	Top-level Design	Data Strategy Planning	[24,33–35]
2			Data Governance System	[24,33–35]
3			Data Strategy Implementation	[24,34,35]
4			Data Governance Team	[24,33–35]
5		Data Standard Management	Data Standard Specifications	[19,35]
6			Data Standard System	[19,35]
7			Data Exchange Standards	[19,35]
8	Flow	Data Collection	Data Abundance	[19,24,35,36]
9			Data Accuracy	[19,24,35,36]
10			Data Timeliness	[19,24,35,36]
11		Data Storage	Data Storage Specifications	[33,36,37]
12			Data Storage Equipment	[33,36,37]
13			Data Storage Effectiveness	[33,36,37]
14	Application	Data Application	Data Operations and Maintenance	[24,33,34,36]
15			Data Analysis	[24,33,34,36]
16			Database Management	[24,33,34,36]
17			Data Exchange	[24,33,34,36]

3. Project Organization Resilience

Organizational resilience refers to an entity's capacity to anticipate, prepare for, respond to, and adapt to both incremental changes and sudden disruptions while maintaining core functions and evolving in dynamic environments [38–41]. It encompasses the ability to absorb external shocks and achieve post-crisis equilibrium through adaptive learning [42,43]. However, despite the increasing attention given to the concept of resilience in the academic literature, the specific definition of project resilience remains largely unclear and ambiguous [39].

Current research in project management is primarily focused on traditional risk management [39]. While both risk management and project resilience share common antecedents such as disturbances, uncertainty, and change [44], risk management focuses on identifying and mitigating predictable threats. In contrast, project resilience acknowledges the presence of unknowns and emphasizes a proactive, adaptive approach to unforeseen challenges [45]. This perspective aligns with the dynamic capabilities theory, which posits that organizations must integrate, reconfigure, and utilize existing resources to maintain competitiveness in uncertain environments [46].

Project, as a temporary organizational entity, exhibit unique characteristics such as distinct lifecycle stages, high temporality, and inherent uncertainties. These features differentiate project organization resilience from traditional organizational resilience, necessitating a tailored approach that considers the transient and project-specific context [47]. Based on the existing representative definitions (Table 4) and the capability-based perspective, project resilience can be conceptualized as a multifaceted capability that includes mitigating risks through anticipation, adapting to changes, and recovering from disruptions [48]. This framework of project organization resilience encompasses three core dimensions: defensive capability, responsive capability, and recovery capability.

Table 4. Definitions of project organization resilience.

Reference	Key Definitions
Geambasu [49] (2011)	The ability of the project system to recover and continuously adapt to changes, as well as the ability to ensure the project operates effectively during disruptive events.
Giezen [50] (2015)	Conceptualization into prevention, reaction, and adaptation.
Turner and Kutsch [51] (2016)	The ability to detect and understand changes in the project environment, plan responses, minimize damage, and adapt to new realities.
Blay et al. [44] (2017)	The ability to respond to, prepare for, and mitigate the impacts of turbulent environments and project complexity, including proactivity, responsiveness, flexibility, and durability.
Zhang et al. [46] (2023)	The dynamic abilities of temporary project organizations to anticipate, respond to changes, adapt, and learn in changing environments to ensure projects can be effectively delivered, including anticipation, coping, and adaptation capabilities.

Defensive capability refers to the proactive measures taken to anticipate and prepare for potential risks, including risk identification and mitigation strategies [49]. Rooted in anticipatory resilience, it aligns with the "sense–seize" phase of dynamic capabilities [52]. Responsive capability concentrates on problem-solving, resource reallocation, participant coordination, and the real-time handling of unexpected events or disruptions [46,53]. The aim is that project organizations can resume normal construction and operation in a timely manner after being disrupted [54]. These coping activities are typically constrained by available budget, resources (such as materials and human resources), and other factors. Recovery capability refers to post-crisis restoration through learning, adaptation, and process optimization [48], encompassing the dynamic ability of project organizations to self-adjust in anticipation of future disruptions.

4. Research Hypotheses

4.1. The Impact Between Dimensions of Data Governance Capabilities

Top-level design capability is fundamental to ensuring data quality and the realization of their value, and it serves as the strategic framework and essential safeguard for an organization's data governance efforts [32]. Mature data governance top-level design capability can effectively address the issues of "what data to collect" and "what data to store", while optimizing the allocation of an organization's data resources. This enables better control over the quality and integrity of data, helping to avoid issues such as redundancy, errors, and inconsistencies. Some scholars argue that accurately understanding the framework and content of data governance is key to ensuring the accuracy, richness, and timeliness of data collection and storage, thus enabling the collected data to accurately reflect the actual situation of project development. This allows the organization to take a targeted approach during the data governance process, improving work efficiency and accelerating project progress.

H1a. *The top-level design capability of data governance has a significant positive effect on data collection capability.*

H1b. *The top-level design capability of data governance has a significant positive effect on data storage capability.*

Data standardization serves as the fundamental pillar of data quality and is crucial for ensuring data consistency, accuracy, and compliance. Well-defined data standards establish appropriate guidelines for data collection and storage, addressing the fundamental questions of "how data should be collected" and "how data should be stored". This capability encompasses the establishment of data standards, including data definitions, classifications, and formatting guidelines. Such a standardized system facilitates the seamless integration, comparison, and analysis of data from diverse sources, formats, and quality levels while enabling the timely identification and regulatory handling of anomalous data. The absence of uniform data standards leads to various data quality issues, such as duplication, inaccuracy, and missing values, thereby compromising data reliability and accuracy and potentially resulting in misguided decision-making [55]. Furthermore, inconsistencies in data standards contribute to fragmented data storage and hinder horizontal data integration, forming "data silos" that obstruct data flow, introduce data risks and accountability challenges, and impede interdepartmental collaboration [55]. The harmonization of data structures, formats, and naming conventions reduces the workload associated with data transformation and cleansing, thereby enhancing data utilization efficiency.

H2a. *Data standard management capability has a significant positive effect on data collection capability.*

H2b. *Data standard management capability has a significant positive effect on data storage capability.*

High-quality data provide accurate information, making data application outcomes more reflective of reality and offering more valuable insights for enterprise decision-making. Enhancing data collection quality improves the efficiency of data analysis and shortens the analytical cycle. Data storage management capability involves establishing standardized data storage protocols, deploying specialized storage infrastructure, and ensuring data quality, security, and traceability. At the same time, it guarantees that various data models are consistently collected and converted into compatible formats for analysis [56]. During project implementation, effective data storage management enhances data utilization, reduces redundant data collection and processing, and consequently saves human, material, and financial resources. Furthermore, optimizing data storage structures and adopting advanced storage technologies can lower the costs associated with acquiring and maintaining data storage infrastructure. Ghasemaghaei and Calic argue that data storage serves as the foundation for ensuring data quality and plays a crucial role in enhancing the effectiveness of data application [57]. Therefore, strengthening data collection and storage management capability contributes to better protection and utilization of data assets.

H3a. *Data collection capability has a significant positive effect on data application capability.*

H3b. *Data storage capability has a significant positive effect on data application capability.*

4.2. The Impact of Data Governance Capability on Project Organization Resilience

Project organization resilience is manifested as the system's adaptive capability to respond to uncertain shocks, a mechanism that can be explained by the dynamic capabilities theory and the resource-based view. As the final stage of data governance, data application capability systematically strengthens project organization resilience across three capability dimensions—defensive, responsive, and recovery—by supporting risk identification, resource allocation, and dynamic adjustments through the deep analysis of governed, high-quality project data [58].

Regarding defensive capability, data application capability enhances the defensive resilience of project organizations through risk anticipation and proactive preparedness [46]. Preparedness is a crucial prerequisite for coping with uncertainty. Before encountering adversity, data application capability is demonstrated by utilizing big data analytics to mine effectively managed historical project data, thereby identifying potential risks and providing preemptive decision-making support to mitigate the impact of adverse external factors [59]. Additionally, data-driven optimization improves information clarity, reduces environmental uncertainty, and sharpens market insights into market dynamics, thereby strengthening the ability to withstand external shocks [60,61].

Regarding responsive capability, data application capability enhances project organization resilience in three ways. First, when facing adversity, data-driven practices endow project organizations with greater flexibility, enabling them to rapidly identify innovation opportunities and adapt to environmental changes [62]. Second, the resource structure of construction projects is often not effectively utilized in risk response [46]; especially in severe crises, an organization lacking sufficient response capability may experience project interruption, cooperation failure, or even negative outcomes [63]. Under resource constraints, data analytics enable precise resource allocation, improving crisis response efficiency [46]. Third, data-sharing mechanisms break down supply chain data barriers by synchronizing cross-team decision-making through real-time dashboards, reducing communication costs, and expediting crisis resolution [64,65].

Regarding recovery capability, data application capability fosters recovery resilience through knowledge integration and innovation. Post-crisis, project organizations leverage data governance to integrate internal and external resources and, through exploratory and exploitative innovation, absorb new knowledge and seek development opportunities [59]. Data-driven decision systems optimize survival capability under resource constraints, for example, by recording crisis response experiences in knowledge repositories to provide a foundation for subsequent project recovery [50]. Such adaptive learning ensures continuous improvement in volatile environments [38].

H4a. *Data application capability has a significant positive effect on project organization defensive capability.*

H4b. *Data application capability has a significant positive effect on project organization responsive capability.*

H4c. *Data application capability has a significant positive effect on project organization recovery capability.*

The model of data governance capabilities and project organization resilience is depicted in Figure 3.

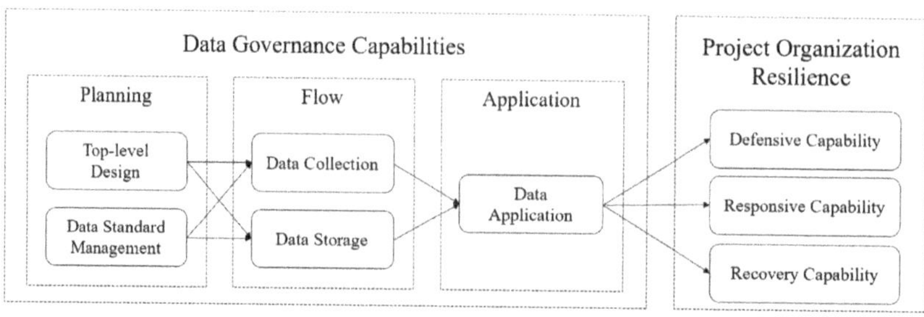

Figure 3. Model of data governance capabilities and project organization resilience.

5. Methods

The methods of this study follow a four-stage workflow, which is depicted in Figure 4. Stage 1 establishes the items of the scale and lays a solid foundation for follow-up research. Stage 2 ensures rigorous data collection through iterative scale refinement and pilot testing. Stage 3 validates the measurement model's reliability and construct validity, while Stage 4 employs PLS-SEM to test hypotheses and control for confounding factors.

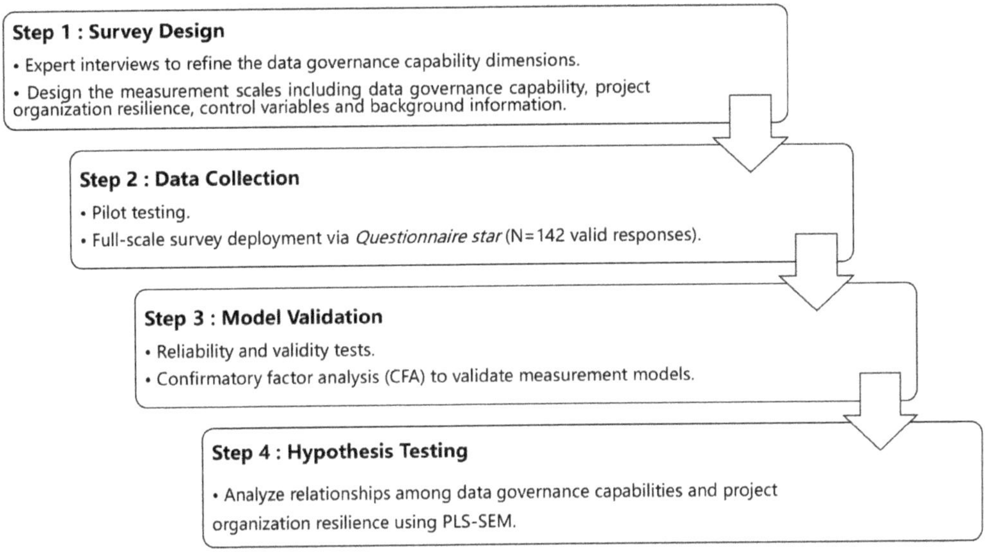

Figure 4. The methodological workflow of the study.

5.1. Sample

This study collected empirical data from professionals in the Chinese construction industry through an online survey distributed via the *Questionnaire Star* platform between December 2023 and March 2024. After a rigorous screening and sorting process, a total of 145 valid questionnaires were collected. To ensure the accuracy and reliability of the research data, two criteria were used to determine invalid responses: (1) selecting the same option for multiple consecutive questions and (2) choosing mutually exclusive options within the same scale. Regarding the selection of respondents' job categories, this study specifically targeted employees of general contracting companies and those involved in general contracting projects to ensure the focus remained on these enterprises or projects. Additionally, the target respondents were required to have a deep understanding of data governance, BIM, and digitalization technologies. The majority of the respondents had over five years of experience in the construction industry, with a strong understanding of construction projects and the ability to independently manage medium-to-large-scale projects.

Based on the above criteria, a final total of 142 valid questionnaires were obtained, resulting in a high validity rate of 97.9%. These data provide a solid foundation for subsequent statistical analysis, ensuring the rigor and reliability of the research results. Table 5 presents detailed background information on the projects and respondents.

Table 5. Description of sample characteristic distribution.

Variable	Option	Frequency	Percentage
Enterprise Type	State-owned enterprise	84	59.20%
	Joint-stock enterprise	4	2.80%
	Private enterprise	39	27.50%
	Other	15	10.60%
Project Type (Multiple Choice)	Housing construction	97	68.20%
	Road and bridge	59	41.50%
	Port/transportation	14	9.90%
	Energy	22	15.50%
	Municipal	55	38.70%
	Telecommunications	10	7.00%
	Industry	10	7.00%
	Other	27	19.00%
Gender	Male	90	63.40%
	Female	52	36.60%
Age	<25	37	26.10%
	25–30	54	38.00%
	31–35	21	14.80%
	36–40	21	14.80%
	>40	9	6.30%
Years of Work Experience	<1	29	20.40%
	1–5	60	42.30%
	6–10	26	18.30%
	>10	27	19.00%
Education	Below undergraduate	2	1.40%
	Undergraduate	70	49.30%
	Master's degree	67	47.20%
	Doctorate	3	2.10%
Position	Senior manager	5	3.50%
	Mid-level manager	34	23.90%
	Junior-level manager	51	35.90%
	Staff member	46	32.40%
	Other	6	4.20%

5.2. Measurement

To obtain a more accurate, scientific, and reliable data governance capability dimension list, the initially identified dimensions of data governance needed further screening and refinement to align with the actual conditions of construction enterprises and projects, as well as meeting the requirements for feasibility studies. Therefore, expert interviews were conducted using the "Expert Survey on Data Governance Capability Dimensions for Construction Enterprises" to solicit professional opinions. Expert interviews are a widely used analytical method in social science research, known for their simplicity and effectiveness in overcoming the limitations of group decision-making, ensuring that important factors are not overlooked during the decision-making process. Based on the preliminary identification of the data governance capability dimensions through literature search, this study used expert interviews to further screen and refine the dimensions, and then revised and categorized all data governance capabilities according to the experts' suggestions. This process resulted in the final list of data governance capability dimensions, which helps to address the gaps and omissions in the literature search, laying a solid foundation for the establishment of the structural equation model in the next chapter.

The study selected 12 experienced industry experts as participants in the survey. These experts have been engaged in the relevant fields for over five years, possessing strong professional backgrounds and extensive practical experience. Among them are 3 university professors with high academic prestige and influence, 5 general contracting project managers with rich hands-on experience and outstanding leadership skills, and 4 professionals from the digital transformation center of general contracting enterprises, who possess forward-thinking approaches and expertise in digital transformation and innovation. The participation of these experts will provide valuable industry insights and professional recommendations for this study.

Based on expert interview feedback, modifications were made to the initial data governance capability dimensions. Since "data standard specification" and "data exchange standard" overlapped, "data standard specification" was revised to "project-level data standard specification", and "data exchange standard" was modified to "cross-enterprise and departmental data exchange standard". Additionally, considering the widespread application of visualization data platforms in construction project management, data visualization capability was deemed essential. Therefore, "data visualization" was added under the data application capability dimension.

The final list of data governance capability dimensions for construction enterprises (Table 6) classifies data governance capabilities into five primary dimensions and sixteen secondary dimensions. The specific categorization is as follows:

- Top-Level Design Capability: Includes three factors—data strategy planning, data governance system, and data governance organization.
- Data Standard Management Capability: Includes three factors—project-level data standard specification, data standard system, and cross-enterprise and departmental data exchange standard.
- Data Collection Capability: Includes three factors—data richness, data accuracy, and data timeliness.
- Data Storage Capability: Includes three factors—data storage specification, data storage equipment, and data storage effectiveness.
- Data Application Capability: Includes four factors—data analysis, database management, data sharing, and data visualization.

Table 6. Final version of the construction enterprise data governance capability dimensions.

Dimension	Factor	Description
Top-Level Design Capability	Data Strategy Planning	Defining the vision and objectives of data management activities; during the project initiation phase, strategic planning documents for data governance are developed.
	Data Governance System	Establishing a corresponding data governance system, with periodic reviews and updates, to ensure standardized execution of data governance tasks.
	Data Governance Organization	Constructing an organizational framework that complements the data governance system; this framework clarifies responsibilities and ensures efficient internal communication.

Table 6. *Cont.*

Dimension	Factor	Description
Data Standard Management Capability	Project-Level Data Standard Specifications	Establishing project-level data standards to ensure consistency across different project phases, such as BIM data standards for design and construction stages.
	Data Standard System	Developing data standard systems, including data definitions, classifications, and formats.
	Cross-Enterprise and Departmental Data Exchange Standards	Adopting industry or national standards to regulate internal and external data exchange.
Data Collection Capability	Data Richness	Collecting data from multiple sources to ensure data comprehensiveness.
	Data Accuracy	Establishing a data quality review system to ensure the accuracy of data reflecting the actual project conditions.
	Data Timeliness	Periodically updating data to maintain data quality and timeliness.
Data Storage Capability	Data Storage Standards	Establishing data storage standards that define data storage processes, methods, and naming conventions.
	Data Storage Equipment	Equipping dedicated data storage devices and ensuring timely project data archiving and backup to facilitate data traceability.
	Effectiveness of Data Storage	Regularly inspecting data storage effectiveness to ensure data quality and security.
Data Application Capability	Data Analysis	Providing data decision support for project implementation by conducting exploratory data analysis and leveraging digital platforms for project management and resilience assessment.
	Database Management	Establishing project databases to facilitate data retrieval and support decision-making.
	Data Exchange	Enabling data sharing through data platforms to enhance project collaboration.
	Data Visualization	Utilizing data platforms to monitor project construction in real time.

After refining the dimensions of data governance capability, a scale for measuring data governance capability in construction enterprises was developed. Based on the theoretical model, the scale items were formulated by extensively referencing the existing literature and the aforementioned data governance capability maturity model classifications [28,33,36,66]. Considering the actual operational conditions and characteristics of enterprises in the construction sector, the relevant scale was carefully adjusted and optimized, leading to the development of a preliminary survey questionnaire for measuring data governance capability. To validate the effectiveness of the questionnaire, a pilot survey was conducted among 18 project managers, confirming its applicability for large-scale research. Respondents were asked to answer the questions based on their most recently completed construction project. The first part of the questionnaire included control variables and background information on the respondents and their projects. Excluding this section, all items were measured using a five-point Likert scale (e.g., 1 = Strongly Disagree; 5 = Strongly Agree).

The measurement items for data governance capability are presented in Table 7.

Table 7. Data governance capability measurement items.

Dimension	Code	Measurement Items
Top-Level Design Capability	DC1	The company has formulated a comprehensive data governance strategy (including vision, goals, and principles).
	DC2	The company has established a sound data governance institutional system to ensure sustainable data management.
	DC3	The company has formed a data governance team, clarifying relevant roles, responsibilities, and workflows.
Data Standard Management Capability	BZ1	The company has established project-level data standards.
	BZ2	The company has formulated a data standard system that includes data definitions, classifications, and formats.
	BZ3	The company adopts industry or national standards to regulate data exchange within and outside the team.
Data Collection Capability	CJ1	The company can collect data from multiple channels to ensure data richness.
	CJ2	The company has established a data quality review system to ensure data accuracy in reflecting project conditions.
	CJ3	The company can ensure timely data updates.
Data Storage Capability	CC1	The company has established data storage specifications.
	CC2	The company is equipped with dedicated data storage devices and ensures timely project data archiving and backup.
	CC3	The company regularly inspects the effectiveness of data storage to ensure data quality and security.
Data Application Capability	YY1	The company can allocate data according to its strategy and business needs.
	YY2	The company has established a database to facilitate data retrieval and support decision-making.
	YY3	Departments within the company can share data through data platforms, enhancing collaboration.
	YY4	The company can observe project construction status in real time through data platforms.

This study identifies three dimensions of project organization resilience through a systematic literature review, namely, defensive capability, responsive capability, and recovery capability, and provides an in-depth analysis of their connotations. Defensive capability refers to an organization's ability to anticipate unexpected events and proactively prepare for risks. The measurement items in this section are primarily based on the theoretical research of Duchek et al. [38]. Responsive capability refers to the ability to acknowledge problems and promptly formulate and implement solutions in response to internal and external changes. The measurement items in this section mainly draw from the research findings of Home and Orr [67], Kendra and Wachtendorf [68], and Lee et al. [7]. Recovery capability refers to an organization's ability to recover from crises in a short period while learning from experiences to lay a solid foundation for future projects. The measurement items in this section are primarily based on the research of Sanchis et al. [69].

The measurement items for project organization resilience are presented in Table 8.

According to previous research, three control variables that are likely to influence project organization resilience are considered to minimize issues related to omitted variable bias [70]. First, human resource, the leadership and competence of core employees, contributes to organizational resilience in crisis response [5]. Human resource was measured by the following single item: "The enterprise possesses leaders with digital strategic vision and employees proficient in applying digital technologies in their work". Second, technical resource may empower project organizational resilience in enabling virtual communication and remote access to resources [71]. Technical resource was measured by the following single item: "The enterprise has comprehensive software resources, hardware facilities, and physical spaces related to the application of digital technologies". Finally, Organization resource, which in some sense reflects the company culture, was measured by the following single item: "The enterprise has established corresponding organizational structures and institutional systems for implementing digital technologies".

Table 8. Project organization resilience measurement items.

Dimension	Code	Measurement Items
Defensive Capability	FY1	The project organization has a clear understanding of potential project risks.
	FY2	The project organization can continuously monitor the project execution process to issue warnings for emerging problems.
	FY3	The project organization has developed an emergency response plan and conducts regular drills for emergency situations.
Responsive Capability	XY1	When a risk occurs, the project organization can coordinate internally to ensure entering a systematic response state.
	XY2	When a risk occurs, the project organization can quickly develop response plans and take action.
	XY3	The project organization can quickly identify and assess different types of crises.
Recovery Capability	HF1	The project organization has established a formal emergency team or group to mitigate crises.
	HF2	The project organization can recover from sudden incidents or risks with minimal time and cost.
	HF3	The project organization can learn from past crises, absorb experience, and lay a foundation for future project construction.

6. Data Analysis and Results

6.1. Reliability and Validity Analysis

Reliability and validity testing ensures the rationality of questionnaire design and the accuracy of sample data in reflecting variable responses.

First, principal component analysis (PCA) was employed to conduct exploratory factor analysis (EFA) to assess the factor structure and item design of the questionnaire. The Kaiser–Meyer–Olkin (KMO) test and Bartlett's sphericity test were used to determine whether each dimension was suitable for factor analysis. The threshold criteria were set as KMO > 0.5 and Bartlett's test $p < 0.05$ to confirm the appropriateness of factor analysis. Cronbach's α coefficient was also examined, with $\alpha > 0.7$ considered indicative of good reliability. The reliability test was performed using SPSS 26.0, yielding an overall KMO value of 0.944, indicating a good model fit for factor analysis. The Bartlett test statistic was significant at the $p < 0.001$ level, further validating the suitability of factor analysis. As shown in Table 9, the Cronbach's α coefficients for DC, CJ, BZ, CC, YY, FY, XY, and HF were 0.878, 0.866, 0.836, 0.878, 0.932, 0.793, 0.889, and 0.863, respectively (all exceeding 0.7), demonstrating the high internal consistency and strong reliability of the questionnaire. After examining the factor loadings for each item, no cross-loading issues were identified, so no items needed to be eliminated.

Table 9. Reliability and validity test results.

Construct and Measure Items	SFL	Cronbach's α	CR	AVE
Top-Level Design Capability		0.878	0.927	0.810
DC1	0.903			
DC2	0.912			
DC3	0.886			
Data Standard Management Capability		0.836	0.902	0.753
BZ1	0.899			
BZ2	0.840			
BZ3	0.864			
Data Collection Capability		0.866	0.918	0.788
CJ1	0.878			
CJ2	0.900			
CJ3	0.884			
Data Storage Capability		0.878	0.925	0.805
CC1	0.892			
CC2	0.861			
CC3	0.937			
Data Application Capability		0.932	0.951	0.830
YY1	0.928			
YY2	0.886			
YY3	0.913			
YY4	0.917			
Defensive Capability		0.793	0.879	0.708
FY1	0.810			
FY2	0.843			
FY3	0.869			
Responsive Capability		0.889	0.931	0.819
XY1	0.926			
XY2	0.889			
XY3	0.898			
Recovery Capability		0.863	0.916	0.784
HF1	0.853			
HF2	0.888			
HF3	0.915			

Second, confirmatory factor analysis (CFA) was conducted using Amos 24.0 to validate the factor structure and item design established in the EFA. Figure 5 shows the first-lever factor analysis results of the scale. The standardized factor loadings (SFLs) for all items exceeded 0.7, the average variance extracted (AVE) for all variables was greater than 0.7, and the construct reliability (CR) values ranged between 0.879 and 0.951, indicating strong convergent validity and composite reliability. Additionally, discriminant validity analysis confirmed significant distinctions among different variables, demonstrating that the indicators effectively differentiate between constructs. As shown in Table 9, all variables exhibited significant correlations, and the square root of each variable's AVE was greater than its off-diagonal correlation coefficients, supporting good discriminant validity. Furthermore, all heterotrait–monotrait (HTMT) values were below the 0.90 threshold [72].

Finally, the fitting indices for the first-level factor analysis are shown in Table 10. According to Table 10, the fit values obtained from the first-level confirmatory factor analysis ($\chi 2/df$ = 1.950 < 3; RMR = 0.031 < 0.05; CFI = 0.928 > 0.9; IFI = 0.930 > 0.9;

TLI = 0.913 > 0.9) indicate an acceptable level of agreement, and the CFA results indicate that the construct validity of the scale is appropriate.

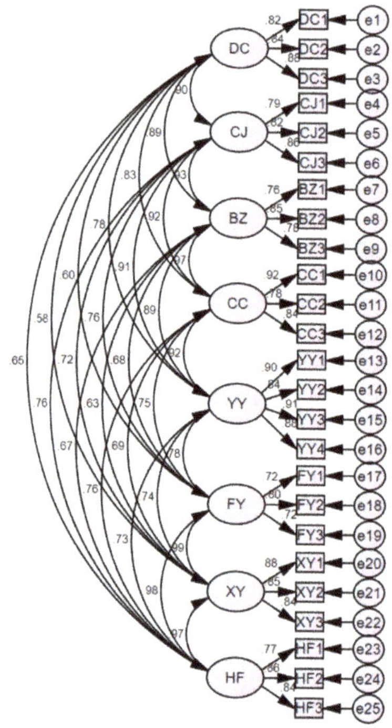

Figure 5. First-level confirmatory factor analysis path (PATH) of the scale.

Table 10. Fitting indices of confirmatory factor analysis.

Index	$\chi 2/df$	RMR	CFI	IFI	TLI
Standard	<3	<0.05	>0.9	>0.9	>0.9
Value	1.950	0.031	0.928	0.930	0.913

6.2. Hypothesis Testing

This study employed Smart PLS 4.0 to test the research hypotheses using the Bootstrapping algorithm with 5000 resamples. As shown in Table 11, the path coefficients (β) from top-level design capability to data collection capability and data storage capability were 0.445 and 0.242, respectively, both significant at the 0.001 level. This indicates that top-level design capability positively influences data collection and storage capabilities, supporting H1a and H1b.

Similarly, the path coefficients (β) from data standard management capability to data collection capability and data storage capability were 0.456 and 0.655, respectively, both significant at the 0.001 level. This confirms the positive impact of data standard management capability on data collection and storage capabilities, supporting H2a and H2b.

The path coefficients (β) from data collection capability and data storage capability to data application capability were 0.378 and 0.538, respectively, both significant at the 0.001 level. This demonstrates that data collection and storage capabilities positively influence data application capability, supporting H3a and H3b.

Table 11. Results for hypothesis testing.

Hypothesis	Path	β	t	CI		p	Correlation
				2.5%	97.5%		
H1a	DC → CJ	0.445	5.108	0.281	0.624	0.000	+
H1b	DC → CC	0.242	3.521	0.110	0.380	0.000	+
H2a	BZ → CJ	0.456	4.952	0.267	0.624	0.000	+
H2b	BZ → CC	0.655	10.096	0.524	0.780	0.000	+
H3a	CJ → YY	0.379	4.070	0.202	0.567	0.000	+
H3b	CC → YY	0.537	5.646	0.343	0.712	0.000	+
H4a	YY → FY	0.677	11.765	0.554	0.779	0.000	+
H4b	YY → XY	0.677	11.433	0.550	0.780	0.000	+
H4c	YY → HF	0.657	9.872	0.513	0.773	0.000	+

Note: DC = top-level design capability; BZ = data standard management capability; CJ = data collection capability; CC = data storage capability; YY = data application capability; FY = defensive capability; XY = responsive capability; HF = recovery capability.

Finally, the path coefficients (β) from data application capability to project organization resilience (defensive capability, responsive capability, and recovery capability) were 0.677, 0.677, and 0.657, also significant at the 0.001 level. This confirms that data application capability has a positive impact on project organization resilience, supporting H4.

With respect to the control variables (human resource, technical resource, and organization resource), our results illustrate that only human resource has a positive effect on project organization resilience (β = 0.199; $p = 0.019$) and other variables do not have a significant effect. Furthermore, when control variables are included in the analysis, they do not significantly alter the main results. This suggests that the primary relationships observed in our study remain robust, regardless of the inclusion of these control variables.

The results of the hypothesis testing are shown in Figure 6.

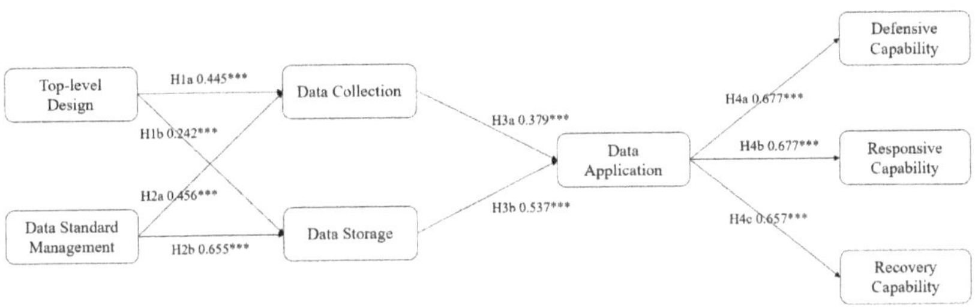

Figure 6. Results of hypothesis testing. Note: *** $p < 0.01$.

7. Discussion

This study aims to explore the influence of data governance capability on the resilience of construction project organizations. To achieve this, we first identified the key dimensions of data governance capability in construction enterprises and developed corresponding measurement scales. The findings not only validate the rationality of these dimensions but also reveal their inter-connections, offering a new theoretical perspective on the application of data governance capability in the field of construction management.

7.1. Dimensions and Hierarchical Relationships of Data Governance Capability

This study introduces five core dimensions of data governance capability in the field of construction management: top-level design, data standard management, data collection, data storage, and data application. More importantly, based on the implementation process of data governance, these dimensions are categorized into three stages, namely, planning, flow, and application, and their inter-connections are verified. The results indicate that these dimensions influence each other, addressing a gap in existing research. Previous studies have primarily enumerated different dimensions of data governance [56,73,74], but have lacked an in-depth discussion of how these dimensions interact with one another.

Consistent with the IBM data governance maturity model and the findings of Mahanti (2021) [66], the validation of H1 and H2 demonstrates that top-level design and data standard management (planning stage) positively influence data collection and data storage capabilities (flow stage). Khatri and Brown (2010) [19] also proposed a similar viewpoint, stating that an effective top-level design in data governance can optimize data collection and storage management processes, thereby improving data quality and usability. However, unlike traditional IT governance frameworks, this study emphasizes the strategic role of data governance, arguing that top-level design not only affects the efficiency of data management but also has profound implications for an enterprise's long-term development and risk resilience, which is consistent with the arguments proposed by Alhassan et al. (2018) [18]. Different from previous studies, this study identifies data governance dimensions more suited to construction projects through interviews with industry experts, thereby making a significant contribution to the existing literature. Furthermore, this study refines the influence pathways of data standard management, emphasizing its role in enhancing data fluidity and integrated management. As the core of the data governance system, data standard management ensures data standardization, accuracy, and consistency, which are crucial for effective data governance.

H3 is supported by the results, indicating that the circulation stage of data governance (data collection and storage) positively influences the application stage. This finding aligns with the study of Alhassan et al. (2018) [18], which suggests that high-quality data collection and storage management are prerequisites for maximizing data application value. Furthermore, this study confirms that, in the context of construction projects, data integrity and traceability enhance the effectiveness of data analysis, providing organizations with more accurate decision-making support.

7.2. The Impact of Data Governance Capability on Project Organization Resilience

This study innovatively establishes the path mechanism through which data governance capability influences project organization resilience. Previous research has primarily explored the potential impact of digital technology [11,71,75], multi-party collaboration [76], data analytics [77], or data management [78] from a singular perspective. Among them, the role of BIM, IoT, and other digital technologies in project organizational resilience is most often mentioned. These technologies provide the necessary tools and infrastructure for data governance, but their effectiveness hinges on alignment with governance strategies. Technology alone cannot address the systemic challenges of data governance—such as stakeholder misalignment or fragmented accountability—which require institutionalized processes and strategic oversight. This study's framework thus prioritizes governance capabilities over tool-specific functionalities, ensuring adaptability across heterogeneous technological environments. By consolidating these fragmented perspectives into a unified data governance capability framework, this study fills this gap and reveals how data governance capability, as a holistic concept, enhances project organization resilience by strengthening its defense, response, and recovery capabilities.

The validation of H4 confirms that data governance capability has a positive impact on project organization resilience. Burnard et al. (2018) emphasized that the core of organization resilience lies in information integration and dynamic adaptability [79]. This study further demonstrates that improving data application capability enhances a construction project organization's ability to manage risks and foster innovation. Particularly in project management, data application capability not only optimizes internal management but also promotes cross-project knowledge sharing, thereby improving a project's ability to recover from crises. This finding offers new managerial insights for construction enterprises seeking to enhance organization resilience in uncertain environments.

8. Conclusions

This study aims to explore the mechanisms through which data governance capability influences organization resilience in construction projects. The findings validate that data governance enhances project organization resilience—including defensive capability, responsive capability, and recovery capability—via strengthened data application capability. Furthermore, the five dimensions of data governance capability do not exist in isolation but follow an inherent logic of data flow, forming an integrated system: top-level design provides a guiding framework for data collection and storage, data standard management regulates data collection processes and storage formats, high-quality data collection ensures the value of stored data, and effective data storage ultimately supports data application. These insights provide a novel pathway for improving construction enterprises' capacity to navigate uncertainties and crises.

8.1. Theoretical Contributions

First, systematizing data governance capability dimensions: This study consolidates fragmented data governance dimensions into an integrated framework with measurable indicators (five dimensions across three stages), thereby addressing conceptual ambiguity in the construction data governance literature and providing theoretical foundations for systematic governance system development.

Second, clarifying hierarchical interdependencies of data governance capability: By investigating dynamic influence mechanisms among capability dimensions through a value-chain logic (planning → flow → application), this study advances beyond static technical interpretations of data governance. These empirically verified sequential relationships offer actionable phased pathways for enhancing capabilities from strategic planning to operational implementation.

Third, establishing the "Data Governance Capability–Project Organization Resilience" theoretical framework: Through theoretical integration, this study positions data governance capability as a critical antecedent of project organization resilience. This framework equips project managers with concrete operational levers while enriching interdisciplinary research at the nexus of data governance and project resilience.

8.2. Managerial Implications

First, establishing a systematic data governance framework: Enterprises should strengthen the top-level design of data governance by defining clear data governance strategies and aligning data management with overall corporate objectives, which involves establishing cross-departmental data governance teams and forming dedicated committees with explicit roles and responsibilities. Additionally, regular reviews should be implemented and efforts made to promote data standardization to develop industry-wide data specifications, enhancing data fluidity and sharing efficiency.

Second, enhancing data quality management to improve data usability: Enterprises should refine data standard management policies, optimize data collection and storage processes, and ensure data integrity, accuracy, and security. For instance, automated data collection tools can be leveraged to reduce human errors, while blockchain and distributed storage technologies can be adopted to enhance data traceability.

Third, developing a resilience management system driven by data application: The true value of data lies in their application. Construction enterprises should increase investment in data analytics and visualization technologies to drive data-informed decision-making, for example, during the defensive stage, using historical data patterns for risk prediction systems; during the responsive stage, establishing real-time monitoring dashboards for crisis management; and during the recovery stage, documenting crisis response lessons in knowledge repositories.

8.3. Limitations and Future Research Directions

First, the data collection in this study is limited to professionals in China's construction industry, without considering governance differences in international projects or multicultural contexts. Data governance policies vary significantly across regions, influencing the capabilities of construction enterprises differently. Future research could enhance sample size and diversity, expanding into cross-national comparative studies. Second, while this study provides a framework for understanding the interplay between data governance and project organization resilience, it does not explicitly incorporate external environmental factors such as market conditions, regulatory policies, and industry competition. Future studies could introduce variables such as environmental uncertainty and project complexity as moderating factors to further examine how external conditions impact the relationship between data governance capability and project organization resilience.

Author Contributions: Conceptualization, Y.H., M.K., Y.F. and H.Y.; methodology, M.K. and H.Y.; formal analysis, M.K.; investigation, Y.H.; resources, Y.H.; writing—original draft preparation, M.K.; writing—review and editing, Y.F. and H.Y.; supervision, Y.F.; project administration, M.K.; funding acquisition, Y.H. All authors have read and agreed to the published version of the manuscript.

Funding: This research was supported by the Beijing Social Science Foundation [No. 22GLC049] and the Project of Cultivation for young top-notch Talents of Beijing Municipal Institutions [No. BPHR202203084].

Data Availability Statement: Some or all data or models that support the findings of this study are available from the corresponding author upon reasonable request.

Acknowledgments: We are sincerely grateful to all the respondents who participated in this research.

Conflicts of Interest: The authors declare no conflicts of interest.

Abbreviations

The following abbreviations are used in this manuscript:

AVE	Average variance extracted
BIM	Building Information Modeling
CFA	Confirmatory factor analysis
CFI	Comparative Fit Index
CMM	Capability maturity model
CMMI	Capability Maturity Model Integration
CR	Construct reliability
DAMA	Data Administration Management Association
DCMM	Data Management Capability Maturity Assessment Model
DGI	Data Governance Institute
DMM	Data Management Maturity
EFA	Exploratory factor analysis

EIM	Enterprise Information Management
HTMT	Heterotrait–monotrait ratio
IBM	International Business Machines Corporation
IFI	Incremental Fit Index
IoT	Internet of Things
KMO	Kaiser–Meyer–Olkin
KPI	Key Performance Indicator
PCA	Principal component analysis
PLS-SEM	Partial Least Squares Structural Equation Modeling
RMR	Root mean square residual
SFL	Standardized factor loading
SU	Syracuse University
TLI	Tucker–Lewis Index

References

1. Lu, J.; Xiang, P. Crisis Process Management: How to Improve Organizational Resilience. *Foreign Econ. Manag.* **2021**, *43*, 3–24. [CrossRef]
2. Bu, W.; Chen, H.; Fu, Y. Development and Validation of a Resilience Scale for Construction Project Organization. *J. Eng. Stud.* **2023**, *15*, 349–358. [CrossRef]
3. Jiao, P.; Bu, W. The Impact of Organizational Learning on Organizational Resilience in Construction Projects. *Buildings* **2024**, *14*, 975. [CrossRef]
4. Hamel, G.; Välikangas, L. The Quest for Resilience. *Harv. Bus. Rev.* **2003**, *81*, 52–63, 131.
5. Lengnick-Hall, C.A.; Beck, T.E.; Lengnick-Hall, M.L. Developing a Capacity for Organizational Resilience through Strategic Human Resource Management. *Hum. Resour. Manag. Rev.* **2011**, *21*, 243–255. [CrossRef]
6. Pal, R.; Torstensson, H.; Mattila, H. Antecedents of Organizational Resilience in Economic Crises—An Empirical Study of Swedish Textile and Clothing SMEs. *Int. J. Prod. Econ.* **2014**, *147*, 410–428. [CrossRef]
7. Lee, A.V.; Vargo, J.; Seville, E. Developing a Tool to Measure and Compare Organizations' Resilience. *Nat. Hazards Rev.* **2013**, *14*, 29–41. [CrossRef]
8. Weick, K.E.; Roberts, K.H. Collective Mind in Organizations: Heedful Interrelating on Flight Decks. *Adm. Sci. Q.* **1993**, *38*, 357. [CrossRef]
9. Gill, T. Building Resilience: Social Capital in Post-Disaster Recovery. *Soc. Sci. Jpn. J.* **2014**, *17*, 118–122. [CrossRef]
10. Jaaron, A.A.M.; Backhouse, C.J. Service Organisations Resilience through the Application of the Vanguard Method of Systems Thinking: A Case Study Approach. *Int. J. Prod. Res.* **2014**, *52*, 2026–2041. [CrossRef]
11. Zhang, B.; Zhang, J.; Chen, C. Digital Technology Innovation and Corporate Resilience. *Glob. Financ. J.* **2024**, *63*, 101042. [CrossRef]
12. He, Z.; Huang, H.; Choi, H.; Bilgihan, A. Building Organizational Resilience with Digital Transformation. *J. Serv. Manag.* **2023**, *34*, 147–171. [CrossRef]
13. Su, W.; Junge, S. Unlocking the Recipe for Organizational Resilience: A Review and Future Research Directions. *Eur. Manag. J.* **2023**, *41*, 1086–1105. [CrossRef]
14. Adewusi, A.O.; Komolafe, A.M.; Ejairu, E.; Aderotoye, I.A.; Abiona, O.O.; Oyeniran, O.C. The role of predictive analytics in optimizing supply chain resilience: A review of techniques and case studies. *Int. J. Manag. Entrep. Res.* **2024**, *6*, 815–837. [CrossRef]
15. Hilkenmeier, F.; Fechtelpeter, C.; Decius, J. How to Foster Innovation in SMEs: Evidence of the Effectiveness of a Project-Based Technology Transfer Approach. *J. Technol. Transf.* **2021**, 1–29. [CrossRef]
16. Gölzer, P.; Fritzsche, A. Data-Driven Operations Management: Organisational Implications of the Digital Transformation in Industrial Practice. *Prod. Plan. Control* **2017**, *28*, 1332–1343. [CrossRef]
17. Chen, Y.-C.; Hsieh, T.-C. Big Data for Digital Government: Opportunities, Challenges, and Strategies. *Int. J. Public Adm. Digit. Age* **2014**, *1*, 14. [CrossRef]
18. Alhassan, I.; Sammon, D.; Daly, M. Data Governance Activities: A Comparison between Scientific and Practice-Oriented Literature. *J. Enterp. Inf. Manag.* **2018**, *31*, 300–316. [CrossRef]
19. Khatri, V.; Brown, C.V. Designing Data Governance. *Commun. ACM* **2010**, *53*, 148–152. [CrossRef]
20. Watson, H.J.; Fuller, C.; Ariyachandra, T. Data Warehouse Governance: Best Practices at Blue Cross and Blue Shield of North Carolina. *Decis. Support Syst.* **2004**, *38*, 435–450. [CrossRef]

21. Abraham, R.; Schneider, J.; Vom Brocke, J. Data Governance: A Conceptual Framework, Structured Review, and Research Agenda. *Int. J. Inf. Manag.* **2019**, *49*, 424–438. [CrossRef]
22. Data Administration Management Association. *The DAMA Guide to the Data Management Body of Knowledge: DAMA-DMBOK Guide*, 1st ed.; Mosley, M., Brackett, M., Earley, S., Henderson, D., Eds.; Technics Publications, LLC: Bradley Beach, NJ, USA, 2010; ISBN 978-0-9771400-8-4.
23. The Data Governance Institute. Definitions of Data Governance. Available online: https://datagovernance.com/the-data-governance-basics/definitions-of-data-governance/ (accessed on 5 April 2025).
24. International Business Machines Corporation (IBM). *The IBM Data Governance Council Maturity Model: Building a Roadmap for Effective Data Governance*; International Business Machines Corporation (IBM): Armonk, NY, USA, 2007.
25. Coleman, S. Foundations of Digital Government. In *Digital Government*; Chen, H., Brandt, L., Gregg, V., Traunmüller, R., Dawes, S., Hovy, E., Macintosh, A., Larson, C.A., Eds.; Integrated Series in Information Systems; Springer: Boston, MA, USA, 2008; Volume 17, pp. 3–19. ISBN 978-0-387-71610-7.
26. Sørensen, E.; Torfing, J. Metagoverning Collaborative Innovation in Governance Networks. *Am. Rev. Public Adm.* **2017**, *47*, 826–839. [CrossRef]
27. Otto, B. A Morphology of the Organisation of Data Governance. In Proceedings of the European Conference on Information Systems, Helsinki, Finland, 9–11 June 2011.
28. Dwivedi, Y.K.; Janssen, M.; Slade, E.L.; Rana, N.P.; Weerakkody, V.; Millard, J.; Hidders, J.; Snijders, D. Driving Innovation through Big Open Linked Data (BOLD): Exploring Antecedents Using Interpretive Structural Modelling. *Inf. Syst. Front.* **2017**, *19*, 197–212. [CrossRef]
29. Benfeldt, O.; Persson, J.S.; Madsen, S. Data Governance as a Collective Action Problem. *Inf. Syst. Front.* **2020**, *22*, 299–313. [CrossRef]
30. Kim, Y.H.; Cho, J.S. Data Governance Framework for Big Data Implementation with NPS Case Analysis in Korea. *J. Bus. Retail Manag. Res.* **2018**, *12*, 36–46. [CrossRef]
31. Sargiotis, D. The Importance of Data Governance: Why It Matters in Today's World. In *Data Governance*; Springer Nature: Cham, Switzerland, 2024; pp. 87–136. ISBN 978-3-031-67267-5.
32. Lin, X. A Model of Big Data-Based Governance: China's National Government Big Data Platform and an Analysis of Its Governance Competence. *Chin. Polit. Sci. Rev.* **2025**. [CrossRef]
33. Begg, C.; Caira, T. Exploring the SME Quandary: Data Governance in Practise in the Small to Medium-Sized Enterprise Sector. *Electron. J. Inf. Syst. Eval.* **2012**, *15*, 3–13.
34. ISACA. *Data Management Maturity (DMM)^{SM} Model*; ISACA: Schaumburg, IL, USA, 2019.
35. The Big Data Standards Working Group of the China Electronics Standardization Institute Data Management Capability Maturity Assessment Model. Available online: http://www.dcmm.org.cn/DCMMbzjs/index.jhtml (accessed on 8 March 2025).
36. Qin, J.; Crowston, K.; Kirkland, A. Pursuing Best Performance in Research Data Management by Using the Capability Maturity Model and Rubrics. *J. EScience Librariansh.* **2017**, *6*, e1113. [CrossRef]
37. Gartner Gartner Introduces the EIM Maturity Model. Available online: https://www.gartner.com/en/documents/846312 (accessed on 8 March 2025).
38. Duchek, S. Organizational Resilience: A Capability-Based Conceptualization. *Bus. Res.* **2020**, *13*, 215–246. [CrossRef]
39. Rahi, K. Project Resilience: A Conceptual Framework. *Int. J. Inf. Syst. Proj. Manag.* **2022**, *7*, 69–83. [CrossRef]
40. Richtnér, A.; Löfsten, H. Managing in Turbulence: How the Capacity for Resilience Influences Creativity. *RD Manag.* **2014**, *44*, 137–151. [CrossRef]
41. Hillmann, J.; Guenther, E. Organizational Resilience: A Valuable Construct for Management Research? *Int. J. Manag. Rev.* **2021**, *23*, 7–44. [CrossRef]
42. Dai, L.; Eden, L.; Beamish, P.W. Caught in the Crossfire: Dimensions of Vulnerability and Foreign Multinationals' Exit from War-afflicted Countries. *Strateg. Manag. J.* **2017**, *38*, 1478–1498. [CrossRef]
43. Meyer, A.D. Adapting to Environmental Jolts. *Adm. Sci. Q.* **1982**, *27*, 515. [CrossRef]
44. Blay, K.B. Resilience in Projects: Definition, Dimensions, Antecedents and Consequences. Ph.D. Thesis, Loughborough University, Leicestershire, UK, 3 November 2017.
45. Pich, M.T.; Loch, C.H.; Meyer, A.D. On Uncertainty, Ambiguity, and Complexity in Project Management. *Manag. Sci.* **2002**, *48*, 1008–1023. [CrossRef]
46. Zhang, S.; Zhang, F.; Xue, B.; Wang, D.; Liu, B. Unpacking Resilience of Project Organizations: A Capability-Based Conceptualization and Measurement of Project Resilience. *Int. J. Proj. Manag.* **2023**, *41*, 102541. [CrossRef]
47. Naderpajouh, N.; Matinheikki, J.; Keeys, L.A.; Aldrich, D.P.; Linkov, I. Resilience and Projects: An Interdisciplinary Crossroad. *Proj. Leadersh. Soc.* **2020**, *1*, 100001. [CrossRef]
48. Madni, A.M.; Jackson, S. Towards a Conceptual Framework for Resilience Engineering. *IEEE Syst. J.* **2009**, *3*, 181–191. [CrossRef]

49. Geambasu, G. Expect the Unexpected: An Exploratory Study on the Conditions and Factors Driving the Resilience of Infrastructure Projects. Ph.D. Thesis, The École Polytechnique Fédérale de Lausanne (EPFL): Lausanne, Switzerland, 2011.
50. Giezen, M.; Salet, W.; Bertolini, L. Adding Value to the Decision-Making Process of Mega Projects: Fostering Strategic Ambiguity, Redundancy, and Resilience. *Transp. Policy* **2015**, *44*, 169–178. [CrossRef]
51. Kutsch, E.; Hall, M. *Project Resilience*; Routledge: London, UK, 2016; ISBN 978-1-317-07498-4.
52. Teece, D.J.; Pisano, G.; Shuen, A. Dynamic Capabilities and Strategic Management. *Strateg. Manag. J.* **1997**, *18*, 509–533. [CrossRef]
53. Varajão, J.; Fernandes, G.; Amaral, A.; Gonçalves, A.M. Team Resilience Model: An Empirical Examination of Information Systems Projects. *Reliab. Eng. Syst. Saf.* **2021**, *206*, 107303. [CrossRef]
54. Liu, L.; Wu, H.; Wang, J.; Yang, T. Research on the Evaluation of the Resilience of Subway Station Projects to Waterlogging Disasters Based on the Projection Pursuit Model. *Math. Biosci. Eng.* **2020**, *17*, 7302–7331. [CrossRef] [PubMed]
55. Duchek, S. Growth in the Face of Crisis: The Role of Organizational Resilience Capabilities. *Acad. Manag. Proc.* **2014**, *2014*, 13487. [CrossRef]
56. Jiang, Z.; Zhu, Z.; Pan, S. Data Governance in Multimodal Behavioral Research. *Multimodal Technol. Interact.* **2024**, *8*, 55. [CrossRef]
57. Ghasemaghaei, M.; Calic, G. Assessing the Impact of Big Data on Firm Innovation Performance: Big Data Is Not Always Better Data. *J. Bus. Res.* **2020**, *108*, 147–162. [CrossRef]
58. Hosseini, S.; Barker, K.; Ramirez-Marquez, J.E. A Review of Definitions and Measures of System Resilience. *Reliab. Eng. Syst. Saf.* **2016**, *145*, 47–61. [CrossRef]
59. Benitez, J.; Arenas, A.; Castillo, A.; Esteves, J. Impact of Digital Leadership Capability on Innovation Performance: The Role of Platform Digitization Capability. *Inf. Manag.* **2022**, *59*, 103590. [CrossRef]
60. Xue, J.; Li, G.; Ivanov, D. Digital Transformation in the Blockchain Era: Balancing Efficiency and Resilience in Operations Management. *Int. J. Prod. Econ.* **2025**, *282*, 109525. [CrossRef]
61. Karsai, I.; Schmickl, T.; Kampis, G. *Resilience and Stability of Ecological and Social Systems*; Springer International Publishing: Cham, Switzerland, 2020; ISBN 978-3-030-54559-8.
62. Vial, G. Understanding Digital Transformation: A Review and a Research Agenda. *J. Strateg. Inf. Syst.* **2019**, *28*, 118–144. [CrossRef]
63. Nachbagauer, A.G.M.; Schirl-Boeck, I. Managing the Unexpected in Megaprojects: Riding the Waves of Resilience. *Int. J. Manag. Proj. Bus.* **2019**, *12*, 694–715. [CrossRef]
64. Pantović, V.; Vidojević, D.; Vujičić, S.; Sofijanić, S.; Jovanović-Milenković, M. Data-Driven Decision Making for Sustainable IT Project Management Excellence. *Sustainability* **2024**, *16*, 3014. [CrossRef]
65. Werner, M.J.E.; Yamada, A.P.L.; Domingos, E.G.N.; Leite, L.R.; Pereira, C.R. Exploring Organizational Resilience Through Key Performance Indicators. *J. Ind. Prod. Eng.* **2021**, *38*, 51–65. [CrossRef]
66. Mahanti, R. Data Governance Maturity Models. In *Data Governance Success*; Springer: Singapore, 2021; pp. 99–126. ISBN 9789811650857.
67. Home, J.F.; Orr, J.E. Assessing Behaviors That Create Resilient Organizations. *Employ. Relat. Today* **1997**, *24*, 29–39. [CrossRef]
68. Kendra, J.M.; Wachtendorf, T. Elements of Resilience After the World Trade Center Disaster: Reconstituting New York City's Emergency Operations Centre. *Disasters* **2003**, *27*, 37–53. [CrossRef]
69. Sanchis, R.; Poler, R. Definition of a Framework to Support Strategic Decisions to Improve Enterprise Resilience. *IFAC Proc. Vol.* **2013**, *46*, 700–705. [CrossRef]
70. Becker, T.E.; Atinc, G.; Breaugh, J.A.; Carlson, K.D.; Edwards, J.R.; Spector, P.E. Statistical Control in Correlational Studies: 10 Essential Recommendations for Organizational Researchers. *J. Organ. Behav.* **2016**, *37*, 157–167. [CrossRef]
71. Mehan, A. The Role of Digital Technologies in Building Resilient Communities. *Bhumi Plan. Res. J.* **2023**, *10*, 33–40. [CrossRef]
72. Hair, J.F.; Risher, J.J.; Sarstedt, M.; Ringle, C.M. When to Use and How to Report the Results of PLS-SEM. *Eur. Bus. Rev.* **2019**, *31*, 2–24. [CrossRef]
73. Bližnák, K.; Munk, M.; Pilková, A. A Systematic Review of Recent Literature on Data Governance (2017–2023). *IEEE Access* **2024**, *12*, 149875–149888. [CrossRef]
74. Chukwurah, N.; Ige, A.B.; Adebayo, V.I.; Eyieyien, O.G. Frameworks for Effective Data Governance: Best Practices, Challenges, and Implementation Strategies across Industries. *Comput. Sci. IT Res. J.* **2024**, *5*, 1666–1679. [CrossRef]
75. Chen, S.; Wang, C.; Yan, K. Assessing Project Resilience Through Reference Class Forecasting and Radial Basis Function Neural Network. *Appl. Sci.* **2024**, *14*, 10433. [CrossRef]
76. Ruijer, E. Designing and Implementing Data Collaboratives: A Governance Perspective. *Gov. Inf. Q.* **2021**, *38*, 101612. [CrossRef]
77. Lu, W.; Lou, J.; Ababio, B.K.; Zhong, R.Y.; Bao, Z.; Li, X.; Xue, F. Digital Technologies for Construction Sustainability: Status Quo, Challenges, and Future Prospects. *npj Mater. Sustain.* **2024**, *2*, 10. [CrossRef]

78. Das, J.; Fisher, A.C.; Hughey, L.; O'Connor, T.F.; Pai, V.; Soto, C.; Wan, J. Considerations for Big Data Management in Pharmaceutical Manufacturing. *Curr. Opin. Chem. Eng.* **2024**, *46*, 101051. [CrossRef]
79. Burnard, K.; Bhamra, R.; Tsinopoulos, C. Building Organizational Resilience: Four Configurations. *IEEE Trans. Eng. Manag.* **2018**, *65*, 351–362. [CrossRef]

buildings

MDPI

Article

Mapping the Impact of Spontaneous Streetscape Features on Social Sensing in the Old City of Quanzhou, China: Based on Multisource Data and Machine Learning

Keran Li and Yan Lin *

School of Architecture and Design, China University of Mining and Technology, Xuzhou 221116, China; ts21190029p31@cumt.edu.cn
* Correspondence: linyan@cumt.edu.cn

Abstract: Streetscapes in old urban areas are not only an important carrier to show regional economies and city style, but also closely correlate to urban residents' everyday life and the hustle and bustle in which they live. Nevertheless, previous studies have either focused on a few examples with low-throughput surveys or have lacked a specific consideration of spontaneous features in the data-driven explorations. Furthermore, the impact of spontaneous streetscape features on diversified social sensing has rarely been examined. This paper combined the mobile collection of street view images (SVIs) and a machine learning algorithm to calculate eight types of spontaneous streetscape elements and integrated two online platforms (Dianping and Sina Weibo) to map the distribution of economic vitality and social media perception, respectively. Then, through comparing multiple regression models, the impacts of the spontaneous streetscape characteristics on social sensing were revealed. The results include the following two aspects: (1) overall, the spontaneous streetscape features have a certain similarity in the impact on both dimensions of social sensing in Quanzhou, with significant clustering and transitional trends and strong spatial heterogeneity; and (2) specifically, the spontaneous streetscape elements can be divided into three categories, given the differentiated roles of significantly positive, negative, and polarizing impacts on the social sensing results. For example, proper use of open-interface storefronts, ads, and banners is consistent with the common suggestions, while the excessive pursuit of interface diversity and the use of cultural elements may bring an ambiguous effect. This paper provides a transferable analytical framework for mixed and data-driven sensing of streetscape regeneration and can potentially inspire related decisionmakers to adopt a more refined and low-cost approach to enhance urban vitality and sustainability.

Keywords: streetscape; spontaneous space; social sensing; multisource data; street view image; machine learning; old urban area; Quanzhou; China

1. Introduction

As the direct reflection of urban images and dynamics, streetscapes in old urban areas are not only an important carrier to show regional economies, local culture, and a city's style, but also closely correlate to residents' everyday life and the hustle and bustle in which they live [1,2]. All these factors make the street space the key area during urban regeneration. Despite the complicated spatial elements in urban environments, the coarse-grained objects along urban streets, such as land use and land type, building interfaces, roadside greening, urban infrastructure systems, and diversified requirements of street section design, attract more focus from urban planners, designers, and administrators [3,4]. This top-down thinking has its advantages, like making it easier to expand to massive street segments and blocks, and seemingly lower financial and management costs, especially for many developing countries and districts [5]. Meanwhile, the universal pattern comes at the cost of sacrificing individual needs and social–spatial diversity, with the numerous spontaneous spatial elements demonstrating this [6,7]. As urban construction patterns shift from incremental development to stock optimization, more refined and humanized objects of urban streetscapes should still be focused on. Furthermore, considering the complexities of property ownership and the difficulties of regeneration in many old city areas, exploiting the multiple spatial features and the impacts of spontaneous streetscape elements can potentially provide a feasible and humanized approach, which can further advance the 2030 United Nations Sustainable Development Goals [5,8,9].

In addition to the direct aspects, more indirect value from spontaneous streetscape components can be created and amplified through the proliferation of online sharing of diverse subjective perceptions [10]. Whether considering historical continuity or economic value, various storefronts comprise the main parts along the streets in old urban areas and become the usual contributors towards spontaneous streetscapes, such as by changing the types and styles of store interfaces and adopting distinctive signboards and attractive advertisements [11]. With the penetration of digital techniques and mobile devices into everyday urban scenarios, lots of store owners have exploited online–offline operation strategies, creating or approaching popular check-in spots to attract consumers and tourists [12]. During the interaction between different media forms, however, the impact mechanism of the spontaneous streetscape on the perception results is still highly obscure, which is mainly restricted by two key problems, including the digital identification of the related spontaneous spatial elements and the fusion analysis of the mixed perception data.

Regarding these two aspects, prior explorations have provided some references. For the former, traditional attention has mainly been paid to the micro-scale analysis of urban form characteristics and related spatial construction methods [6,13]. Usually by using low-throughput traditional surveys, typical spontaneous spatial elements, including door and window interfaces, awning and sunshade, ground steps, ads and banners, roadside greening, seating facilities and other urban furniture, have been studied [1,13]. Meanwhile, some of the latest works have utilized computer vision algorithms and all kinds of SVIs to identify and map the spatial distribution, although their perspectives dominantly concentrate on the all-factor-segmentation of streetscape photos [14–19]. As the nature of complexity and diversity of the elements is quite obvious, however, one of the potentially vital problems concerns what kinds of spontaneous streetscape elements should be considered, which is also an important factor that limits the regeneration strategies integrating spontaneity into old urban areas. In this respect, therefore, the latter may provide valuable inspirations by incorporating social-sensing-related objective elements into the research vision [20–23]. In other words, two steps, including a direct choice of the spatial elements themselves and another filtering judgement, are combined to target what really matters. As for the specific methods of the perception-related studies, while traditional on-site interviews and questionnaires are still important data sources, some of the latest attempts using public online perception platforms (e.g., Dianping, Yelp, Sina Weibo, Flickr, Panoramio, Facebook, etc.) or crowdsourced datasets, like the well-known Place Pulse datasets, have provided a cluster of approaches to analyze diversified social sensing data, including economic vitality, social media perception, etc. [24–30]. These methods are also believed to require lower labor, time, and financial costs.

Despite these efforts, few studies have focused on the spontaneous streetscape features and the impact on social sensing, especially with the latest techniques and tools. Specifically, there are at least three aspects of research gaps to be bridged and optimized further. First, most of the studies either focus on only a few street segments or lack specific considerations that are suitable for spontaneous streetscape features. While various costs may have restricted the spatial area in the studies with traditional investigation methods, the data-driven explorations with the latest perception techniques are far from sufficient. As an example of the latter, SVIs can deliver rich details of the micro-scale objects in streetscapes, even at the pixel level [31]. However, the mainstream of the application has paid much attention to sky, universal greenery, roads, and building facades, which are relatively coarse-grained streetscape elements compared with the precision level of human visual perception [11]. Accordingly, a fine-grained identification method is needed. Second, studies adopting SVIs often lack consistent categorizations of spontaneous streetscape components, causing ambiguity among different explorations. For example, one of the representative problems is that both public- and private-oriented spatial elements are chosen, such as confusing urban streetlights and public greening with store interfaces and private potted greenery. This is a natural processing if viewed solely from the perspective of visual classification, as in the frequently-used datasets based on autonomous driving scenarios, but there is a fundamental difference between humanized urban regeneration and driving perception. Third, the digital explorations of the impact of spontaneous streetscape features on diverse social sensing are still lacking, which is partly due to the unclear studies of the former problems. On the other hand, most of the related literature has taken the views of pedestrians, riders, and drivers, which can be categorized into the demand end of the street space use. From the perspective of the supply end, numerous ordinary socioeconomic entities, like storefronts, also play a significant role in providing and maintaining a vibrant economic environment and urban streetscape [11]. This is also one of the bases that supports the classical theory of Jane Jacobs [32]. Therefore, revealing the spontaneous streetscape features that are significantly correlated with social sensing can be helpful to deepening our knowledge of the spatial and social value of the street components.

In view of the gaps, this paper can contribute to the existing literature by using multisource data and the latest algorithms to reveal the spatial features of spontaneous streetscapes and the impact on diverse social sensing in an area containing hundreds of street segments. To the best of our knowledge, it is one of the few attempts to integrate a data-driven workflow into streetscape studies from a supply-end perspective. Specifically, there are three main questions in our research, as follows:

(1) How can we combine the latest practices of visual perception methods and the nature of spontaneous streetscape elements?
(2) What are the spatial distribution characteristics of the spontaneous streetscape elements at an intraurban scale?
(3) What are the impacts of the features on diversified social sensing?

In order to tackle the problems, this paper first categorized the spontaneous streetscape elements through a literature review, taking the private-oriented elements into consideration. Taking the Old City in Quanzhou as a case, the study then integrated mobile collection of SVIs and a machine learning algorithm to establish a set of indexes for the analysis of spontaneous streetscape features, followed by mapping the distribution of economic vitality and social media perception from two large open platforms, namely Dianping and Sina Weibo. Finally, the impacts of the spontaneous streetscape characteristics on the sensing dimensions were revealed after adopting and comparing multiple regression models. The main results comprise the following two aspects.

(1) Overall, the spontaneous streetscape features have a certain similarity in the impact on both dimensions of social sensing in Quanzhou, with significant clustering and transitional trends and strong spatial heterogeneity.
(2) Specifically, the spontaneous streetscape elements can be divided into three types, given the differentiated roles of significantly positive, negative, and polarizing impacts on both types of social sensing.

2. Literature Review

2.1. Nature of the Spontaneous Streetscape and the Elements

Extracting the element-level representations along urban streets has a close relationship with the nature of spontaneous streetscapes. As opposed to point or areal space, the linear streets in old urban districts feature flat, dense, and continuous distributions of roadside elements. Storefronts, as important units of urban streetscapes, are usually controlled to avoid obvious concave–convex changes along the roads. Instead, diversified spontaneous spatial elements are preferred by them to create uniqueness [13].

The classification of spontaneous spatial elements along urban streets provides the basis for fine-grained analysis of the composition, distribution, and perceptual characteristics. As a prerequisite, understanding the meaning of spontaneity is of great help [7]. With the etymology of "spontaneity" traced back to the Latin root "sponte", several key features of the concept can be extracted from the interpretations given by the Cambridge Dictionary website, i.e., being natural rather than planned, feeling natural, and internal stimulus [33].

Based on this, spontaneous spatial elements inherently possess a bottom-up and demand-oriented style in urban environments. However, there are still many kinds of elements involved, with diversified taxonomies. For example, some of them totally belong to personal choices, while others are chosen or advised by designers and local officials; some are adopted by storefronts (e.g., store interfaces and awnings), while others are used by temporary and mobile vendors or shared between storefronts and vendors (e.g., stalls and umbrellas); some have relatively large scales, while others are possibly minor changes; some are legal, while others, which encroach upon the street, look disagreeable. All the above are just examples. Actually, it seems impossible to list all the spontaneous spatial elements. Nevertheless, there are still some filter conditions to be chosen for this specific study. First, the research scope along with the selected technological characteristics is a vital approach. On-site observation can reach many seemingly inaccessible places and achieve high precision, but is usually low-throughput, labor-intensive, and limited by a small area. On the other hand, the existing automatic or semi-automatic image recognition techniques can be utilized in a much larger district but probably sacrifice some accuracy [19]. Second, as some spontaneous elements look similar to public items (e.g., signs), the division of public- or private-oriented is a feasible approach to classify spontaneous streetscape elements. This has potential value for urban planning and governance, as making the distinction of ownership of related elements is a fundamental prerequisite to some extent. Third, filtering and focusing on some elements with special value can narrow down the scope of urban regeneration movements. In this regard, this paper attempts to focus on those with potentially significant impacts on socioeconomic sensing.

2.2. Image Identification of Micro-Scale Elements in Street Studies

With the development of digital technologies, visual perception media represented by SVIs as well as the related processing of image identification and classification have been frequently examined in recent years. Given the advantages, micro-scale image identification methods, as well as data, like SVIs, have been utilized in multiple areas, including urban planning and design, geography, and psychology [34]. The data-driven workflow can be used to detect the visual features of spontaneous spatial elements at an (intra)urban scale.

As different meanings apply to varied resolutions for image identification, establishing a realistic connection with urban settings for SVIs requires human knowledge, which acts as a prefabricated impetus. Influenced by the works of early exploration with driving scenario datasets, pursuing all-factor segmentation and identification has predominantly become a trend. As briefly listed in Table 1, the explorations with SVIs and multiple kinds of advanced computer vision algorithms are capable of identifying more than a hundred types of objects, with the research areas varying from (intra)urban to global scales [11,16,21,35–40]. The elements that are more suitable for the spontaneous streetscapes mainly include basic interfaces, commercial, greenery, cultural features, etc. However, most of the selected micro-scale elements of the studies have focused on general scenarios, coarse-grained building components, or public urban furniture. The results are twofold. On the one hand, the universal object identification and segmentation workflows make large-scale detection and comparative studies possible, although the problem of lacking consideration of local characteristics exists to some extent. On the other hand, compared to the relatively more explored streetscape elements from the perspective of the demand end, the one that is related to the supply-end streetscape component, namely ordinary roadside storefronts, has gradually become the weak side and needs to be examined further. Actually, it should be noted that the so-called fine-grained attempt is still a relative concept.

Table 1. A brief summary of the types and algorithms of micro-scale streetscape elements.

Study Area	Image Data	Selected Streetscape Element	Algorithm	Source
56 global cities	Google Street View	Wall, building, road, sidewalk, fence, signboard, path, stairs, door, bench, awning, streetlight, pole, fountain, sculpture, traffic light, stairway, tree, grass, plant, water, waterfall, lake, mountain, rock	PSPNet	[16]
Shenzhen, China	Tencent Street View	Greenspace vegetation (tree, forest, greenbelt, lawn), building, sky, wall, road, pavement	SegNet	[21]
Guangzhou, China	Tencent Street View	Street furniture (fences, streetlights, traffic lights, cameras, windows), pedestrian, bicycle, booth, trade name, signboard, pavement, roadway, motor vehicles	FCN	[35]
Shanghai, China	Baidu Street View	Sky, building, tree, road, pavement, fence, sign, vehicle, pedestrian, bike	SegNet	[36]
Xuzhou, China	Baidu Street View	Signboard, frame door, glass interface, open store, closed store	YOLO	[11]
Atlanta, GA, USA	Google Street View	Building, house, sidewalk, road, car, tree, plant, grass, walk signal, crosswalk, sidewalk, buffer, streetlight	PSPNet	[37]
Seoul, Korea	Naver Street View	Building, tree, plant, sky, road, sidewalk, street furniture	High-resolution networks; Canny Edge	[38]
Seoul, Korea	Google Street View	Vegetation, building, road, sidewalk	Deeplab V3+	[39]
ZG city, China	Baidu Street View	Wall, building, tree, road, window, grass, plant, sidewalk, fence	Faster R-CNN; PSPNet	[40]

Based on the current techniques, a few studies have chosen to exploit some characteristic and flexible applications of SVIs in the area, rather than focusing on excessive all-factor analysis. For example, Liu and Liu (2022) combined manually labeled images of four kinds of street vendors and a deep learning detection framework to map the spatial pattern of the informal sector in Shenzhen, China [41]. Furthermore, they found that the agglomeration of street vendors lay in the proximity of low-level roads and far away from the central business districts. Focusing on bus shelters, Kim et al. (2024) only collected SVIs near the geolocations of bus stops and put forward their bus shelter scores in 20 cities of the United States [42]. Cinnamon and Gaffney (2021) indicated the problems of absence, fragmentation, and obsolescence of commercial SVI platforms, like Google Street View, and explored the notion of do-it-yourself (DIY) street view production with the latest panoramic imaging technologies [43]. Similarly, Williams et al. (2019) also examined DIY sensor technologies and evaluated their performance and potential in urban public spaces [44]. Overall, these explorations can provide inspiration about customized data acquisition and the flexible use of streetscape image data.

2.3. Data-Driven Social Sensing and Its Methods

Using geospatial big data to reveal the distribution and dynamics of socioeconomic elements has become a prevalent branch in geography and urban science studies in the latest decade. As a special academic discourse, however, the concept was first put forward by Yu Liu in 2015 to indicate the counterpart to remote sensing [45]. Taking human beings as units of sensors, social sensing utilizes the collected high-resolution data to reveal the spatiotemporal distribution, dynamics and processes of socioeconomic phenomenon [46,47].

Given the diversity of the involved data sources and methods in social sensing scenarios, related practices are multifaceted, from relatively rarely used records of urban complaint hotlines to the datasets of trajectories, consumption, and social media records [24,47–50]. As examples, the commonly used crowdsourced platforms of Dianping and Sina Weibo not only possess a large number of stable users but also fit well in Chinese urban contexts. Considering the characteristics and user preferences of the applications, the two sources can deliver economic vitality and social perception liveliness, respectively. Still, it should be noted that the ordinary scenarios of the platforms usually agglomerate in popular urban areas to form uneven spatial distribution, which may cause bias in large-scale studies [47]. Therefore, the study area has to consider the influence or focus on the data concentration areas.

Furthermore, as recommended in Liu's book, fusion analysis of SVIs and social sensing data represents a promising trend as well, which is exactly in line with this paper [51]. The similar traits of the point data or point-based data make the integration feasible, whether at an individual level or aggregate level.

3. Materials and Methods

3.1. Study Area

This paper takes the Old City located in Licheng District, Quanzhou City, Fujian Province of China as the study area (see Figure 1). The whole area covers approximately 6.4 square kilometers. There are basically two aspects of consideration for this paper. First, due to the profound historical background and a long history of economic development in Quanzhou, it has both bottom-up commercial traditions and pretty good environments for numerous small and medium businesses along the roads, which can support the development of a spontaneous streetscape. Additionaly, the distinct phenomenon and culture in this Hokkien region is highly consistent with the ordinary experience of on-site investigation. Second, after Quanzhou was successfully listed in the World Cultural Heritage List in 2021, the Old City, as one of the core districts, has attracted a large number of residents and tourists [52]. Particularly, more and more youngsters are visiting and posting their experience through various social media platforms, which caused Quanzhou to be selected as the one of the vibrant cities at the China Vitality City Conference [53]. Given the abundant spontaneous streetscape elements and the relatively concentrated social sensing use, the Old City in Quanzhou can be an appropriate case for the study.

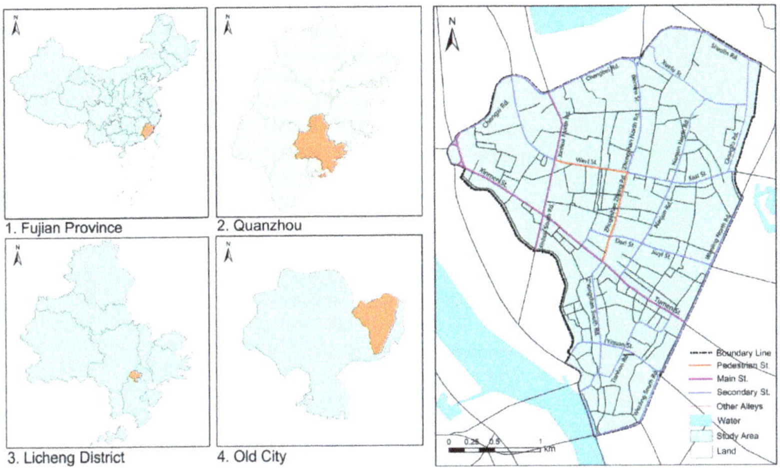

Figure 1. Study area.

3.2. *Data Acquisition and Methods*

As is illustrated in Figure 2, the overall analytical framework comprises three main parts, to study the distribution features of spontaneous streetscape elements, distribution of social sensing results, and impact of spontaneous streetscape features on social sensing, respectively. Among them, the first part has further been divided into three subsectors in the following text, considering the fundamental focus of spontaneous streetscape in this paper. Furthermore, the physical environment orientation can also better support urban regeneration.

3.2.1. Mobile Collection of Street View Images

As the basis of the subsequent analysis, the road network was first downloaded from OpenStreetMap (OSM) and we made some necessary typology corrections compared to Baidu Map. After that, 179 streets and alleys were obtained and divided into street segments based on road intersections. Then, the segments were used to generate sampling points of SVIs with an interval of 20 m in ArcGIS (version 10.6, Esri) [11,54]. In total, 3447 points were attained.

Despite the usual workflow of using the points to gain SVIs, the timeliness of these data from prevailing open platforms may have a significant influence on the results. After checking the Baidu Map and Tencent Map platforms, we found that there is an evident time lag in the provided SVI data products in Quanzhou. For example, the latest data for most of the SVIs from the Baidu Street View platform have remained the record of 2017, with only a few of 2021. Given the significant context of the area's successful inclusion in the World Heritage List in 2021, these data can reflect neither the recent intervention of urban regeneration nor the current streetscape. In fact, the problem of spatiotemporal accuracy has become one of the major limitations on geospatial big data. As a reflection and response, some scholars have put forward the concept of mobile sensing technology [55,56]. As an emerging exploration direction, it can integrate portable sensors to achieve low-cost collection and high temporal granularity while reaching more places [56].

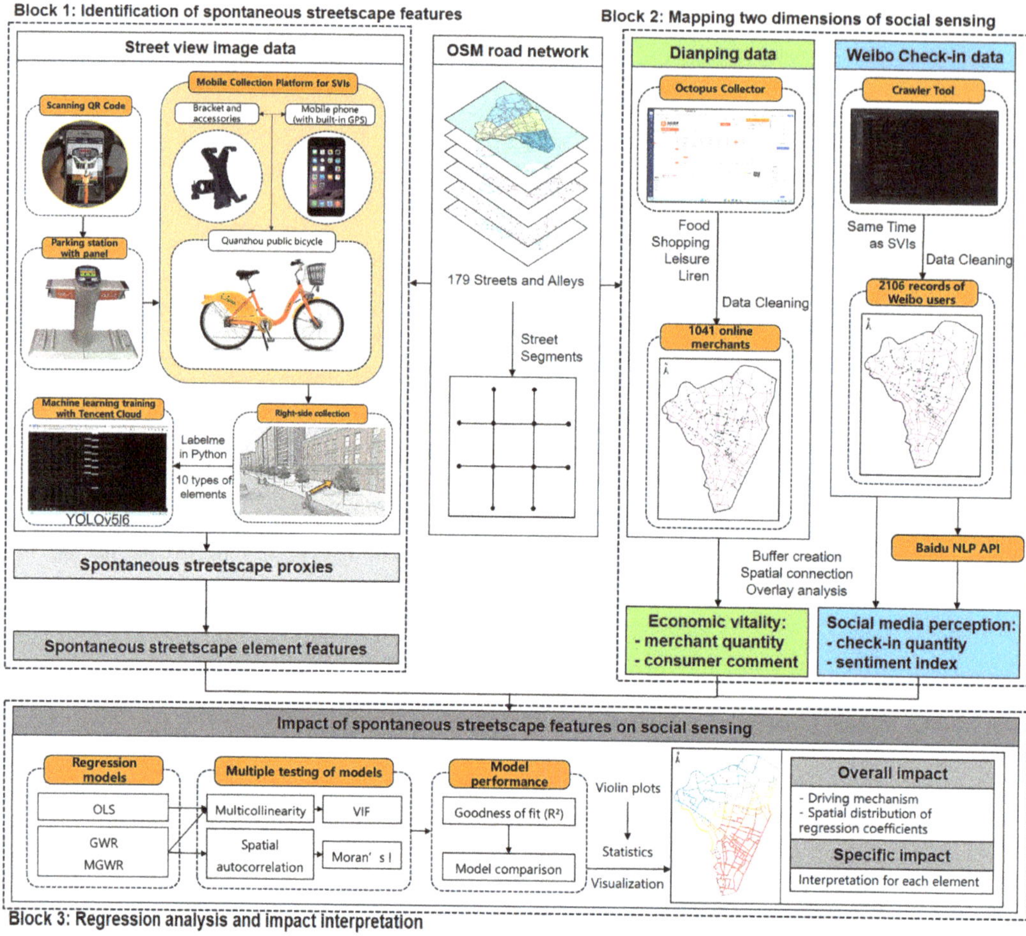

Figure 2. Analytical framework.

Likewise, the SVIs in the study area were finally collected using a mobile and detachable sensing tool, which mainly consists of a public bicycle, a mobile phone with built-in GPS sensors, a bracket, and accessories (see Figure 2). Before on-site collection, a pre-survey was arranged to divide the whole area into several task modules and conduct route planning, which aimed to ensure a balanced workload between task modules and to avoid duplicate routes. Recording the images on the left and right sides of the street segments, the in situ mobile collection took place from the 3rd to 8th of July 2023. During the process, 274 sampling points were deleted due to temporary traffic blockage of urban regeneration and pipeline laying projects. Furthermore, the collection task was arranged between 9:30 and 18:30 each day. This was an essential but passive choice, because the selected period had to be consistent with the normal business hours of the storefronts. Finally, we obtained 6346 SVIs in total, with the image size of 1600 × 1200 pixels. Among them, 3624 images were selected as the training dataset and 910 images were used as the validation dataset.

3.2.2. Classification of Spontaneous Streetscape Elements

As mentioned in Section 2.2, the examined diverse micro-scale streetscape elements mainly focus on some common functions, like basic interfaces, commercial, greenery, and cultural elements. However, the lack of consideration on the divergent public- or private-orientation and fine-grained identification of the storefronts as the supply end has restricted the humanized and complete perception of streetscape element system. According to the filtering conditions in Section 2.1, the spontaneous streetscape elements in the study area were categorized into four types, including basic, commercial, movable, and cultural elements. They could be further divided ten subtypes (see Figure 3). All the classification and definitions of the related elements are listed in Table 2. All the elements were labelled manually using the Labelme software (version 5.1.1) with Python (version 3.9). It should be noted that the selected spontaneous streetscape elements are the representative components of an orderly and typical street scenario and are not intended to include all the spontaneous counterparts in urban streets.

Figure 3. Illustrations of the spontaneous streetscape elements.

Table 2. Classification and related definitions of spontaneous streetscape elements.

Type	Subtype	Label Code	Definition
Basic	Framed door/window	Framedoor	(1) There are clear horizontal and vertical frames inside the outline of doors and windows; (2) an interface mainly used by the storefronts operating indoors, with relatively poor display quality.
	Glass gateway/window	Glassinterface	(1) Glass occupying the main part of the gateways or windows, with no frame inside; (2) an interface mainly used by the storefronts needing to be displayed externally.
	Open entrance	Openentry	(1) Only the outlines of entrances can be seen, with shadows or various goods in them; (2) an interface of the storefronts that require direct and constant outdoor activities.
	Closed entrance	Closedstore	(1) Storefronts that are shut down or not in business; (2) a special component of the streetscape that can reflect urban economies and the experience of street vitality.
Commercial	Store signboard	Signboard	(1) A reflection of street economic vitality and the number of storefronts, whether in business or not; (2) usually fixed above entrance interfaces.
	Banner/ad	Banner	(1) Larger quantity and smaller scale compared to store signboards; (2) generally used at the height of the visual level or lower.
Movable	Sunshade/awning	Sunshade	(1) It mainly serves to shield against wind, rain, and sunlight, and also indicates a certain outdoor usage range; (2) usually located at a higher position than doors and windows and close to the vertical interfaces of storefronts.
	Outdoor umbrella	Umbrella	(1) Commonly used by street vendors, or storefronts to set up stalls and outdoor businesses along the street.
	Potted plant	Pottedplant	(1) A kind of private greenery for store decoration and embellishment of natural elements; (2) usually placed at entrances.
Cultural	Wall decoration painting/sculpture	Decoration	(1) Generally, hard materials, such as bricks and stones, are used with clear outlines; (2) a cultural element usually placed at vertical interfaces or independently near entrances.

3.2.3. Machine Learning Model Training

As summarized in Section 2.2, many kinds of machine learning algorithms have been explored, among which YOLOv5 (version 6.0) was selected for further model training. YOLO is an example of a one-stage object detection algorithm. Compared to two-stage algorithms, like R-CNN, YOLO takes a detection task as the solution of regression models and considers global features simultaneously during the detection process, resulting in a significant speed improvement while paying a small cost in terms of accuracy [57]. After subsequent iterative optimization, YOLOv5 has been widely used due to its maturity and stability, with further improvements arriving in the latest 6.0 version. Given the different characteristics of the involved n, s, m, l, and x models, the l model was chosen to obtain a balance of data volume, accuracy, and speed.

To avoid the impact of long-term training on other parallel workflow, we built a server environment based on Tencent Cloud for model training (see Figure 4a,b). Specifically, the adopted type of instance is a GPU Computing GN7, with eight cores and 32GB configuration and the Windows Server 2019 system. The GPU model is an NVIDIA Tesla T4 16GB. During the training process, the batch-size, epoch, and IOU (intersection over union) were set to 16, 300, and 0.5, respectively, with other hyperparameters as the default. As the selection of IOU complies with the normal range (e.g., from 0.5 to 0.75), the setting of the batch-size and epoch has to consider many factors, like hardware resources, calculation period, and the model convergence speed. Given the selected model, we have began a pre-experiment with a smaller epoch (e.g., 50) and batch-size (e.g., 8) and adjusted them to obtain a relatively suitable value for the allowed GPU memory capacity. To avoid the influence of the randomly selected images during a training process, we further divided the training dataset into four sub-datasets. Each sub-dataset was taken as a test set and the others were used for training in turn, along with the validation dataset. The average performance of the models was then calculated to provide a relatively robust result in this study.

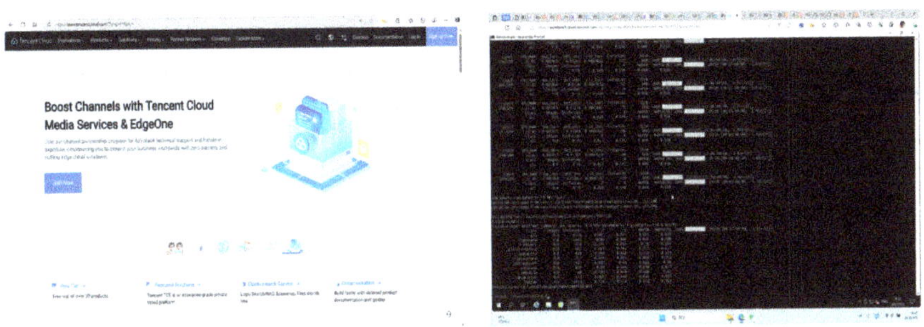

(a) Interface of Tencent Cloud

(b) Model training process

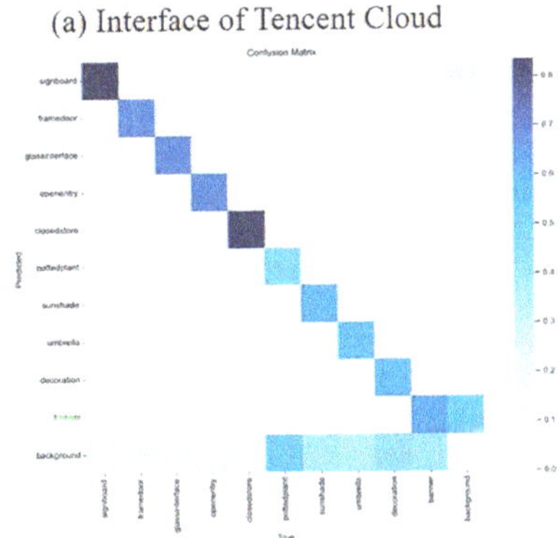

Type	Average	Min	Max	Std.
signboard	0.837	0.828	0.852	0.0105
framedoor	0.751	0.738	0.763	0.0121
glassinterface	0.708	0.704	0.713	0.0039
openentry	0.726	0.715	0.752	0.0177
closedstore	0.864	0.848	0.872	0.0110
pottedplant	0.486	0.467	0.523	0.0262
sunshade	0.640	0.599	0.658	0.0277
umbrella	0.571	0.548	0.604	0.0271
decoration	0.603	0.520	0.643	0.0524
banner	0.592	0.482	0.635	0.0734
All	**0.678**	**0.655**	**0.694**	**0.0163**

(c) Confusion matrix

(d) Model performance statistics

Figure 4. Machine learning model training processes and performance: (**a**) interface of Tencent Cloud; (**b**) model training process; (**c**) confusion matrix; (**d**) model performance statistics.

The performance of the model can be seen in Figure 4c,d. As is illustrated in the table, the mAP (mean average precision) value of the 10 subtypes reached 0.678, showing a good comprehensive performance. The small standard deviations for the specific subtypes and the all-factor approach reflects a relatively stable performance. In terms of the specific elements, the type of basic interfaces has high and relatively balanced results, with most of the mAP values over 0.75. While commercial signboards achieve the second highest record of all, the difference value between them and banners and ads is 0.217. This may correlate with the usual lower precision of identification of small-scale objects, as well as an amplified effect in the mixed recognition of multiscale elements. The performance of the movable and cultural elements mainly lies between 0.6 and 0.7, slightly lower than the average. However, the mAP of the potted plants has the lowest value of 0.484. From the confusion matrix, we can see that all the basic interface elements and the commercial signboards have higher values of the 10 subtypes, while the relatively low values of the others are basically due to the distraction of the image's background. In other words, the model has a lower recognition ability to correctly distinguish the small-scale elements from the background of the sample images at a higher probability. The phenomenon can be caused due to the quality of the collected images, imbalanced quantities of different tags, etc. Although directly feeding machine learning models with SVIs is a common process, the lack of multiscale training inputs and specific control of the tag numbers of different elements may potentially affect the model performance in our study. Therefore, it should be noted that the small-scale elements with a relatively lower identification precision are likely to be overestimated in the related spatial analysis.

Given the obviously smaller number of tags of the umbrella element identified by YOLO, the distribution of it is likely to be zero in most of the street segments, resulting an in incomparable quantity compared to the other elements. Consequently, the other nine kinds of elements, as well as the interface diversity of storefronts, were adopted in the following text.

3.2.4. Social Sensing Data Processing

Two social sensing platforms, namely Dianping (https://www.dianping.com/, accessed on 12 October 2023) and Sina Weibo, were utilized to indicate online economic vitality and social media perception, respectively. These data platforms have demonstrated satisfactory support value due to their high user quantity and stickiness [27,58,59]. Based on the latest reports released by the related companies, youngsters and females are the main users of these platforms, which may reflect the characteristics of the population.

For the online economic vitality, we used Octopus Collector in July 2023 to obtain four types of businesses in the Old City area of Quanzhou, including food, shopping, leisure and entertainment, and liren (namely beauty issues). The collected information includes merchant type, location, per capita consumption, rating score, number of reviews, etc. Among these tags, the per capita consumption is affected by multiple factors, such as merchant type and individual consumption level, and the rating scores from different businesses have both differentiated standards and missing values of a whole type (e.g., liren). However, the number of reviews can avoid the abovementioned drawbacks. Thus, this factor was adopted in this study. In fact, the four types also contain different numbers of lower-level tags. For example, the type of food comprises 40 subtypes from the website, including buffet, bread/drinks, Western cuisine, coffee shops, barbecue skewers, snack noodles, and various local and foreign cuisines. These types are selected due to their preferences for offline operation and the experience during the on-site investigation in the Old City area. After inspecting the duplicated and missing items, 1041 pieces of store information were obtained in total (see Figure 5). Among them, however, the type of food accounts for over 80% of all the businesses. Given the obviously uneven quantities of the four types of points of interest (POIs), the related data were combined together to calculate the Dianping merchant quantities and consumer comments, which can, thus, indicate the online economic vitality in this paper.

For the social media perception, to maintain the temporal consistency with other data, the Weibo check-in data of the study area were crawled, with their timeframe limited to July 2023. There were 2106 pieces of data as a results. These records possess approximately 10 types of tags, such as user identity, spatiotemporal information, text content, and online interaction data. In this paper, we focus on both the locations and contents of the Weibo texts to calculate check-in quantity and sentiment index. Although the former was easy to obtain, the latter was calculated using the Natural Language Processing (NLP) API of the Baidu Cloud platform (https://cloud.baidu.com/product/nlp_basic, accessed on 10 January 2024). Given the rich corpus resources and the multiple deep learning training from Baidu, the sentiment analysis tool can obtain a high overall precision and a good generalization ability when faced with diversified text content. By requesting the URL, it can return a probability value of a positive/negative judgement, a confidence level (with the interval from 0 to 1), and a predicted sentimental classification code (e.g., 0, 1, or 2, for negative, neutral, and positive results, respectively). As the sum of the positive and negative probabilities of a post text is 1, a higher positive probability is likely to indicate a better feeling. However, only considering the predicted emotional judgement result from the deep learning model may overlook the impact of the confidence level. Accordingly, this paper multiplied the positive probability of a text with its confidence level to obtain the sentiment index for a Weibo record. The average of all the nearby records was then deemed as the corresponding sentiment index of a street unit.

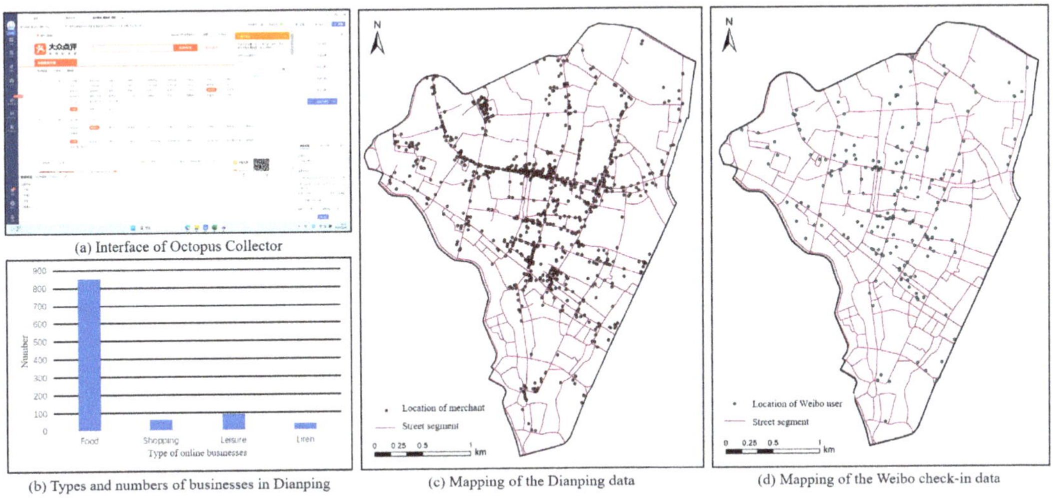

Figure 5. Data collection and description of social sensing: (**a**) interface of Octopus Collector (a Chinese website); (**b**) types and numbers of businesses in Dianping; (**c**) mapping of the Dianping data; (**d**) mapping of the Weibo check-in data.

Taking street segments as units, their centerlines were adopted to expand by 50 m on both sides to form buffer zones. The distance of about 50 m has become a common practice for the visual recognition of SVIs [60] and is suitable for the street scale in the Old City of Quanzhou. The quantities of online businesses and Weibo check-in records were then used to conduct overlay statistics and visualization in ArcGIS.

3.2.5. Multiple Regression Models

Based on the handling of the multisource data, a set of independent and dependent variables was established (see Table 3). Then, multiple regression models were used to reveal the impact of spontaneous streetscape features on social sensing, including the ordinary linear squares (OLS) model, the geographically weighted regression (GWR) model, and the multiscale geographically weighted regression (MGWR) model. Among them, the OLS model is used as a benchmark model reference because its insufficient consideration of spatial heterogeneity often leads to poor model performance. In this regard, the GWR and MGWR models highlight spatial heterogeneity by automatically selecting the optimal bandwidth and calculating the respective bandwidths for different variables to avoid this weakness. Meanwhile, the workflow of the MGWR model usually makes the results more robust [61]. The software for the GWR and MGWR calculation can be downloaded directly from the website of the School of Geographical Sciences and Urban Planning at Arizona State University (https://sgsup.asu.edu/form/windows-sparc-mgwr (accessed on 15 May 2023)). The selection of the spatial kernel function, the model function, and the convergence threshold of the model were all consistent with the settings from one of our prior papers [11]. Nonetheless, to avoid a potential interrupted calculation due to a zero problem for certain elements, we replaced 0 with 1×10^{-5}, a very small value without a significant impact on the result [62]. However, it should be noted that this minor adjustment is relatively reasonable, partly because we have focused on the common elements of streetscape features and removed the umbrella category with a very low quantity.

After the construction of the models, the best one can be obtained by comparing the values of the goodness of fit. It can then be used for further analysis. To avoid the influence of different variable dimensions and value ranges on calculation robustness, all proxies were calculated to the [0,1] interval according to min–max normalization method.

Table 3. Proxies and calculation methods of the spontaneous streetscape features and social sensing.

Variables	Abbr.	Description	Calculation Method
Dependent variables			
Online economic vitality index	DMQ	The quantity of the Dianping merchants in a unit	Buffer creation and the spatial connection tool in overlay analysis in ArcGIS
	DCC	The sum of the Dianping consumer comments in a unit	Buffer creation and the spatial connection tool in overlay analysis in ArcGIS
Social media perception index	WCQ	The quantity of the Weibo check-in records in a unit	Buffer creation and the spatial connection tool in overlay analysis in ArcGIS
	WSI	The average sentiment index of Weibo texts in a unit	$WSI_i = \sum_{WCQ_i}(conf_{iw} \times prob\_p_{iw})/WCQ_i$ where $conf_{iw}$ and $prob\_p_{iw}$ are obtained from the Baidu Natural Language Processing (NLP) API, and stand for the confidence level and positive sentiment probability of the w-th Weibo text in the i-th unit, respectively.
Independent variables			
Basic interface usage	SFD	The sum of the framed doors and windows in a unit	$SFD_i = FD_{li} + FD_{ri}$ where FD_{li} and FD_{ri} are the sums of the framed doors and windows of the i-th unit from the L- and R-side, respectively.
	SGI	The sum of the glass gateways and windows in a unit	$SGI_i = GI_{li} + GI_{ri}$ where GI_{li} and GI_{ri} are the sums of the glass gateways and windows of the i-th unit from the L- and R-side, respectively.
	SOE	The sum of the open entrances in a unit	$SOE_i = OE_{li} + OE_{ri}$ where OE_{li} and OE_{ri} are the sums of the open entrances of the i-th unit from the L- and R-side, respectively.
	SCS	The sum of the closed entrances in a unit	$SCS_i = CS_{li} + CS_{ri}$ where CS_{li} and CS_{ri} are the sums of the closed entrances of the i-th unit from the L- and R-side, respectively.

Table 3. *Cont.*

Variables	Abbr.	Description	Calculation Method
Interface diversity index	SDI	The interface diversity degree of the storefronts in business calculated by Shannon entropy	$SDI_i = -\sum(P_j \times \ln P_j)$ where Pj is the proportion of the j-th type of the interfaces from operating storefronts within the i-th unit.
Commercial element usage	SSB	The sum of the store signboards in a unit	$SSB_i = SB_{li} + SB_{ri}$ where SB_{li} and SB_{ri} are the sums of the store signboards of the i-th unit from the L- and R-side, respectively.
	SBA	The sum of the banners and ads in a unit	$SBA_i = BA_{li} + BA_{ri}$ where BA_{li} and BA_{ri} are the sums of the banners and ads of the i-th unit from the L- and R-side, respectively.
Movable element usage	SSS	The sum of the sunshades and awnings in a unit	$SSS_i = SS_{li} + SS_{ri}$ where SS_{li} and SS_{ri} are the sums of the sunshades and awnings of the i-th unit from the L- and R-side, respectively.
	SPP	The sum of the potted plants in a unit	$SPP_i = PP_{li} + PP_{ri}$ where PP_{li} and PP_{ri} are the sums of the potted plants of the i-th unit from the L- and R-side, respectively.
Cultural element usage	SDE	The sum of the wall decoration paintings and sculptures in a unit	$SDE_i = DE_{li} + DE_{ri}$ where DE_{li} and DE_{ri} are the sums of the cultural decoration paintings and sculptures of the i-th unit from the L- and R-side, respectively.

4. Results

4.1. Features of Spontaneous Streetscape Element Distribution

The spatial distribution of the spontaneous streetscape elemental proxies is shown in Figure 6. Overall, various spontaneous streetscape elements exhibit a similar characteristic of concentrated distribution along the high-level roads of the Old City in Quanzhou, although the differentiated distribution results of different elements vary.

Specifically, the spatial characteristics of the basic interface categories are relatively consistent. The high-value areas of the three types of interfaces from operating storefronts and the low-value area of the closed store entrances are concentrated along the main roads north of Xinmen Street and Tumen Street. The interface diversity index exhibits a high value in most of the street segments, indicating a diversified trend of the streetscapes. As for the commercial type of elements, the spatial distributions of store signboards and banners and advertisements show a high similarity. While the commercial elements mainly agglomerate along the main and secondary roads at the edge of and inside the Old City, some of them also exist in some residential streets and alleys. In terms of the movable type, the spontaneous greenery from potted plants basically shows a higher quantity in the north part of Xinmen Street and Tumen Street, including most of the high-level roads and some famous alleys in the central and western parts. Meanwhile, the southern area maintains a stable and sparse distribution of the spontaneous plants. The sunshades and awnings possess both low level of spatial distribution continuity and high degree of fragmentation. They mainly concentrate along the roads near the northwest and southeast boundaries of the Old City. On the other hand, the scattered clusters in some important roads located at the center parts of the study area, such as Xinmen Street, Tumen Street, Zhongshan Middle Road, and Zhongshan South Road, are probably due to the continuous arcade forms of these street segments. In terms of the cultural elements, they are generally located in a V-shape along the two lines of Chengxi Road—Chengbei Road and Xinmen Street–Tumen Street. Actually, the area of their distribution is rather limited, which presents a high degree of aggregation in the Old City.

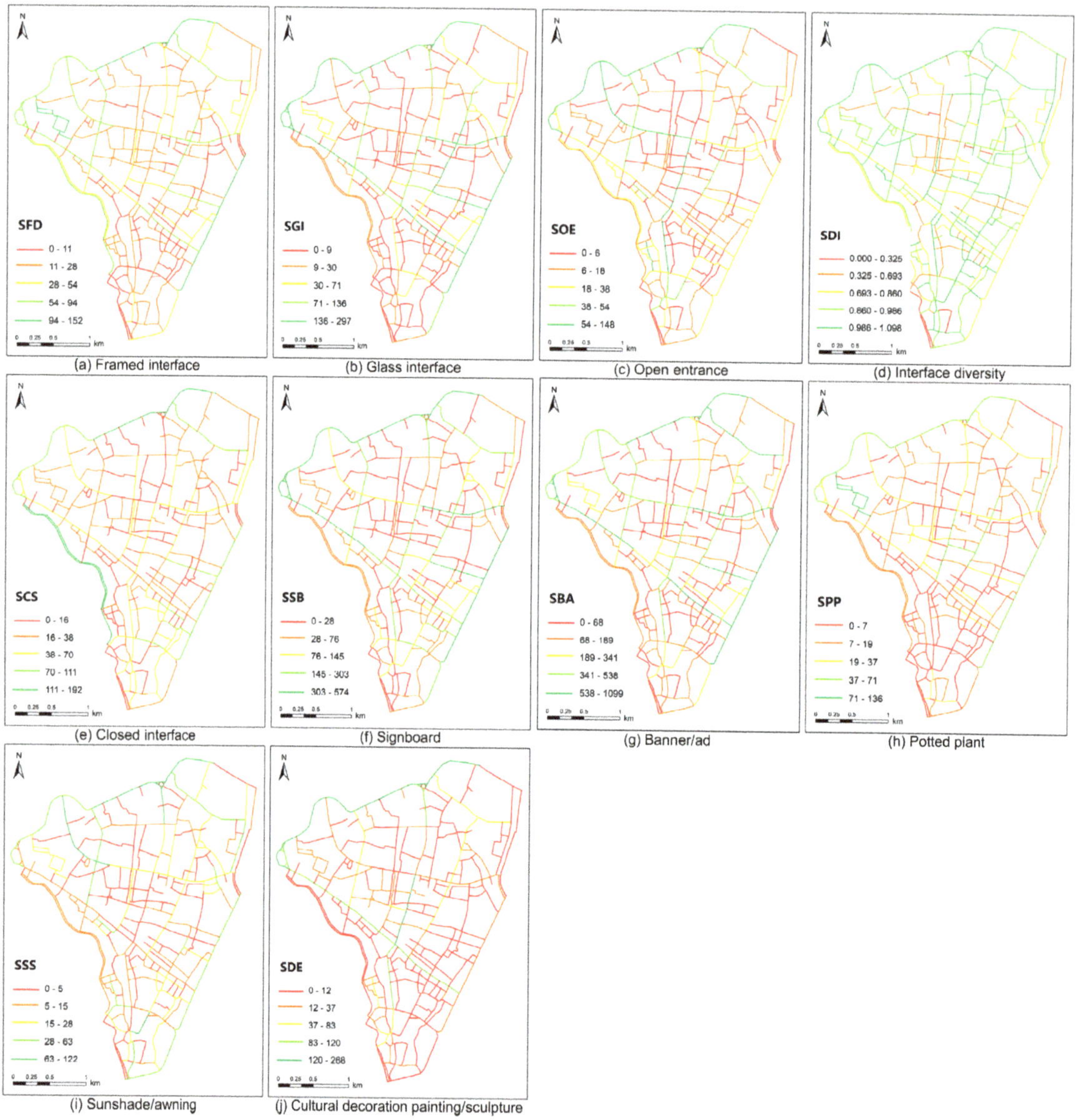

Figure 6. Distribution of the spontaneous streetscape elements in the Old City of Quanzhou.

4.2. Spatial Distribution of Social Sensing

Two dimensions of social sensing, namely economic vitality and social media perception indexes, have been illustrated in Figure 7. In terms of the overall characteristics, the high-value areas of them agglomerate approximately at the central district of the Old City.

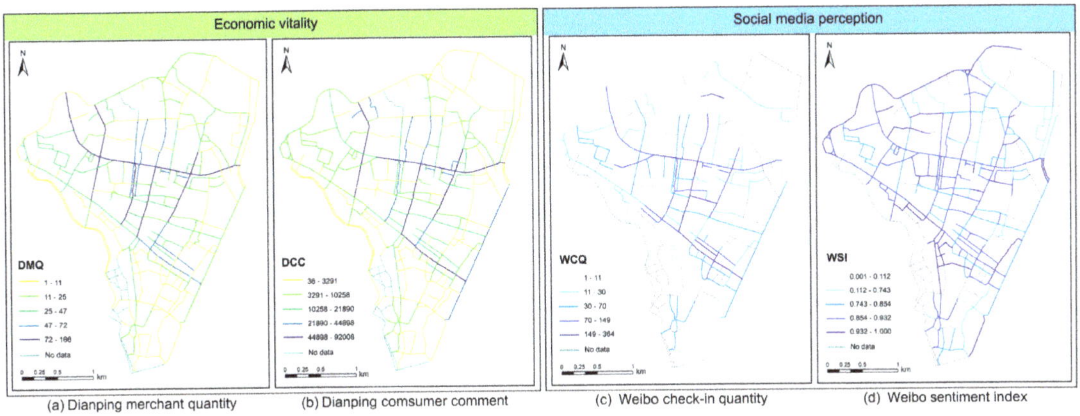

Figure 7. Distribution of economic vitality and social media perception.

Specifically, for the economic vitality, DMQ generally exhibits a fishbone-shaped distribution pattern. Along the axis of West Street and East Street, several connected roads, including Xinhua Road, Haogouqian, Zhongshan Middle Road, and Nanjun Road with a west–east sequence, form a core area of a merchant gathering that extends north to Aiguo Road and south to Xinmen Street Tumen Street. In addition, other street segments, such as the back alleys near Tumen Street, the surrounding roads of 1916 Creative Industry Park, and the northern segments of Zhongshan South Road, have exhibited a moderate level of DMQ. As some of them have already become well-known check-in spots for residents and tourists, potential future regeneration can stimulate a larger area of online economic vitality to connect with the existing patches. Similarly, the pattern of DCC shows a fishbone-shaped and agglomerated distribution pattern in the central area, with a shorter axis of West Street and East Street. This reflects a more compact core consumption area, compared to the distribution of DMQ. Considering the huge distinction in the number of comments from different segments, the attractiveness to consumers of the core area is overwhelming.

From the perspective of social media perception, WCQ extends in a zigzag pattern to the surrounding road network. In the area between West Street–East Street and Xinmen Street–Tumen Street, some contiguous high-value clusters appear. In addition, moderate levels of check-in numbers are found in the street segments near Tianhou Road, Zhongshan South Road, Wanshou Road, and north of West Street in the study area. As opposed to the limited concentration area of check-in records, streets with high sentiment indexes show a wide distribution across the Old City. Specifically, those with a sentiment index of over 0.8 account for 77.4% of all the effective measured street segments, which indicates an overall positive perception in the whole area. Beyond the central part, the segments with high indexes also exist in the boundaries of the City, like Chengxi Road and Chengbei Road, which may reflect a differential motivation or social media attraction between them.

4.3. *Impact of the Spontaneous Streetscape Features on Social Sensing*

4.3.1. Performance Comparison of Multiple Regression Models

Given the features of linear regression models, the variance inflation factors (VIFs) of all the variables are usually adopted to test the multicollinearity. While most of the VIFs of the variables are lower than 10 [63], the VIF value of the store signboard type is not. As such, it was excluded from further analysis. The OLS model was established, with the performance summarized in Table 4. Although the explanatory power of the independent variable for changes in the dependent variable of economic vitality is higher than that of social media perception, the overall performance of the OLS models still remains to be improved.

Table 4. Performance of the OLS model.

Social Sensing Index	R^2	Adjusted R^2	Durbin-Watson Value
DMQ	0.439	0.406	1.976
DCC	0.463	0.431	2.119
WCQ	0.255	0.201	2.076
WSI	0.183	0.124	1.949

Moran's I index was then used in ArcGIS to test the spatial autocorrelation of the dependent variables [63]. The calculation results showed that at a confidence level of 0.001, the Moran's I indexes for the four types of social sensing proxies were 0.242, 0.167, 0.213 and 0.275, respectively, indicating a positive spatial autocorrelation for the social sensing indexes that can be further analyzed through the GWR and MGWR models. The model performance of the two kinds of models can be seen in Table 5. The overall model performance of the economic vitality is higher than that of the social media perception, showing a stronger spatial heterogeneity of the latter. Furthermore, the adjusted R^2 values of the spatial heterogeneity models are higher than those of the OLS model, with the best performance of the MGWR model. Therefore, the next explorative analyses are based on that.

Table 5. Performance of the GWR and MGWR models.

Social Sensing Index	GWR			MGWR		
	R^2	Adjusted R^2	AIC	R^2	Adjusted R^2	AIC
DMQ	0.786	0.709	300.633	0.800	0.744	274.549
DCC	0.819	0.746	281.044	0.812	0.755	268.980
WCQ	0.377	0.271	357.567	0.502	0.374	343.209
WSI	0.245	0.134	378.724	0.364	0.242	364.564

4.3.2. Impact of Spontaneous Streetscape Features on Economic Vitality

The local parameter estimates of the spontaneous streetscape on the economic vitality and the significance performance are shown in Figures 8 and 9. Overall, it can be interpreted through the spatial distribution and variation of regression coefficients, and the driving mechanisms of each element. First, spontaneous streetscape elements generally exhibit clustering and transitional features, reflecting the strong spatial heterogeneity of the impact. Second, the influence tendency between spontaneous spatial elements and the economic vitality can be divided into three categories, except for the insignificant estimates for some elements. The first category, such as open entrances, banners and ads, can exert a positive effect, while the second type, like closed store interfaces, sunshades and awnings, shows significantly negative correlations. Furthermore, some elements may exhibit polarizing characteristics to some extent, which means that their coefficients are significantly positive in some street segments and significantly negative in others. Still, the regression coefficients of varied spontaneous streetscape features have some differences for the two proxies of economic vitality; thus, the two sets of results are interpreted separately, as follows.

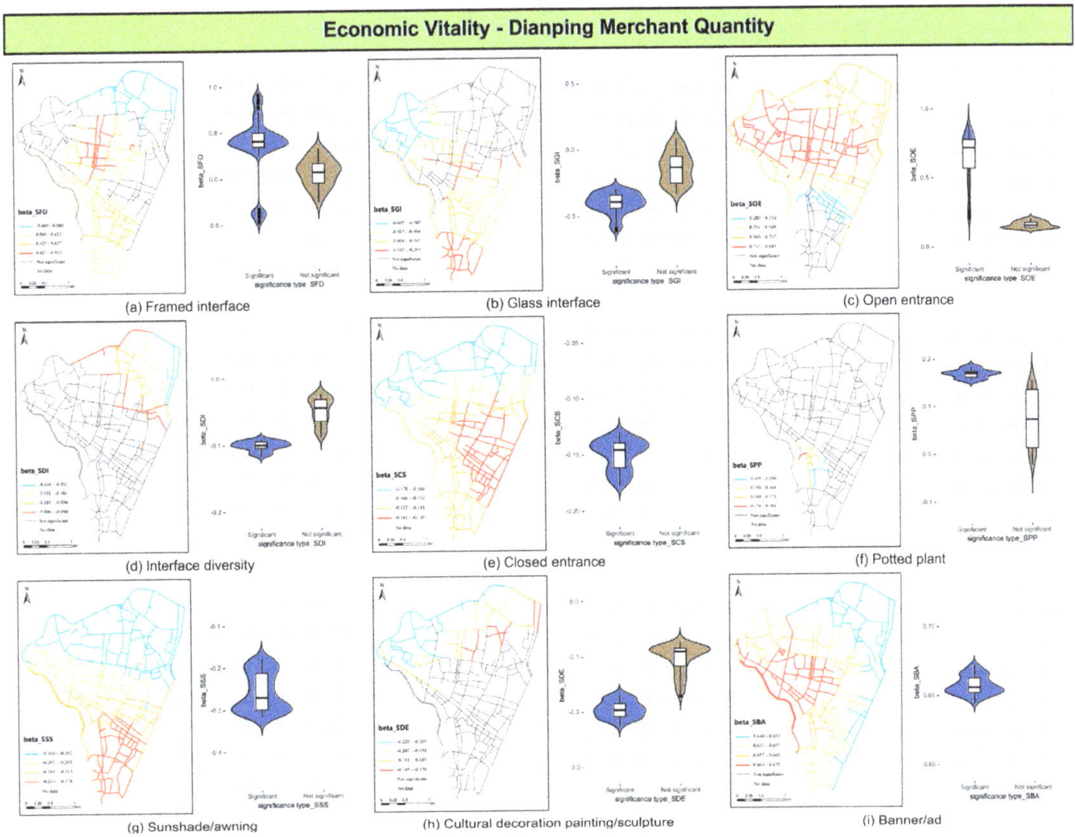

Figure 8. Regression coefficient distribution with Dianping merchant quantity as the dependent variable.

Dianping Merchant Quantity as the Proxy of Economic Vitality

Specifically, the above three significant categories exist for all the independent variable when taking DMQ as the dependent variable. In terms of the elements that exert a positive influence, SOE, SPP, and SBA are included. Among them, the commercial type of elements, namely SBA, has the smallest standard deviation, as the upper and lower quartiles are 0.66 and 0.65, respectively. Its relatively stable positive driving force reflects the broad adaptability in promoting the perception of online economic vitality. On the other hand, the open entrance interface (i.e., SOE) has demonstrated larger variations, with the upper and lower quartiles of 0.77 and 0.58. While it possesses relatively low positive intensity in most of the area near Tumen Street, a higher driving intensity is reflected in the streets and alleys along the axis of West Street and East Street. Furthermore, the proxy of SPP also shows a significantly positive impact in limited segments along the north part of Zhognshan South Road.

The spontaneous streetscape elements that are negatively correlated with DMQ can be ranked from high to low according to the absolute value of the significantly median regression coefficient in parentheses as follows: SGI (−0.391), SSS (−0.270), SDE (−0.197), SCS (−0.146), and SDI (−0.099). Among them, the glass interface shows a high and negative driving force in the northwest of the Old City, and the diffusion gradients weaken towards the southeast. Similarly, while sunshades and awnings have a lower driving intensity in the streets located south of Tumen Street, they exhibit a relatively higher negative driving intensity in the large areas north of West Street and East Street. The intensity attenuation of the cultural elements from northwest to southeast is mainly influenced by the relatively concentrated distribution of the cultural elements. Although the node-like cultural buildings, such as Linzhang Gate and Chaotian Gate along the north outline of the Old City, act as good indications of the city's image and urban history, the adoption range is too limited and non-systematic. The negative impact presented by the closed interface is consistent with our general understanding, which can be partly reflected by the significant results in all the effectively measured street segments. Its spatial variation mainly indicates the sensitivity of different regions' economic vitality to the negative interfaces. The negative driving intensity of the interface diversity index in the northeast of the Old City is high, which may be related to the relatively limited daily consumption scenarios dominated by the surrounding large-scale communities.

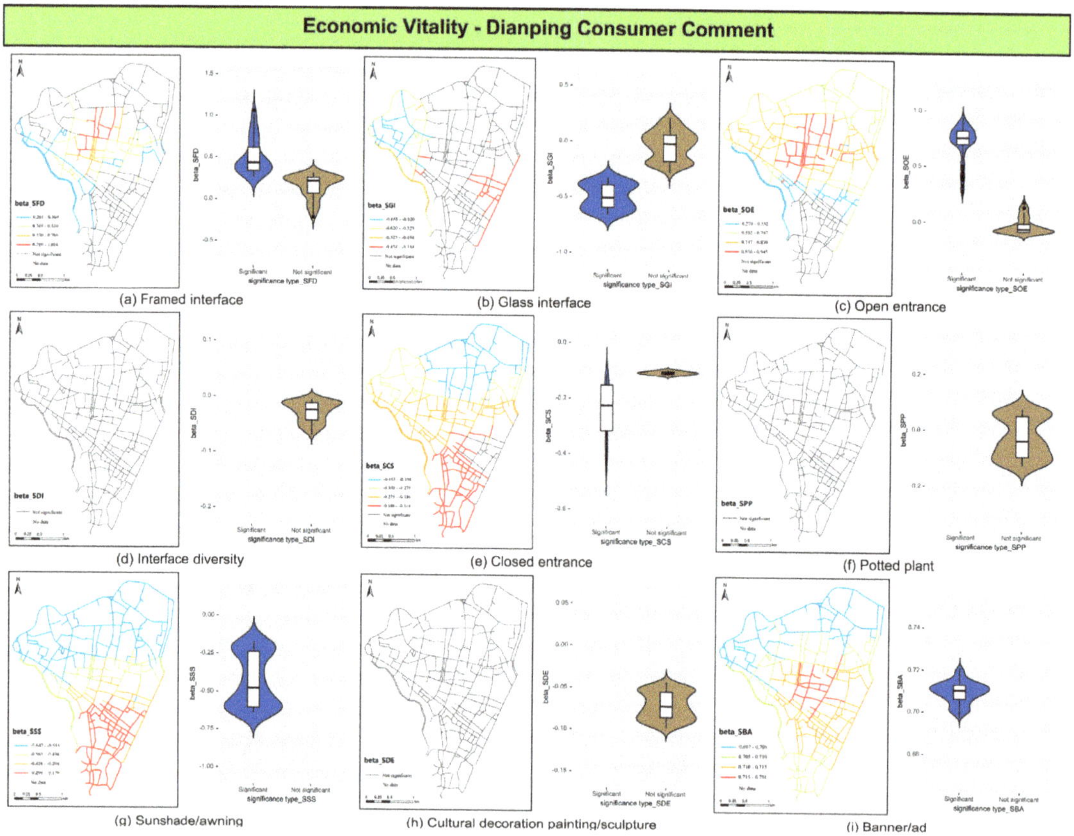

Figure 9. Regression coefficient distribution with Dianping consumer comments as the dependent variable.

The interfaces of framed doors and windows exhibit a polarizing correlation across the entire area, demonstrating both a strong positive impact in the adjacent alleys of the eastern section of West Street, and a strong negative driving intensity in the northern area of the Old City. This can also be reflected in the two concentration areas in the violin plot. With a relatively traditional and historical charm, the pedestrian street blocks represented by the eastern section of West Street have gathered a large number of residents and tourists, which attracts more categories of small businesses, like catering, drinking, and specialty snacks.

Therefore, the businesses that tend to operate indoors and use the framed interfaces can not only provide a distinctive place for people to relax and enjoy the traditional flavor, but also avoid the adverse effects on the historical style of the streetscape. The small measures can help to obtain a balance between the multifaceted needs of related stakeholders. On the other hand, as the northern region has relatively scattered flow of people and the greatly changing street environment, the framed interfaces in this area are negatively correlated to DMQ.

Dianping Consumer Comments as the Proxy of Economic Vitality

When taking DCC as the dependent variable, categories of streetscape elements with significantly positive and negative impact exist. Furthermore, there are also some proxies with totally insignificant performance in the study area, such as SDI, SPP, and SDE. Compared to the results of DMQ, the difference may indicate the varied influence between the supply and consumer ends, although they were both chosen as proxies of economic vitality.

Specifically, the proxies of SFD, SOE, and SBA all present positive correlations with DCC. In the whole area, the regression coefficients of SBA are close to 0.71. The street network in the central part of the Old City has a slightly higher value, with a small interval between 0.697 and 0.721. All these indicate that the commercial elements, including advertisements and banners, have shown a stable and significantly positive impact on DCC, as with the effect on DMQ. On the other hand, the value ranges of the estimates of SFD and SOE are similar and share a larger variance. The high-value areas of them are in the central area, near the eastern part of West Street. Although both of them have some insignificant segments, the significant group of SOE reflects an obviously higher driving intensity than the insignificant one. Furthermore, the median value of SOE (0.751) is also higher than that of SFD (0.431).

The proxies of SCS, SSS, and SGI all appear to have a negative influence on DCC. Among them, the former two basically show a significantly negative effect from every street segment. The distributions of their estimates also share a similar structure of spatial gradient, with a stronger negative performance in the northern part and a smaller absolute value in the southern area located south of Tumen Street. In terms of their numerical performance, the median of SCS (−0.226) is larger than that of SSS (−0.478), indicating a weaker negative effect on DCC. Actually, as the negative association of SCS is understandable, the variation of its esti-mates may be influenced by the types of storefronts, with different sensitivities. This might partly explain the smaller absolute values in the southern part of the Old City. Wholesalers of daily necessities and hardware processing industry are the dominant store types in that area; both are demand-oriented and not usually deemed to be businesses that have strong dependence on the conditions of nearby economic entities. Similarly, the effect from the SSS also has the relatively smaller absolute value in the southern area. In addition, unlike SCS and SSS, which have a wide distribution of significant results, SGI only shows a scattered pattern in limited segments of the eastern and western areas. The higher negative driving force in the eastern area may correlate with the insufficient demand for the display consumption characteristics, as usually realized through large and transparent glass interfaces.

4.3.3. Impact of Spontaneous Streetscape Features on Social Media Perception

The impact of the spontaneous streetscape elements on the social media perception is illustrated in Figures 10 and 11. Likewise, the spatial distribution of regression coefficients of the independent factors basically exhibits significant clustering and transitional characteristics, as well as strong spatial heterogeneity. From the perspective of driving mechanisms, the spontaneous streetscape elements with differentiated impacts on the social media perception can also be divided into the following three types: the first category, like SPP, shows a clear positive influence; the second one, like SCS, can exert negative impacts; the third one presents polarizing features. Besides, although the regression estimates of some elements are insignificant, their statistical characteristics are illustrated in the violin plots as well. Similarly, interpretations based on different dependent variables are explained separately.

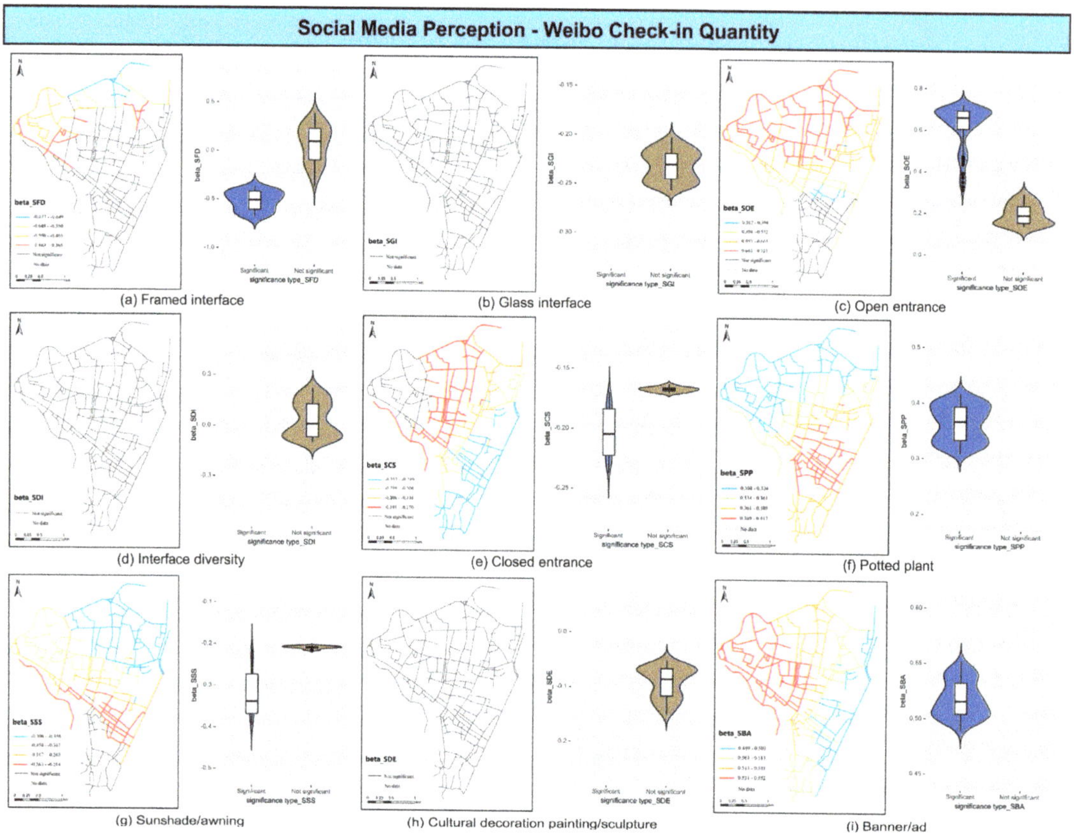

Figure 10. Regression coefficient distribution with Weibo check-in quantity as the dependent variable.

Weibo Check-In Quantity as a Proxy of Social Media Perception

In the case of WCQ as the dependent variable, the elements of SOE, SPP, and SBA show significantly positive correlation in almost every street segment of the Old City. Among them, the regression coefficients of open entrance interfaces, banners, and advertisements have a smaller range of values ([0.308,0.417] for SPP, and [0.489,0.552] for SBA), and the median of SBA (0.516) is higher than that of SPP (0.367), indicating a stronger positive impact. On the other hand, as the estimate of SOE has a greater range ([0.312,0.721]), its median (0.661) is the largest of the three. In terms of the distribution, the high-value segments of SPP mainly agglomerate near Tumen Street, while the western part of the Old City shows a stronger driving force of SBA. With a larger spatial gradient, SOE has a higher performance in the central pedestrian streets and the northwestern area.

The elements with significantly negative impacts can be sorted by the absolute value of the median regression coefficients as SFD (−0.508), SSS (−0.338), and SCS (−0.205). Among them, the segments with significant estimates of SFD are rather limited and are mainly located in the northern part of the Old City. The proxy of SSS exhibits a high negative driving intensity in a large area located north of West Street and East Street. This may be not only related to the lack of distinctive check-in attractions near the high-level traffic streets, such as Chengbei Road, but also to people's short-term consumption experience and "lagging" check-in behavioral habits in the main streets represented by West Street. On the other hand, the significantly negative driving intensity of SCS decreases from Wenling South Road and Wenling north Road to the central area. Considering that the southeastern boundary is near several clusters of commercial centers and squares, the gradient may be correlated with the massive influx of pedestrians and the requested continuous social places.

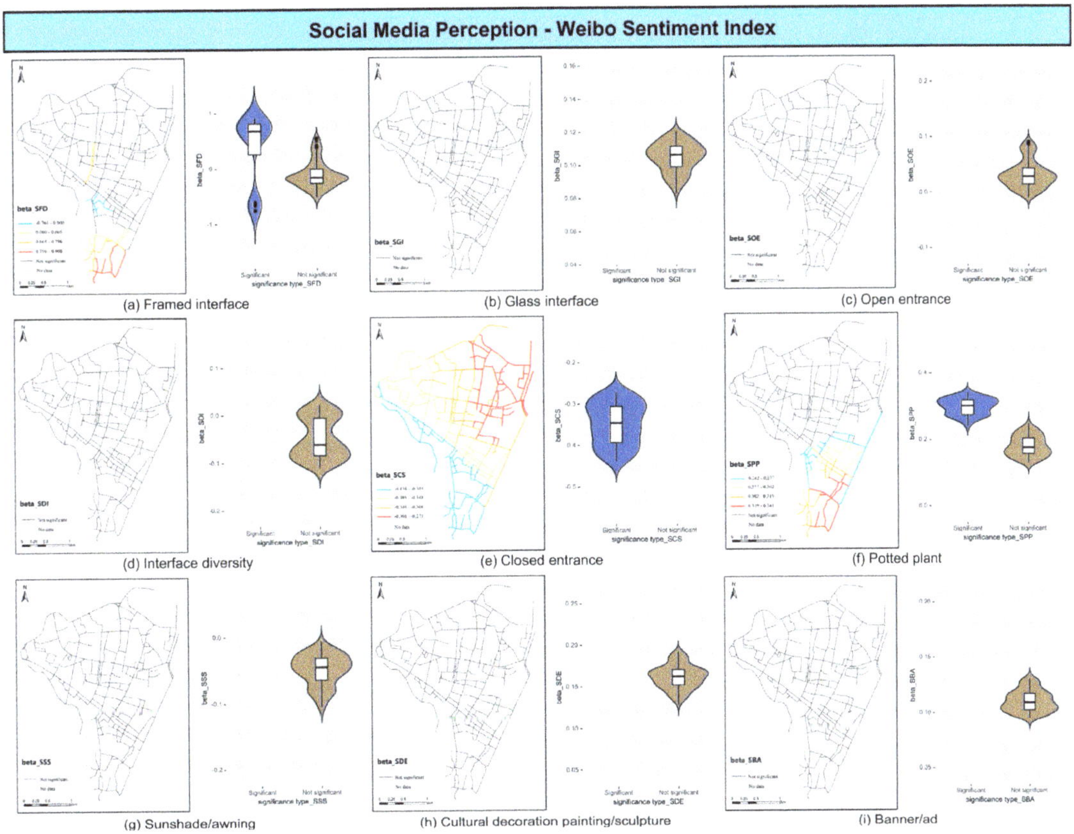

Figure 11. Regression coefficient distribution with the Weibo sentiment index as the dependent variable.

Weibo Sentiment Index as the Proxy of Social Media Perception

When taking WSI as the dependent variable, only three independent proxies show significant regression estimates. The other six categories of elements, including SGI, SOE, SDI, SSS, SDE, and SBA, present insignificant results over the whole area.

The proxy of SPP has shown a significantly positive impact on WSI in the southern and southeastern parts, with a range between 0.242 and 0.341 and a median of 0.301. Like the barbell-shaped violin plot in Figure 8, SFD also shows a polarizing correlation. On the one hand, it exhibits a significantly negative driving force in several segments, such as Yanzhi Alley and Yanggong Alley, with coefficients of about −0.7. On the other hand, it presents significantly positive estimates in Haogouqian Alley and in the southern area of the Old City, with a median of about 0.7. As opposed to the limited distribution of effective results of the other two proxies, SCS has a significantly negative impact on WSI in nearly every segment. With the upper and lower quartiles of −0.31 and −0.39, respectively, it shows a relatively stronger negative driving force in the areas located south of Xinmen Street and Tumen Street, and a weaker negative intensity in the northeastern part of the Old City.

5. Discussion and Conclusions

As a complement to top-down streetscape creation, the spontaneous streetscape is closely related to individual subjective initiatives. If these approaches are integrated together, we can strive for a better balance between societal efficiency and order and humanized tenderness and vitality. To fully understand how the spontaneous streetscape influences diversified social sensing, both a fine-grained identification of the spontaneous streetscape elements from the perspective of the supply end of streetscapes, and a data fusion framework at a relatively large scale is required. However, previous studies have either focused on few examples with low-throughput surveys or lacked specific consideration of spontaneous features in the data-driven explorations. Furthermore, the impact of spontaneous streetscape features on diversified social sensing has rarely been examined. Given the context, this paper first categorized the spontaneous streetscape elements, by taking the existing micro-scale visual identification method, functions and ownership of the spontaneous elements into consideration. Taking the Old City in Quanzhou as a case, the study then adopted the mobile collection of SVIs and a machine learning algorithm to establish a set of indexes of spontaneous streetscape features, followed by mapping the distribution of economic vitality and social media perception from two large open platforms, namely Dianping and Sina Weibo. Finally, the impacts of the spontaneous streetscape features on both dimensions of social sensing were revealed using multiple regression models. The main results comprise the following two aspects:

(1) Overall, the spontaneous streetscape features have a certain similarity in terms of their impact on both dimensions of social sensing in Quanzhou, indicating significant clustering and transitional trends and strong spatial heterogeneity.
(2) Specifically, the spontaneous streetscape elements can be divided into three categories, given their differentiated roles of significantly positive, negative, and polarizing impacts on the social sensing results.

The idea of using computer vision techniques to identify and map the spontaneous streetscape elements at an (intra)urban scale is one of our main contributions. To the best of our knowledge, this is the first paper to conduct such explorations. Compared to the other related literature on urban street studies, an important difference is to distinguish micro-scale spatial elements from their spontaneous counterparts in the study. Although many scholars have exploited multifaceted applications of SVIs to reveal some vital implications concerning refined physical environments, socioeconomic development, and urban perception [30,64,65], the universal framework is primarily used as a tool for all-factor identification scenarios. However, it should be noted that so-called visual precision is still a relative concept under the existing technological context. Although it has been proven to work reasonably well for micro-scale streetscape components, like streetlights, poles, signboards, etc., the ownership details of these elements are complex, and, as such, they can hardly be used to support urban regeneration practices [16,35]. This is particularly true with the transformation from large-scale incremental construction to stock optimization. In this regard, therefore, we explored a feasible approach to visually identify the spontaneous streetscape-related elements that are usually involved in a localized area and used by roadside storefronts. The workflow is not limited to the coarse classification of "building" and "sign" types, nor just store interfaces, as we have examined previously [11]. Drawing on more elements from the spontaneous spatial studies can make this field of study more systematic.

As the main aim of the paper, we further revealed the impact of spontaneous streetscape features on two dimensions of social sensing, namely economic vitality and social media perception. On the one hand, as we investigated the storefronts as the supply end of streetscapes, exploring the correlation with economic vitality is both a natural association and a beneficial attempt to enrich the interpretation of urban vitality. In Jane Jacobs' discourse about Manhattan, the seemingly common retail sector and catering businesses that were under decay were closely correlated with the depression of street life [32]. A similar viewpoint about the significance and vitality component of various individual-owned shops can be found in Maas' doctoral thesis [66]. Despite these classic discussions, however, investigating the economic vitality from the supply end of streetscape components is still scarcely conducted. Using the latest social sensing data from third-party platforms, like Dianping, our methodology is consistent with the related existing research in terms of data sources [58,67]. Besides, we also utilized another dimension of social sensing, namely social media perception, to explore the correlation with spontaneous streetscape features. Similarly, both types of data were also adopted by Long and Huang (2019) to indicate economic vitality [59]. Given the initial aim and common use scenarios of Sina Weibo, we consider whether social media data are a direct source for large-scale economic assessment. In addition, scholars have found an obvious agglomeration tendency with social media data, such as in limited tourist attractions and scenic spots [47]. Therefore, we decided to use social media perception to demonstrate a kind of online popularity and vitality for the statistical unit of street segments.

Apart from the theoretical discussions, the results of this study can perhaps provide some inspiration for urban street planning and design practices. For example, due to the positive impacts on social sensing, areas for open entrance interfaces, spontaneous potted plants, and commercial elements can be purposefully reserved and added to storefronts. This is also consistent with some proposals on vibrant street construction in some guidelines of Chinese cities, like the Shanghai Street Design Guidelines [68]. However, there are still some differences. Under the target of comfortable activities, for instance, while the movable elements including sunshades and awnings can create more grey space and bring a richer offline spatial experience, there may be a reduction in the marginal benefits for some popular streets and alleys driven by online social sensing, and a significantly negative impact is even likely to appear. Besides, our results also reveal that there is probably a significantly negative or insignificant impact of the interface diversity index and the use of cultural elements on social sensing. We do believe in the significance of the proposals, such as diverse streetscape and highlighting regional and cultural symbols, as general guidelines of street design. However, planners and designers should be more cautious, especially from an (intra)urban scale. In addition, if a street is mainly dominated by online users, the disconnection between offline experience and online sensing may be different, and the influence should be considered carefully by urban planners and designers. With the flourishing development of agglomerated online retail sectors in the digital era, the corresponding streetscapes with the abovementioned transformed demand cannot be ignored.

The study also has some limitations that should be highlighted. First, the classification of streetscape elements in the study does not include all the related spontaneous spatial counterparts. Through on-site surveys and observation, for example, some behaviors causing changes in the material conditions and microtopography (e.g., steps) are the result of spontaneous spatial practices in some prior literature. They are excluded from our classification mainly because they are both seldom and difficult to consider in the mainstream computer vision techniques based on SVIs. However, further related spontaneous elements can be added in the future. Second, although the social sensing dimensions have been taken into consideration, the related exploitation was not very deep, due to us placing more focus on spontaneous streetscape digital identification and interpretation. Given the rich tags and high spatiotemporal resolution of the social sensing data, more detailed dimensions can be further explored.

In addition to the further expansion of spontaneous elements and multifaceted dimensions of social sensing, we also attempt to highlight some future directions from the perspective of mobile collection and data fusion. First, portable and low-cost sensors can help to perceive the pulses of diverse socio-spatial aspects. As many scholars have noticed, despite the limitations of large commercial platforms in terms of data availability, quality, and timeliness, humans as sensors can collect the requested data wherever they walk, ride, or drive. To gather numerous lively samples of bottom-up practices, like spontaneous space surveys, this approach could be quite useful. As the elements selected in the study basically belong to the common streetscape components, future explorations can be conducted in the urban street space of more regions. Second, from the perspective of data-aided planning and design practices, we also suggest that future studies should explore a flexible framework to integrate the multisource data and their specific implications for socioeconomic sensing. If the results in the socioeconomic dimension are different from or contradictory to the other, this can be useful, especially for urban planners, designers, and decisionmakers.

Author Contributions: Conceptualization, K.L.; methodology, K.L.; software, K.L.; validation, K.L.; formal analysis, K.L.; investigation, K.L.; resources, K.L.; data curation, K.L.; writing—original draft preparation, K.L.; writing—review and editing, K.L. and Y.L.; visualization, K.L.; supervision, Y.L.; project administration, K.L. and Y.L.; funding acquisition, Y.L. All authors have read and agreed to the published version of the manuscript.

Funding: This research was funded by the National Natural Science Foundation of China, grant number 52308040.

Data Availability Statement: The data presented in this study are unavailable due to privacy restrictions.

Conflicts of Interest: The authors declare no conflicts of interest.

References

1. Shang, W.; He, C.; Lv, S. Sustainable Renewal of Spontaneous Spatial Characteristics of a Historical–Cultural District: A Case Study of Tanhualin, Wuhan, China. *Sustainability* **2023**, *15*, 2038. [CrossRef]
2. Zhang, W.; Xin, J. Green Spaces and the Spontaneous Renewal of Historic Neighborhoods: A Case Study of Beijing's Dashilar Community. *Sustainability* **2023**, *15*, 16566. [CrossRef]
3. Hartmann, T.; Jehling, M. From diversity to justice-Unraveling pluralistic rationalities in urban design. *Cities* **2019**, *91*, 58–63. [CrossRef]
4. Baharoglu, D.; Leitmann, J. Coping Strategies for Infrastructure: How Turkey's Spontaneous Settlements Operate in the Absence of Formal Rules. *Habitat Int.* **1998**, *22*, 115–135. [CrossRef]
5. Itma, M.; Monna, S. Sustainable Housing Design Potentials in Palestine: A Focus on the Spontaneous Urban Design Form. *J. Urban Plan. Dev.* **2024**, *150*, 04023050. [CrossRef]
6. Li, J.; Wang, F.; Wuzhati, S.; Wen, B. Urban or village residents? A case study of the spontaneous space transformation of the forced upstairs farmers' community in Beijing. *Habitat Int.* **2016**, *56*, 136–146. [CrossRef]
7. Cozzolino, S. On the spontaneous beauty of cities: Neither design nor chaos. *Urban Des. Int.* **2022**, *27*, 43–52. [CrossRef]
8. Wang, F.; Liu, X.; Zhang, Y. Spatial landscape transformation of Beijing compounds under residents' willingness. *Habitat Int.* **2016**, *55*, 167–179. [CrossRef]
9. Shen, J.; Lin, L. State-sponsored and spontaneous urbanization in Fujian province of China, 1982–2010. *Cities* **2017**, *60*, 420–427. [CrossRef]
10. Zhou, L.; Wang, T. Social media: A new vehicle for city marketing in China. *Cities* **2014**, *37*, 27–32. [CrossRef]

11. Li, K.; Lin, Y. Exploring the Correlation between Streetscape and Economic Vitality Using Machine Learning: A Case Study in the Old Urban District of Xuzhou, China. *ISPRS Int. J. Geo-Inf.* **2023**, *12*, 267. [CrossRef]
12. Ali Qalati, S.; Li, W.; Ahmed, N.; Ali Mirani, M.; Khan, A. Examining the Factors Affecting SME Performance: The Mediating Role of Social Media Adoption. *Sustainability* **2021**, *13*, 75. [CrossRef]
13. Zhang, Z.; Wang, X.; Zhu, G.; Zhang, W.; Chen, L.; Fang, K.; Xie, Y.; Shen, M.; Furuya, N. A micro-scale study on the spontaneous spatial improvement of in-between spaces in Chinese traditional districts considering the relationship between modifications and encroachment. *J. Asian Archit. Build.* **2023**, *22*, 783–801. [CrossRef]
14. Jeon, J.; Woo, A. Deep learning analysis of street panorama images to evaluate the streetscape walkability of neighborhoods for subsidized families in Seoul, Korea. *Landsc. Urban Plan.* **2023**, *230*, 104631. [CrossRef]
15. Chen, S.; Biljecki, F. Automatic assessment of public open spaces using street view imagery. *Cities* **2023**, *137*, 104329. [CrossRef]
16. Larkin, A.; Gu, X.; Chen, L.; Hystad, P. Predicting perceptions of the built environment using GIS, satellite and street view image approaches. *Landsc. Urban Plan.* **2021**, *216*, 104257. [CrossRef]
17. Koo, B.W.; Guhathakurta, S.; Botchwey, N.; Hipp, A. Can good microscale pedestrian streetscapes enhance the benefits of macroscale accessible urban form? An automated audit approach using Google street view images. *Landsc. Urban Plan.* **2023**, *237*, 104816. [CrossRef]
18. Gong, Z.; Ma, Q.; Kan, C.; Qi, Q. Classifying Street Spaces with Street View Images for a Spatial Indicator of Urban Functions. *Sustainability* **2019**, *11*, 6424. [CrossRef]
19. Huang, Y.; Zhang, F.; Gao, Y.; Tu, W.; Duarte, F.; Ratti, C.; Guo, D.; Liu, Y. Comprehensive urban space representation with varying numbers of street-level images. *Comput. Environ. Urban Syst.* **2023**, *106*, 102043. [CrossRef]
20. He, J.; Zhang, J.; Yao, Y.; Li, X. Extracting human perceptions from street view images for better assessing urban renewal potential. *Cities* **2023**, *134*, 104189. [CrossRef]
21. Ma, X.; Ma, C.; Wu, C.; Xi, Y.; Yang, R.; Peng, N.; Zhang, C.; Ren, F. Measuring human perceptions of streetscapes to better inform urban renewal: A perspective of scene semantic parsing. *Cities* **2021**, *110*, 103086. [CrossRef]
22. Rui, J. Measuring streetscape perceptions from driveways and sidewalks to inform pedestrian-oriented street renewal in Düsseldorf. *Cities* **2023**, *141*, 104472. [CrossRef]
23. Lynch, K. *The Image of the City*; The MIT Press: Cambridge, MA, USA, 1960.
24. Dunkel, A. Visualizing the perceived environment using crowdsourced photo geodata. *Landsc. Urban Plan.* **2015**, *142*, 173–186. [CrossRef]
25. Huang, J.; Obracht-Prondzynska, H.; Kamrowska-Zaluska, D.; Sun, Y.; Li, L. The image of the City on social media: A comparative study using "Big Data" and "Small Data" methods in the Tri-City Region in Poland. *Landsc. Urban Plan.* **2021**, *206*, 103977. [CrossRef]
26. Zahnow, R.; Verrier, J.; Hames, S.; Corcoran, J. Mapping and measuring neighbourhood social media groups. The case of facebook. *Appl. Geogr.* **2024**, *172*, 103415. [CrossRef]
27. Jia, J.; Zhang, X.; Zhang, W. Between place attachment and urban planning in Jinan: Does environmental quality affect human perception in a developing country context? *Land Use Pol.* **2025**, *148*, 107384. [CrossRef]
28. Kim, S.; Woo, A. Streetscape and business survival: Examining the impact of walkable environments on the survival of restaurant businesses in commercial areas based on street view images. *J. Transp. Geogr.* **2022**, *105*, 103480. [CrossRef]
29. Zhang, A.; Li, W.; Wu, J.; Lin, J.; Chu, J.; Xia, C. How can the urban landscape affect urban vitality at the street block level? A case study of 15 metropolises in China. *Environ. Plan. B Urban Anal. City Sci.* **2021**, *48*, 1245–1262. [CrossRef]
30. Zhang, F.; Zhou, B.; Liu, L.; Liu, Y.; Fung, H.H.; Lin, H.; Ratti, C. Measuring human perceptions of a large-scale urban region using machine learning. *Landsc. Urban Plan.* **2018**, *180*, 148–160. [CrossRef]
31. Danish, M.; Labib, S.M.; Ricker, B.; Helbich, M. A citizen science toolkit to collect human perceptions of urban environments using open street view images. *Comput. Environ. Urban Syst.* **2025**, *116*, 102207. [CrossRef]
32. Jacobs, J. *The Death and Life of Great American Cities*; Vintage Book Company: New York, NY, USA, 1961.
33. Cambridge Dictionary. Available online: https://dictionary.cambridge.org/dictionary/english/spontaneity (accessed on 19 March 2025).
34. Yuan, Y.; Wang, R.; Niu, T.; Liu, Y. Using street view images and a geographical detector to understand how street-level built environment is associated with urban poverty: A case study in Guangzhou. *Appl. Geogr.* **2023**, *156*, 102980. [CrossRef]
35. Wang, X.; Liu, Y.; Yao, Y.; Zhou, S.; Zhu, Q.; Liu, M.; Helbich, M. Adolescents' environmental perceptions mediate associations between streetscape environments and active school travel. *Transp. Res. Part D Transp. Environ.* **2023**, *114*, 103549. [CrossRef]
36. Ye, Y.; Zeng, W.; Shen, Q.; Zhang, X.; Lu, Y. The visual quality of streets: A human-centred continuous measurement based on machine learning algorithms and street view images. *Environ. Plan. B Urban Anal. City Sci.* **2019**, *46*, 1439–1457. [CrossRef]
37. Koo, B.W.; Hwang, U.; Guhathakurta, S. Streetscapes as part of servicescapes: Can walkable streetscapes make local businesses more attractive? *Comput. Environ. Urban Syst.* **2023**, *106*, 102030. [CrossRef]

38. Lee, J.; Kim, D.; Park, J. A Machine Learning and Computer Vision Study of the Environmental Characteristics of Streetscapes That Affect Pedestrian Satisfaction. *Sustainability* **2022**, *14*, 5730. [CrossRef]
39. Woo, A.; Han, J.; Shin, H.; Lee, S. Economic benefits of urban streetscapes: Analyzing the interrelationships between visual street environments and single-family property values in Seoul, Korea. *Appl. Geogr.* **2024**, *163*, 103182. [CrossRef]
40. Yue, H.; Liu, L.; Xu, C.; Song, G.; Chen, J.; He, L.; Duan, L. Investigating the diurnal effects of on-street population and streetscape physical environment on street theft crime: A machine learning and negative binomial regression approach using street view images. *Appl. Geogr.* **2024**, *163*, 103194. [CrossRef]
41. Liu, Y.; Liu, Y. Detecting the city-scale spatial pattern of the urban informal sector by using the street view images: A street vendor massive investigation case. *Cities* **2022**, *131*, 103959. [CrossRef]
42. Kim, J.; Park, J.; Lee, J.; Jang, K.M. Examining the socio-spatial patterns of bus shelters with deep learning analysis of street-view images: A case study of 20 cities in the U.S. *Cities* **2024**, *132*, 104852. [CrossRef]
43. Cinnamon, J.; Gaffney, A. Do-It-Yourself Street Views and the Urban Imaginary of Google Street View. *J. Urban Technol.* **2021**, *29*, 95–116. [CrossRef]
44. Williams, S.; Ahn, C.; Gunc, H.; Ozgirin, E.; Pearce, M.; Xiong, Z. Evaluating sensors for the measurement of public life: A future in image processing. *Environ. Plan. B Urban Anal. City Sci.* **2019**, *46*, 1534–1548. [CrossRef]
45. Liu, Y.; Liu, X.; Gao, S.; Gong, L.; Kang, C.; Zhi, Y.; Chi, G.; Shi, L. Social Sensing: A New Approach to Understanding Our Socioeconomic Environments. *Ann. Assoc. Am. Geogr.* **2015**, *105*, 512–530. [CrossRef]
46. Liao, C.; Brown, D.; Fei, D.; Long, X.; Chen, D.; Che, S. Big data-enabled social sensing in spatial analysis: Potentials and pitfalls. *Trans. GIS* **2018**, *22*, 1351–1371. [CrossRef]
47. Liu, Y. *Geographical Big Data and Social Sensing*; Science Press: Beijing, China, 2023.
48. Peng, X.; Li, Y.; Si, Y.; Xu, L.; Liu, X.; Li, D.; Liu, Y. A social sensing approach for everyday urban problem-handling with the 12345-complaint hotline data. *Comput. Environ. Urban Syst.* **2022**, *94*, 101790. [CrossRef]
49. Li, C.; Huang, Y.; Shen, Y.; Xu, L. Spatiotemporal patterns and mechanisms of street vending from the social sensing perspective: A comparison between law-enforcement reported and residents complain events. *Cities* **2022**, *124*, 103597. [CrossRef]
50. Ghermandi, A.; Sinclair, M. Passive crowdsourcing of social media in environmental research: A systematic map. *Glob. Environ. Change* **2019**, *55*, 36–47. [CrossRef]
51. Zhang, Y.; Kwan, M.-P.; Yu, B.; Liu, Y.; Song, L.; Chen, N. Quantitative Identification of Mixed Urban Functions: A Probabilistic Approach Based on Physical and Social Sensing Data. *Trans. GIS* **2025**, *29*, e13272. [CrossRef]
52. UNESCO World Heritage Center. Quanzhou: Emporium of the World in Song-Yuan China. Available online: https://whc.unesco.org/en/list/1561 (accessed on 22 March 2025).
53. Quanzhou Municipal People's Government. Available online: https://www.quanzhou.gov.cn/zfb/xxgk/zfxxgkzl/qzdt/qzyw/202112/t20211205_2664783.htm (accessed on 22 March 2025).
54. Li, Y.; Yabuki, N.; Fukuda, T. Exploring the association between street built environment and street vitality using deep learning methods. *Sustain. Cities Soc.* **2022**, *79*, 103656. [CrossRef]
55. Jiang, J.; Wang, J.; Yao, H.; Zhang, Z.; Xu, Z. Revealing the mobile sensing powers of semi-random and deterministic "drive-by" vehicle fleets. *Trans. GIS* **2024**, *28*, 2–22. [CrossRef]
56. Birenboim, A.; Helbich, M.; Kwan, M.-P. Advances in portable sensing for urban environments: Understanding cities from a mobility perspective. *Comput. Environ. Urban Syst.* **2021**, *88*, 101650. [CrossRef]
57. Redmon, J.; Divvala, S.; Girshick, R.; Farhadi, A. You Only Look Once: Unified, Real-Time Object Detection. In Proceedings of the 2016 IEEE Conference on Computer Vision and Pattern Recognition (CVPR), Las Vegas, NV, USA, 27–30 June 2016. [CrossRef]
58. Ye, Y.; Li, D.; Liu, X. How block density and typology affect urban vitality: An exploratory analysis in Shenzhen. *China. Urban Geogr.* **2018**, *39*, 631–652. [CrossRef]
59. Long, Y.; Huang, C. Does block size matter? The impact of urban design on economic vitality for Chinese cities. *Environ. Plan. B Urban Anal. City Sci.* **2019**, *46*, 406–422. [CrossRef]
60. Yang, D.; Wang, X.; Han, R. Nonlinear and Synergistic Effects of the Built Environment on Street Vitality: The Case of Shenyang. *Urban Plan. Forum* **2023**, *5*, 93–1002. [CrossRef]
61. Fotheringham, A.S.; Yang, W.; Kang, W. Multiscale Geographically Weighted Regression (MGWR). *Ann. Assoc. Am. Geogr.* **2017**, *107*, 1247–1265. [CrossRef]
62. Xia, C.; Zhang, A.; Yeh, A.G.O. The Varying Relationships between Multidimensional Urban Form and Urban Vitality in Chinese Megacities: Insights from a Comparative Analysis. *Ann. Assoc. Am. Geogr.* **2022**, *112*, 141–166. [CrossRef]
63. Chen, Y.; Yu, B.; Shu, B.; Yang, L.; Wang, R. Exploring the spatiotemporal patterns and correlates of urban vitality: Temporal and spatial heterogeneity. *Sustain. Cities Soc.* **2023**, *91*, 104440. [CrossRef]
64. Hou, C.; Zhang, F.; Kang, Y.; Gao, S.; Li, Y.; Duarte, F.; Li, S. Transferred Bias Uncovers the Balance Between the Development of Physical and Socioeconomic Environments of Cities. *Ann. Assoc. Am. Geogr.* **2025**, *115*, 148–166. [CrossRef]

65. Zhang, F.; Salazar-Miranda, A.; Duarte, F.; Vale, L.; Hack, G.; Chen, M.; Liu, Y.; Batty, M.; Ratti, C. Urban Visual Intelligence: Studying Cities with Artificial Intelligence and Street-Level Imagery. *Ann. Assoc. Am. Geogr.* **2024**, *114*, 876–897. [CrossRef]
66. Maas, P.R. Towards a Theory of Urban Vitality. Doctor's Thesis, University of British Columbia, Vancouver, BC, Canada, 1984.
67. Zikirya, B.; He, X.; Li, M.; Zhou, C. Urban Food Takeaway Vitality: A New Technique to Assess Urban Vitality. *Int. J. Environ. Res. Public Health* **2021**, *18*, 3578. [CrossRef]
68. Shanghai Municipal Bureau of Planning and Land Resources Management, Shanghai Municipal Commission of Transport, Shanghai Urban Planning and Design Institute. *Shanghai Street Design Guidelines*; Tongji University Press: Shanghai, China, 2016.

buildings

MDPI

Article

Analysis of the Urban Form of Bechar through the Attributes of Space Syntax "for a More Sustainable City"

Mohammed El Fatih Meknaci [1,*], Xingping Wang [1], Ratiba Wided Biara [2] and Wiem Zerouati [3]

1 School of Architecture, Southeast University, Nanjing 210094, China; wxpsx@seu.edu.cn
2 Department of Architecture, Tahri Mohamed University, Bechar 08000, Algeria; biara.wided@univ-bechar.dz
3 Department of Architecture, Ferhat Abbas University, Setif 19000, Algeria; wiemzerouati@univ-setif.dz
* Correspondence: 233217083@seu.edu.cn

Abstract: Located in the vast Algerian Sahara, Bechar has experienced over the years uncontrolled urban growth. It has evolved from a simple oasis and ksar (traditional settlement) to a city with a contrasting center and segregated outskirts. This growth has led to significant urban fragmentation that affected residents' quality of life. This research uses space syntax theory to analyze Bechar's urban configurations and their impact on functionality and sustainability. The study examines integration, connectivity, control, and intelligibility through axial maps and syntactic measures. Findings reveal that areas with high integration show better accessibility and reduced vehicle reliance, while segregated areas face accessibility challenges and limited services. Global integration values range from −1 to 1.192, averaging at 0.767, which indicates different accessibility among neighborhoods. These insights are crucial to identify areas where sustainable urban planning can mitigate adverse effects and enhance city efficiency. We created an axial map using DepthmapX v0.8.0 software and a modified digital QGIS map to collect data and measure Bechar's urban properties. The methodology we used translates social behaviors into spatial representations, generating interpretative models for social and spatial phenomena. This article provides valuable information on the infrastructure and service needs, guiding strategic resource allocation. The proposed urban design strategies include urban canopies and optimized building ratios to reduce energy consumption and promote sustainability. To conclude, we recommend practical solutions to enhance the sustainability and functioning of urban areas similar to Bechar.

Keywords: Bechar; city/oasis; urban fragmentation; spatial syntax; spatial configurations; urban sustainability

1. Introduction

Urban sustainability has become a concerning subject due to the negative impact of the global increasing urbanization [1]. This increase requires sustainable urban planning to mitigate adverse environmental effects and promote social equity and economic prosperity [2]. For instance, sustainable urban spaces promote green spaces, help reduce carbon footprints, and help improve the quality of life.

Despite efforts showcasing the effectiveness of sustainable urban planning in promoting economic growth while reducing congestion and pollution, a significant gap remains in understanding the impact of sustainable characteristics of urban spaces on urban quality of life and environmental sustainability. The Saharan context, where oases were once synonymous with resilience, is an example of this case [3]. This study explores the relationship between space syntax variables and sustainability parameters integrating into the Saharan urban space, an area largely unexplored in architecture and urban planning [3]. We address the following research questions to examine Bechar city:

How does the urban structure impact the sustainability and functionality of Bechar?

In what way does the current urban form influence socioeconomic and environmental dynamics in Bechar?

How do spatial configurations within Bechar enhance or hinder urban integration and interaction?

Given its challenging living conditions, Bechar faces issues of urban coherence and effective management of public spaces. These spaces are essential for fostering a livable and sustainable urban environment. The contemporary urban fabric of Bechar, heavily influenced by unsustainable construction driven by residential demands [4], has led to spontaneous and unstructured development patterns. This deviation from the city's historical growth trajectory, exacerbated by land appropriation (notably agricultural) after independence, has resulted in a diverse and antagonistic architectural landscape [5]. Consequently, uncontrolled expansion has led to fragmented and dysfunctional urban extensions, disrupting harmonious integration with the existing urban structure. This result highlights the importance of integrating urban sustainability indicators with the analysis of landscape patterns [6].

Urban fragmentation risks lead to new urban fragments and a drop in the average parcel size. This fragmentation indicates discontinuous growth and a trend toward compaction in the urban fabric [7]. Therefore, an in-depth analysis of urban spatial structures using measurable spatial attributes (such as space syntax) is crucial to promote sustainable urban development. This approach helps identify and prevent sustainability challenges caused by various land allocations, thus minimizing their effects on the city's functionality.

The research on Bechar's urban sustainability has the potential to make a substantial impact on multiple fronts. By applying spatial syntax analysis to evaluate the city's spatial configurations, this study offers critical insights that can significantly enhance urban planning strategies aimed at improving functionality and sustainability. The research provides actionable recommendations for urban planners to optimize spatial layouts in Bechar, offering a clear understanding of the city's spatial dynamics. This supports the creation of policies that promote sustainable urban development and efficient resource allocation.

1.1. Presentation of the Research Area

Bechar is situated strategically in the western region of the Algerian Sahara. Once known as the historical department of Saoura, this city is positioned approximately 950 km southwest of Algiers (the capital of Algeria) (Figure 1). Bechar expands at a substantial area of 5050 km^2, making it an urban entity in terms of geographical size (Figure 2), with approximately 279,851 inhabitants.

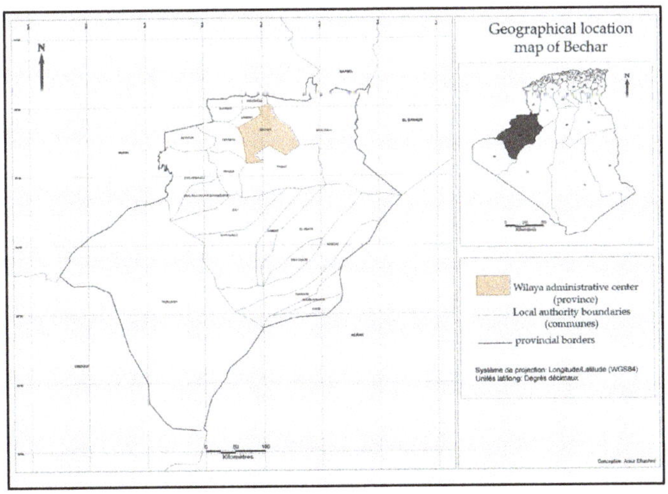

Figure 1. Geographical situation of Bechar.

Figure 2. The urban space of Bechar city.

1.2. The Evolution of Urban Space in Bechar over Time

The evolution of urban space in Bechar has resulted in a unique urban landscape resembling a patchwork of distinct fabric fragments. This mosaic-like urban form, shaped through successive contributions over time, has not been obliterated but integrated with the pre-existing urban space, allowing historical forms to persist [8]. Bechar's urbanization reflects the influences of various civilizations across its extensive history. The ancient Ksour ruins and the Old Ksar advocate its long-lasting habitation. Characterized by oasis habitats and agriculture, Bechar was historically a hub of Saharan civilization for its surrounding nomadic cultures and the bustling caravan trade [9].

Following the creation of the Trans-Saharan route and the CITROEN expedition, the city's development took a new trajectory. This change marked a significant shift by effectively replacing the traditional caravan routes and initiating several urban transformations. Bechar experienced substantial restructuring and densification in this phase, especially in the colonial village. This period of transformation led to what can be called a second duplication, which confined the barracks and facilitated the city's expansion towards the east [8].

Before 1903

Historically, Bechar was already a significant crossroads of civilizations. The remnants of ancient Ksour and the Old Ksar testify to its rich past. The city was known for its Saharan civilization (evident in its traditional dwellings, oasis agriculture, and nomadism in its outskirts) and caravan trade [8].

French Colonial Intervention, 1903–1917

The arrival of the French in 1903 marked a transformative period for Bechar as it began evolving into a colonial military post. The establishment of the French Foreign Legion greatly influenced its development [8]. The construction of the Trans-Sahara Railway, linking Bechar with Oran and other cities, was instrumental in its economic expansion. During this period, new infrastructures were established, including administrative buildings and military installations. The colonial administration's imposition of a grid-like urban pattern (reflective of European preferences for order and control) introduced segregation between European, military, administrative zones, and traditional living areas.

The Period 1917–1940

The city's growing role is exploiting coal mines and constructing the Trans-Saharan railway. The most significant urban expansion occurred between the two world wars, during which Bechar became a military base and a communication hub towards the Saharan South [8]. Large companies (such as the Trans-Saharan) exploited the riches of the

Saoura and Touat regions and built the road that extended to Gao in Niger. A significant part of the road would later be used for National Road n°06.

The economic and administrative role of Bechar had a clear impact on the organization of the territory in general and on the urban development of the city in particular. However, the city's growth coincided with the natural and social environment. A significant event later disrupted this harmony and, consequently, the city's urban structure when the French administration encouraged locals to acquire individual property titles for tribal and communal lands and assets to facilitate their sale to French settlers [9].

The Period 1940–1958

Bechar underwent significant development during this period, affirming its position as a central military and administrative hub. A vast residential area emerged west of the town, near the crest of the 'la Barga' mountain (Figure 3). This expansion stemmed primarily from the influx of residents from the Ksour in the Saoura and the settlement of nomads. This period marked a transformative phase in Bechar's urban evolution, establishing the groundwork for its present urban structure and social dynamics [9].

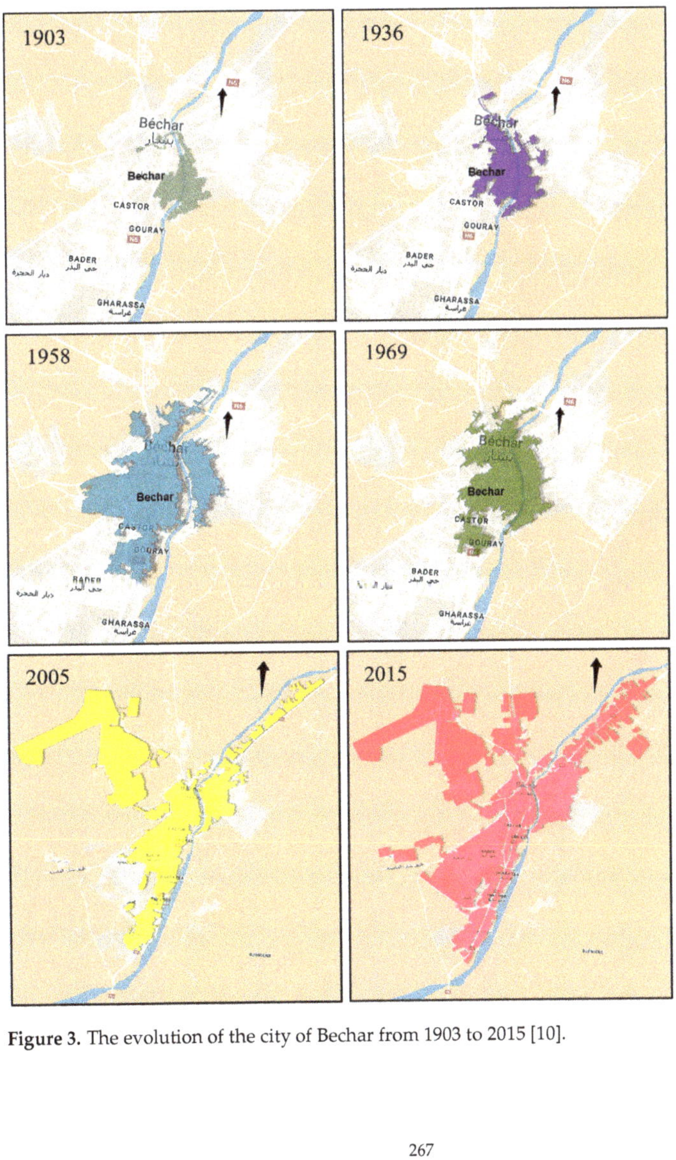

Figure 3. The evolution of the city of Bechar from 1903 to 2015 [10].

After Independence

Following Algeria's independence, Bechar embarked on a period of rapid development. The city capitalized on government investments, shifting towards industrialization with a focus on mining and energy sectors [11]. This economic upswing attracted a diverse population seeking employment opportunities, leading to unplanned urban sprawl (Figure 3). As a result, residential areas and informal settlements emerged, mirroring the socioeconomic stratifications of the growing population.

Since 1988

The urban development in Bechar progressed at an accelerated rate. However, this development often lacked a cohesive planning strategy, further exacerbating urban segregation (Figure 3). The city's expansion was sporadic, responding more to immediate demands than long-term urban planning objectives [8]. This period witnessed accelerated development in the part that extends from Bechar Djedid to Ouakda through the construction of residences, taking on a linear configuration in its primary form. This period of development was accompanied by an irrational consumption of space, making the management of the city very complex. Consequently, Bechar's cityscape evolved into a mosaic of varied urban styles and developmental stages (Figure 4). The absence of an integrated urban policy has led to noticeable disparities, with affluent areas sharply contrasted against underdeveloped neighborhoods, highlighting the socioeconomic divisions within the city.

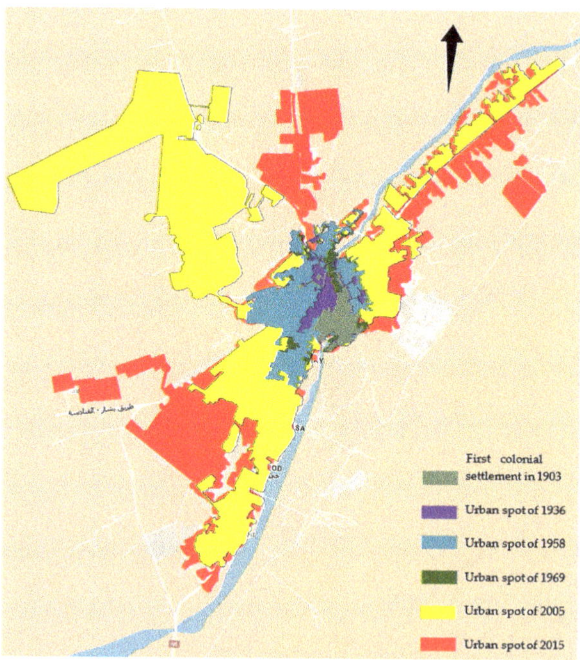

Figure 4. Urban evolution of the city of Bechar [10].

Bechar has evolved into a mosaic-like urban landscape, blending historical elements with modern developments. Key transformations began with French colonialism, establishing Bechar as a military post and introducing the Trans-Saharan Railway. This spurred infrastructural growth and a European-style grid urban pattern. The city mainly expanded due to coal mining and railway construction, making it a military and communication hub. Bechar's role grew progressively, leading to significant urban expansion. After gaining independence, rapid industrialization and urban sprawl reflected socioeconomic stratifications. Recent uncoordinated development has led to urban segregation and management challenges, creating noticeable disparities between affluent and underdeveloped areas. As a result, Bechar's urban evolution is a dynamic blend of historical integration and modern expansion amid socioeconomic divisions.

1.3. Literature Review

The urban evolution of Bechar has reached a critical stage, focusing on spatial and functional growth driven by the concentration of 99% of the population in the central area. The city's current urban structure, influenced by its historical formation, consists of five main entities: the city center, Debdaba, Ouakda, Bechar Djedid, and a new zone. Except for the city center, these areas lack complete autonomy. A new urban area is emerging along CW03 (provincial road) towards Lahmar, outside the old master plan perimeter. The urban fabric is imbalanced in terms of population distribution and activity allocation.

Urban planners have always prioritized rational space use, yet Bechar's residential density is very low (less than 15 dwellings/ha), indicating significant spatial disparity. The master plan revision emphasizes densifying spaces to maximize urban surface use while maintaining an integrated, dense, and well-equipped environment. The linear expansion affects infrastructure costs adversely, extending them disproportionately to the population and causing space shortages by limiting constructible areas to a narrow strip. It is thus crucial to halt the linear expansion and avoid densifying Ouakda to preserve its agricultural role. Instead of meeting the 2013 master plan for urban extension, the city's expansion followed an unplanned path, reflecting chaotic land choices that disregard urban and land plans. Despite numerous urban plans from 1966 to 2007, Bechar still faces the same issues highlighted in past studies. The city's development remains uncoordinated and fragmented, as observed in historical urban planning documents.

1.3.1. Sustainable Urban Development: Global Insights and Applications

The literature review helps us define our problem, strengthen existing research knowledge, and explore how different urban areas have addressed sustainability challenges (focusing primarily on Saharan regions). It also identifies research methods used to approach Bechar's unique challenges.

Bechar experienced significant development since 1948, marked by the construction of the Trans-Saharan railway and coal mining [8]. This led to rapid urban expansion that overshadowed previous development patterns. Post-independence land appropriation further influenced the formation and transformation of various neighborhoods, shaping the city's heterogeneous urban fabric [5]. Over a century, spatial production in Bechar resulted in a patchwork of distinct urban fabric fragments [8]. This unique urban fabric results from the demographic explosion, causing structural and functional transformations in the city's urban composition [12].

Despite the critical need to understand the dynamics of urban expansion in Bechar, most studies focus on the built environment and urban planning [5], sustainability [13–15], and other aspects of urban planning [5]. The impact of spatial structure on Bechar's urban form and planning strategies remains unexplored and requires further investigation to fully grasp the complexities of Bechar's urban evolution.

Methodologies and solutions that adapt to Bechar's unique challenges, such as uncontrolled urban growth, fragmentation, and socioeconomic disparities, will rely on lessons from global case studies on sustainable development in arid regions and developing countries. The collective research from Talpur MAH, Khahro SH, and Chandio IA offers significant insights into urban sustainability in developing countries. These studies emphasize the importance of integrated urban planning that incorporates socioeconomic, environmental, and spatial dimensions to mitigate the adverse impacts of rapid urbanization. Additionally, the research underscores the socioeconomic benefits of sustainable planning, which improves residents' quality of life through better access to amenities and economic development. Studies include the socioeconomic impacts of urbanization [16–18], green infrastructure in urban planning [19,20], and sustainable green buildings in Hyderabad [21].

Furthermore, the integration of sustainable practices in urban planning is supported by case studies on energy-efficient practices [22,23] and the role of public transportation in enhancing urban livability [18,24]. By providing actionable policy recommendations, these works advocate for comprehensive planning approaches that balance economic, social, and environmental objectives to achieve sustainable urban growth.

By 2050 (Figure 5), approximately 68% of the global population will live in urban areas [25], which are responsible for about 70% of global carbon dioxide emissions and consume between 60% and 80% of the world's energy [26]. An exemplary model of sustainable urban development in an arid region is Masdar City in Abu Dhabi, which employs innovative technologies and design principles to create a zero-carbon, zero-waste city. Passive cooling techniques, solar energy, and sustainable building materials are used to mitigate the harsh desert climate [27]. Masdar City also addresses the urban heat island effect by strategically placing vegetation and shaded walkways, reducing ambient temperatures and enhancing pedestrian comfort. Ouarzazate, Morocco, is home to one of the world's largest solar power plants. The Noor Solar Power Station uses concentrated solar power technology to generate clean energy, significantly reducing the region's carbon footprint and providing a sustainable energy source [28]. Copenhagen, Denmark, has invested substantially in cycling infrastructure (Figure 6), making it a leading example of sustainable urban mobility. They successfully reduced traffic congestion and pollution by prioritizing bike lanes (approximately 41% of trips are made by bicycle) and pedestrian pathways [29].

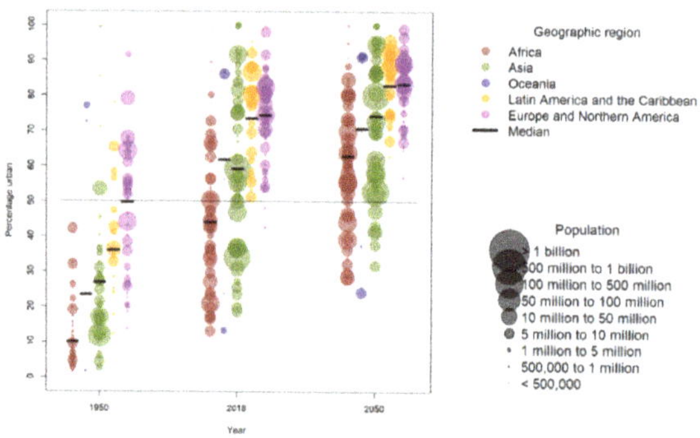

Figure 5. Urban population percentage by region and population size [25].

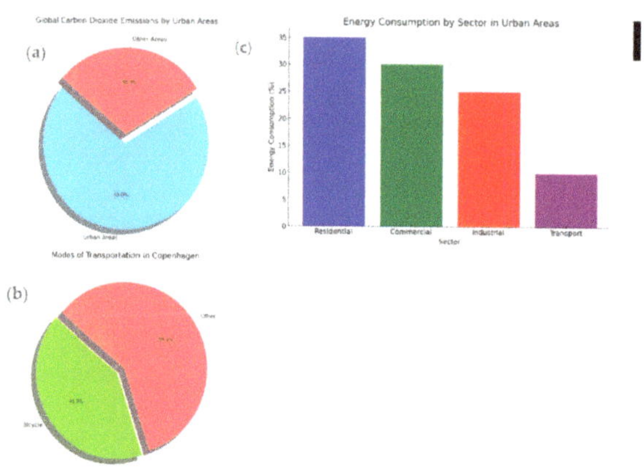

Figure 6. (**a**) Carbon dioxide emissions by urban areas. (**b**) Bicycle usage in Copenhagen. (**c**) Energy consumption by sector.

In New York, the High Line Park is a notable example of adaptive reuse and urban green space development [30]. An abandoned elevated railway was transformed into a public park, attracting millions of visitors annually and spurring over $2 billion in real estate development nearby. Bogotá, Colombia's TransMilenio bus rapid transit (BRT) system, demonstrates how efficient public transport can address urban congestion and reduce carbon emissions. The city integrates transport planning with urban development, improving accessibility and reducing traffic issues [31]. Singapore's garden city vision successfully integrates nature into urban planning. Initiatives like the Park Connector Network and vertical greening of buildings aim to maximize green space and enhance biodiversity in a densely urbanized environment [32].

The methodologies used in these case studies include space syntax, urban canopy, green roofs, and integrated public transport systems. Space syntax helps understand spatial configurations and their impact on social, economic, and environmental outcomes, optimizing urban layouts for better functionality and sustainability. Techniques like urban canopies and green roofs are critical in reducing urban heat islands and improving energy efficiency. Examples from Bogotá and Copenhagen demonstrate the benefits of integrating public transport with urban planning, significantly improving mobility, reducing emissions, and enhancing urban livability.

Urban heat islands are a significant issue in many arid regions, where high temperatures can severely impact urban livability. For instance, urban heat islands have been mitigated in Phoenix, Arizona, USA, through reflective roofing materials, increased vegetation, and green spaces [33]. These measures have helped lower surface temperatures and improve overall urban comfort. Accessibility issues are prevalent in many rapidly urbanizing areas. In Mexico City, accessibility challenges have been addressed by developing an extensive metro and bus system, improving mobility for millions of residents, and reducing reliance on private vehicles [33]. These solutions have significantly reduced traffic congestion and improved air quality in one of the world's largest cities. Urban fragmentation and segregation are critical challenges impacting cities' social and economic fabric. Johannesburg, South Africa, exemplifies how historical segregation policies resulted in a fragmented urban landscape. Efforts to reintegrate these areas include developing mixed-use zones and improving public transportation links to enhance connectivity and reduce socioeconomic disparities [34].

Reducing high temperatures is a significant focus of urban sustainability efforts in arid regions. Urban planners have incorporated extensive green spaces and water features in Riyadh, Saudi Arabia, to cool the urban environment. Traditional architecture with features like wind towers and shaded courtyards also helps naturally reduce temperatures and improve comfort [35].

A historical overview of urban development helps contextualize current challenges and solutions. For instance, Talpur MAH, Khahro SH, and Chandio IA emphasize integrated urban planning in developing countries. They highlight green infrastructure and sustainable practices to enhance energy efficiency, reduce urban heat islands, and improve quality of life through balanced socioeconomic and environmental objectives. The transformation of New York's High Line from an industrial relic to a green urban space reflects broader trends in urban renewal and adaptive reuse. The study emphasizes need-based solutions tailored to the specific challenges of congested urban centers. In Bogotá, for example, the BRT system addresses severe traffic congestion and air pollution. In Singapore, vertical greening responds to limited horizontal green space. In Saharan regions, traditional architecture and water management systems, such as qanats and foggara, address water scarcity and extreme temperatures.

The study uses global case studies, including those from arid regions, to guide sustainable urban development strategies in Bechar. It adapts methodologies and solutions to address Bechar's unique challenges, such as uncontrolled growth, fragmentation, and socioeconomic disparities. This need-based approach tackles urgent urban issues in Bechar. A comprehensive review of global and regional case studies provides a robust framework for targeted solutions, enhancing urban sustainability and functionality. By incorporating these elements, the literature review offers a solid foundation, demonstrating relevance and applicability to urban centers worldwide (including Bechar) and ensuring practical solutions informed by successful global examples.

1.3.2. Expanding Urban Sustainability: From Core Pillars to Spatial Integration

The urban sustainability topic is diverse and multifaceted, with various works exploring its definitions, implications, and foundational concepts. For instance, Mondal [36] emphasizes the need to understand urban sustainability through the Brundtland report's tripartite model, which includes environmental, economic, and social pillars. Mazzoncini, Somaschini, and Longo [37] expand this model by adding cultural aspects as a fourth pillar. The Brundtland report's definition of sustainable development, focusing on meeting present needs without compromising future generations' abilities, is widely acknowledged. This definition underscores the integrated approach necessary for achieving sustainable urban development, embracing economic growth, environmental safeguards, and social justice [38]. The inclusion of social sustainability through Universal Design highlights urban planners' role in fostering social interaction, community cohesion, and cultural dialogue, emphasizing the social dimension's critical importance in sustainability [38].

Despite a common understanding of the three core pillars of sustainability, ongoing debates and efforts address their practical implications and adaptability. José G. Vargas-Hernández [39] explores the educational aspects, creating courses for civil engineering students on sustainable design and evaluation methods, enhancing future professionals' understanding of sustainability principles. Framing urban sustainability as a normative choice highlights the critical challenges linked to urban environments, such as traffic congestion, social tensions, and environmental impacts [40,41]. Research spans various areas, from ecological protection and infrastructure development to urban metabolism, reflecting the multifaceted nature of urban sustainability.

Zhang and Li [42] enhance our understanding by categorizing existing studies on urban sustainability into five distinct clusters (Table 1). This classification highlights the complexity of the issue and helps navigate the extensive body of research. Organizing studies into clusters provides a structured framework that encapsulates diverse sustainability challenges, allowing researchers and practitioners to focus on specific areas of interest.

Table 1. Key clusters of research on urban sustainability and their focus areas.

Cluster	Focus Area	Key Aspects
Ecological systems	Relationship between urban areas and environmental systems	Maintaining environmental balance, utilizing ecological resources sustainably, the impact of urban development on the ecology
Climate change	Adaptation to and mitigation of climate change effects in urban areas	Developing resilient infrastructure, managing risks of rising sea levels, increased natural disasters
Infrastructure systems	Sustainability of urban infrastructure systems	Sustainable design and maintenance of transportation networks, green buildings, and energy systems
Emergency operations	Preparedness and response to urban emergencies	Robust emergency response systems and resilient infrastructure to handle natural disasters and terrorist attacks
Urban recovery	Strategies for recovery of cities after disruptive events	Social, economic, and infrastructural recovery processes, building back better and more resilient urban environments

Exploring spatial aspects within urban sustainability, such as spatial integration and city form, is a critical yet underexplored area that significantly impacts sustainability outcomes. This research gap signals a shift from traditional and functionally focused approaches to more spatially oriented perspectives. The city's form is now seen as crucial in addressing sustainability challenges, recognizing that spatial dynamics and urban form play a key role beyond mere functional and environmental considerations [43,44].

Bibri and Krogstie [45], Beatley and Collins [46], and Sev [47] emphasize the critical role of spatial dimensions in urban sustainability, advocating for the integration of spatial considerations into sustainability planning and policy-making. Space syntax analysis was used widely to understand urban spatial configurations and their implications for urban sustainability. Hillier and Iida [48] demonstrated the relationship between street network configuration and pedestrian movement patterns, showing how urban design influences social interaction. Karimi [49] applied space syntax to the analytical design of urban spaces, focusing on how spatial configurations impact social and economic activities. Baran et al. [50] compared walkability in new urbanist and suburban neighborhoods, highlighting the role of spatial configurations in walking behavior. Vaughan and Geddes [48] explored the relationship between urban form and social deprivation in London, while Chiaradia et al. [51] analyzed the economic implications of spatial design on residential property values in London. Hillier [52] and space syntax theory provide a foundational framework for optimizing urban spatial structures to enhance sustainability outcomes.

These studies support our methodology by demonstrating the use of space syntax to assess urban configurations and their impacts on functionality and sustainability.

Further investigations into configurational theories, like the space syntax approach, reveal the potential for understanding spatial relations within neighborhoods. This understanding can inform interventions to address segregation and improve urban design, fostering more sustainable and integrated environments [53]. The application of space syntax theory in the Metropolitan Region of Porto Alegre, Brazil, demonstrates its effectiveness in spatial analysis for large urban areas, offering insights into spatial navigation and urban intelligibility [44].

These studies support our methodology by highlighting the use of space syntax to assess urban configurations and their impacts on functionality and sustainability. This approach suggests considering spatial configurations for more effective resource allocation and policy-making, thus improving urban sustainability.

1.3.3. Urban Fragmentation

Urban fragmentation is a complex challenge in urban studies, involving physical and social segregation leading to isolated and disconnected communities. The United Nations Human Settlements Program [54] highlights how urban development that fails to integrate community spaces often results in socioeconomic disparities exacerbated by physical barriers like highways and walls.

The concept extends beyond physical segregation to social fragmentation, as Mebirouk [55] and Santos [56] explored. They describe urban fragmentation as disrupting the social fabric, leading to segregated communities in isolated enclaves. This segregation is evident among immobile urban populations, resulting in a fragmented urban landscape characterized by ghettos.

The literature on urban fragmentation identifies a typological and morphological heterogeneity within urban landscapes, resulting in disorderly urban sprawl [57]. This heterogeneity is known for the proliferation of boundaries and the privatization of public spaces. The fragmentation creates strong contrasts and accessibility issues within cities [58,59] and weakens political, social, and economic relations across neighborhoods [60].

Vidal [61] explored socio-spatial fragmentation, which examines the interplay between social and spatial dimensions in urban environments. It highlights how morphological differences and socioeconomic transformations contribute to urban disparities. Political or managerial fragmentation addresses urban governance complexities, with public–private partnerships and segmented urban management posing challenges to metropolitan governance and functionality [55,60].

In contrast, Paquot [62] discussed that urban integration emphasizes the importance of fostering solidarity and interdependence among urban segments. Chouadra [63] highlights the adverse effects of fragmented urban policies in Algerian cities, showcasing the need for a more integrated urban development approach.

The literature on urban development, sustainability, and fragmentation comprehensively explores city dynamics, focusing on Bechar's evolution. Significant developmental milestones, such as the construction of the Trans-Sahara railway and coal mining, spurred urban expansion and shaped Bechar's heterogeneous urban fabric [5,8]. While urban sustainability discourse has evolved to include broader considerations beyond the traditional pillars, emphasizing the need for spatial integration remains a notable gap in addressing the impact of spatial structures on urban form in Bechar [36,37]. Urban fragmentation, characterized by physical and social segregation, is a critical challenge, and the literature lacks a cohesive approach to addressing its implications on urban cohesion and functionality. This synthesis underscores the need for further research into the spatial dynamics influencing urban development and planning strategies and calls for a more integrated approach to tackling urban fragmentation to foster sustainable and cohesive urban environments.

Despite extensive research on urban sustainability, there is a significant gap in studies focusing on the unique urban context of Bechar, an arid region facing distinct socioeconomic and environmental challenges. Previous urban plans have not effectively addressed critical issues such as spatial fragmentation, infrastructure costs, and socioeconomic disparities. Although there is a critical need to understand the dynamics of urban expansion in Bechar, most studies have concentrated on the built environment, covering aspects such as urban planning [5], sustainability [13–15], architecture [64], and further aspects of urban planning [5], but they do not adequately address the spatial configurations complexities specific to Bechar.

This study fills the existing gap by applying spatial syntax analysis to Bechar's urban landscape, offering novel insights and methodologies for sustainable urban planning in arid regions. The research provides innovative strategies tailored to Bechar's specific conditions, such as optimizing spatial layouts to improve functionality and reduce socioeconomic disparities. By identifying and addressing unique spatial dynamics, this study contributes to a better understanding of sustainable urban development in Bechar and offers valuable methodologies that can be applied to other arid regions facing similar challenges.

The literature review highlights the critical stage of Bechar's urban evolution, focusing on the city's spatial and functional growth, and the significant role of urban planners in addressing the imbalanced population distribution and activity allocation. Despite efforts to rationalize space use, Bechar's low residential density and unplanned expansion underscore the need for integrated urban planning to enhance sustainability and functionality. Lessons from global case studies, such as Masdar City, Noor Solar Power Station, and sustainable urban strategies from Copenhagen, New York, Bogotá, and Singapore, provide valuable insights for addressing Bechar's unique challenges. The review emphasizes the importance of understanding spatial dynamics and integrating sustainable methodologies to foster cohesive and sustainable urban environments. Addressing uncontrolled growth, fragmentation, and socioeconomic disparities through targeted solutions will be crucial for Bechar's sustainable urban development.

2. Material and Methods

2.1. Description of the Variable

Evaluating urban sustainability is a complex but essential process for improving urban planning and management [65]. This study employs spatial syntax methodology (Figure 7) to examine the influence of urban spatial configurations on the functionality of Bechar, significantly impacting urban sustainability. The relationship between urban sustainability and spatial syntax lies in analyzing and optimizing spatial configurations to enhance urban functionality and viability.

Spatial syntax provides a methodological framework to understand the impact of urban form on social, economic, and environmental aspects [66]. The urban form plays a vital role in urban sustainability, as the land influences travel patterns, fuel consumption, and air quality [67]. The spatial configuration of cities and metropolitan regions profoundly impacts sustainability, shaping ecological and socioeconomic processes.

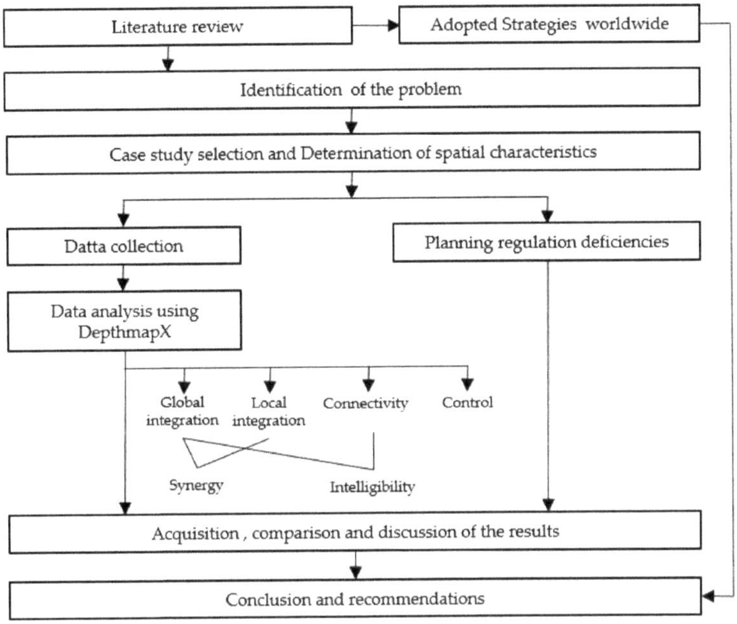

Figure 7. Methodology flow chart used in this study.

The analysis of spatial segregation in Bechar will highlight its significant impact on social vulnerability and the persistence of societal divisions. A comprehensive analytical framework is essential to effectively explore the complex interaction of integration and segregation in this urban landscape. In urban public space studies, axial line analysis has produced significant relationships [68,69]. Spatial syntax theory stands out as a crucial tool, offering unique insights into the nuances of this case study.

The development of this study, which examines the impact of urban spatial configurations on the functionality of the city of Bechar, relies on the city map provided by the urban planning department to perform spatial syntax analysis using DepthmapX software.

2.1.1. Space Syntax

Developed in the late 1970s and early 1980s by Bill Hillier and colleagues at the Bartlett University College London, space syntax theory originated as a morphological approach in response to urban and social changes in major British cities during that period [70]. Evolving over the years, it now focuses on analyzing spatial configurations of architectural and urban spaces. The theory translates social behaviors into spatial representations, employing theories and techniques to generate interpretative models for social and spatial phenomena, such as social segregation, crime, business locations, and urban movement [71]. Its applications have expanded into diverse fields, including urban planning, architecture, transportation, human geography, anthropology, and archaeology [69].

Space syntax views urban space as a complex single entity but distinguishes two levels of duality. First, it recognizes the urban space as having local and global dimensions, linking it to the broader urban context [72–74]. Second, it identifies two distinct components: the built and communicative spaces within and between these built areas. Focusing on the continuous space, especially its topological configuration, space syntax analyses the relationships between spaces, considering how each space is influenced by its relation to others. The interplay of these two components, the 'urban full' and the 'urban empty', results in specific spatial configurations that correlate with different urban operations, including mobility, a key element in syntactic analysis [72–74]. The configuration of space is seen as a movement generator, with a multiplier effect on the attraction of economic activities linked to pedestrian flows [70,72,73,75].

2.1.2. Axial Line Analysis

In this research, the city of Bechar is modeled using Space Syntax techniques and DepthmapX v0.8.0 software to create an axial map based on a modified digital QGIS map suitable for export to Depthmap©10. The axial analysis conducted with this software reveals distinct syntactic characteristics of Bechar by measuring its global and local urban properties. This map consisted of the minimum number of continuous axial lines, each representing the longest straight line of sight needed to cover the entire spatial layout [70]. The axial map is designed to represent the urban area based on its street network structure. For numerical evaluations, the axial map is converted into an axial graph where each node corresponds to an axial line, and each edge represents an intersection. This formalization allows the decomposition of the continuous spatial layout into component units, enabling the evaluation of their interrelationships using graph-based methods.

The axial map is a fundamental technique in modeling and analyzing urban forms as spatial configurations, widely recognized in space syntax representations for characterizing urban forms. This topological map simplifies spatial morphology into a system of relationships, connecting people to their social and spatial environments. Based on the axial lines (the longest lines of uninterrupted visibility connecting spaces), the axial map is formed by the minimum number of lines required to cover an urban area's free space [76,77]. These lines accurately represent the accessible line of sight in each street, mirroring the human perception of the built environment. They are instrumental in modeling and analyzing urban configurations, aiding in predicting movement patterns and activity locations [78]. In the graph, lines represent connections, while intersections represent nodes. This approach allows for a comprehensive understanding of the urban system and the analysis of spatial configurations, with its design initially focused on human movement within space. Computational algorithms efficiently manage the complexity of calculating the axial map, determining values such as connectivity and integration through intricate calculations [77].

Integration

Integration classifies as a global static measure and is key in measuring spatial syntax and assesses any spatial system's relative depth or shallowness from a specific point [48]. It evaluates the capacity of a space to integrate, making it a crucial space syntax measure and a primary indicator of urban movement [74].

Topological measurements were performed on the graph-based representation of the axial map to understand the spatial environment's behavioral characteristics. The integration index for each axial segment (portions of axial lines between their intersections) is calculated [48]. Higher integration values for a segment signify a better integration within the spatial network, on average. In other words, it has a shorter topological distance from all segments. In the integration map, blue indicates the lowest integration value, gradually increasing to red for the highest value. We define high-integration areas as integrated, and low-integration areas as segregated [79].

Previous studies have demonstrated that integration and segregation effectively describe the social use of urban environments. Typically, areas with high integration are active economic zones, for instance, main streets with retail shops and large businesses. Conversely, segregated areas are usually underutilized spaces associated with neglect or crime [80]. Empirical findings have shown a significant correlation between local integration and pedestrian flow rates in urban settings [81]. Across various geographic locations and cultures, natural pedestrian movement patterns closely align with the integration index map [82]. On an average scale, pedestrian destinations can be predicted accurately based on the integration index. This evidence supports Hillier et al.'s theory. Hillier suggests that pedestrian distribution in a street network is mainly influenced by the spatial system's inherent morphological and geometrical characteristics [83]. Additional studies have backed these findings, highlighting the role of integration in economic vitality and social interaction. For example, research has shown that areas with high integration attract more economic activities and pedestrian traffic, leading to vibrant public spaces [84]. Moreover, the spatial distribution of integration values has been linked to variations in land use and social cohesion, further underscoring the importance of spatial configuration in urban planning [85].

Connectivity

Connectivity is a local static measure that indicates the number of directly connected streets [86]. It serves as a key metric in understanding the comprehensibility of the global network [78]. In detail, connectivity measures the number of immediate neighbors directly connected to a node, providing a local perspective on network interconnections [48,73,87]. Analyzing the relationship between connectivity and overall integration within a spatial system leads to the concept of intelligibility [88]. Its value equals the R^2 coefficient of determination from the regression between connectivity values and global integration. A subset with a high R^2 is interpreted as a group of spaces with good intelligibility [89].

Control Value

The control value is a dynamic local measure that quantifies the extent to which space influences access to its neighboring areas. It assesses the number of alternative connections available to each neighboring space, thus evaluating the relative influence of a particular space on accessibility and movement patterns within its immediate vicinity [88].

2.1.3. Second Degree Measures

Intelligibility

As defined by Hillier [88], intelligence is the capacity of a structure to provide a clear local indicator to understand spaces at a global level, allowing the study of different parts of urban areas and their relationship [83]. This intelligence is calculated by analyzing the correlation between Connectivity (a local measure) and Integration (a global measure). A high coefficient value, close to 1, suggests a strong correlation and indicates that the urban system is easily understandable. This concept is also closely linked to cognitive mapping and Lynch's notion of legibility, focusing on the readability of an urban environment [66,90]. Intelligibility measures the integration of axial lines within the system, reflecting how easily a pedestrian can understand the overall urban layout from a specific location. It indicates how well the local environment helps navigate the entire area. A low intelligibility value suggests a labyrinthine spatial organization, while a high intelligibility value means the area is easy to navigate [83].

Synergy

As described by Dalton [91] and Hillier [75] in 'Space is the Machine', synergy is the correlation between local integration (radius 3) and global integration (radius n). This measure assesses how well a space's local structure relates to the larger-scale system it is part of. A diagram typically visualizes this relationship, with the degree of correlation indicated by the coefficient (R^2).

The analysis of spatial segregation in Bechar reveals its significant impact on social vulnerability and societal divisions. Using space syntax theory, we explore urban fragmentation through spatial accessibility indicators, showing that a territory is fragmented when its parts are poorly connected. Space syntax, developed by Bill Hillier, translates social behaviors into spatial representations, aiding in analyzing phenomena like segregation and urban movement. Key measures such as integration, connectivity, control value, intelligibility, and synergy provide insights into the city's morphological continuity, urban fabric heterogeneity, and the articulation between centralities and peripheries.

These measures are crucial for optimizing urban spatial configurations, directly linking to urban sustainability by promoting accessibility, enhancing social interaction, and ensuring cohesive development. Integration and connectivity improve mobility and accessibility, reducing reliance on private vehicles. Intelligibility and synergy facilitate coherent urban planning, making spaces more legible and functional. The control value helps identify areas for interventions to enhance accessibility and equity, contributing to a balanced and sustainable urban development.

3. Results

The axial map shows a color gradient shifting from blue to red, effectively highlighting areas of segregation and integration within the city (as depicted in Figures 8 and 9). Figure 10 illustrates the city's connectivity, identifying highly and poorly connected areas. Additionally, Figure 11 displays the analyses of control values across Bechar, indicating the extent of spatial influence. These visual and quantitative assessments offer an overall understanding of Bechar's spatial dynamics, underlining the importance of syntactic measures in urban analysis.

Figure 8. Results of the axial analysis for the global integration measurement. Scale: Red = most integrated areas; blue = least integrated areas.

Figure 9. Results of the axial analysis for the local integration measurement. Scale: Red = most integrated areas; blue = least integrated areas.

Figure 10. Results of the axial analysis for the connectivity measurement.

Figure 11. Results of the axial analysis for the control measurement.

3.1. Axial Map Analysis Results

The results of the axial map analysis (Table 2) for Bechar reveal valuable insights into the spatial dynamics and their impact on sustainability, socioeconomic structures, and urban interaction. The axial map analysis, encompassing 3195 axial lines over a 45.5 km^2 area, shows intricate patterns of connectivity and integration. The global integration (HH) values, ranging from −1 to 1.192 with an average of 0.767, highlight areas where high integration supports sustainable urban living by promoting accessibility and reducing reliance on private vehicles. Specifically, neighborhoods such as Hay El-Karma, Hay Badr, Hay Inara, and Hay Gharassa, which exhibit high integration values, demonstrate the importance of integrated urban design in enhancing the city's functionality (Figure 8). Higher foot traffic is due to increased levels of integration. Subsequent research by Ozbil, Peponis, and Stone [90] has confirmed that this correlation remains robust and significant, even when we consider the development densities on the observed streets and their surrounding areas. Similarly, the distribution of land uses, especially commercial land uses, follows this pattern.

Table 2. Summary attributes of the axial map.

	Minimum	Maximum	Mean	Std. Deviation
Integration HH	−1.00	1.19	0.7674	0.16134
Connectivity	0	67	3.72	3.005
Control	−1	21	1.00	0.921

The map highlights several marginalized and segregated neighborhoods, primarily located on the city's outskirts. These include informal settlements such as Hay Boutlis and the Bechar Djedid neighborhood to the south. Hay Teghlyine and the new extensions in the Debdaba neighborhood are other examples in the northeast. The 198 Housing Units neighborhood also exhibits signs of segregation in the north.

Most integrated axes in red correspond to the city's main thoroughfares. Road n°06 (Saoura Road), linking the city center to southern neighborhoods, notably records the highest integration values. Its extensions, including Boulevard Toumi Tayeb, Boulevard Razni Chikh, Rue Menouni, and Rue Maalem Mohamed, also exhibit substantial integration. These main roads and interconnected streets create vibrant urban public spaces, hosting various social and commercial activities, with Road n°06 serving as a focal point for neighborhood life.

The map identifies several marginalized areas with poor accessibility, resulting in a lack of vibrant social and commercial activities. These areas typically feature predominantly residential land use with limited commercial or recreational facilities, resulting in a lack of diversity. Minimal pedestrian traffic is due to these neighborhoods' peripheral location and poor connectivity. Additionally, infrastructure in marginalized areas is often inadequate, with limited access to transportation, utilities, and public services. Economic disadvantages are expected, with fewer businesses and services available to residents. The limited number of roads effectively connecting various urban fragments highlights a significant lack of inter-neighborhood connectivity. Suggesting, therefore, that urban planning has not adequately addressed the need for thoroughfares that facilitate better integration across different parts of the city, impacting overall urban functionality and social cohesion.

The local integration analysis within Bechar city reveals significant disparities in road system integration (Figure 9). Road n°06 emerges as the most integrated, marked by its red color on the map. This street is dynamic and heavily trafficked, playing a vital role in the structural organization of the neighborhood. Boulevard Razni Cheikh also displays significant integration values, though marginally less than Road n°06. These well-integrated axes are distinguished by their ease of access, high permeability, and visibility, contributing to their popularity for movement and transportation. They also enhance urban functionality and promote sustainable urban living by easing movement and reducing reliance on private vehicles.

Contrarily, less integrated axes (highlighted in blue) are characterized by reduced accessibility and lower integration levels. This is particularly evident in the peripheral areas of the neighborhood, such as the southwest regions, northeast regions, and around critical areas (i.e., the main neighborhood entrance and interstitial spaces between houses). These areas exhibit minimal integration values that constrain movement and interaction, especially for visitors and newcomers. The lack of integration in these areas contributes to socioeconomic differentiation, vehicle reliance, and environmental degradation, thus underscoring the urban segregation within the city. This analysis aligns with identifying and addressing spatial disparities to enhance urban sustainability and functionality.

In the analysis of Bechar city's connectivity map (Figure 10), roads connecting neighborhoods (represented in shades from red to light blue) indicate the highest connectivity values.

These roads have a regular and near-orthogonal urban layout, enhancing permeability and fostering the emergence of new urban centers. Areas with spontaneous and irregular housing patterns display strong segregation from the broader urban system, as indicated by the blue-colored streets in these regions.

The map also reveals that the number of roads effectively connecting the various parts of Bechar to the overall system is quite limited. Road n°06 stands out as the most connected axis, followed by Rue Hamlili Brahim and Rue Menouni. These areas are known for their high integration values, central location, proximity to major thoroughfares, and dense street network, allowing vibrant social and commercial activities, diverse land uses, and well-developed infrastructure. Hay El Badr, Debdaba, and the city center are examples of these areas. In contrast, disconnected areas exhibit low-integration values, peripheral location, distance from major roads, and a sparse street network. This lack of connectivity between neighborhoods causes limited social and commercial activities, homogeneous land uses, underdeveloped infrastructure, and economic disadvantages. As a result, we can suggest that urban planning may not have sufficiently addressed the need for thoroughfares that facilitate better integration across different parts of the city.

The control map analysis reveals significant navigational challenges in Bechar's urban layout. Aside from Road n°06, none of the tertiary or secondary roads exhibit high control values (Figure 11), indicating their relative disconnection and lack of clear navigational routes.

The overall spatial control within the city is inadequate, with few road intersections, alleys, and inter-district passages showing better navigability. Mostly, this is due to their regular and near-orthogonal structure in Hay El Badr, some parts of Debdaba, the city center, and Hay Tnkerod. This disconnection primarily benefits residents familiar with the city's layout, who can navigate its fragmented routes. For outsiders or newcomers, however, the city's layout is confusing and difficult to navigate, leading to disorientation and unfamiliarity. The lack of clear routes can also decrease perceptions of safety, particularly in complex and poorly visible areas. These proprieties are showcased in Hay Fajr, Bleu Zone, Hay Inara, and recent urban extensions of Debdaba neighborhood.

3.2. Second Degree Measures Results

Intelligibility measures how well local properties of a spatial configuration reflect its global structure. This is assessed by the correlation between Connectivity (a local measure) and Integration (a global measure). A strong correlation indicates higher intelligibility. This measure is crucial since studies have shown that higher intelligibility correlates with different behaviors [88,92]. To assess the intelligibility of Bechar, a scatter plot of Integration and Connectivity was examined. Figure 12 shows no significant relationship between these two measures in the city. The low intelligibility value ($R^2 = 0.048$) means that high local connectivity does not necessarily translate to global integration. This conclusion suggests that Bechar's urban layout is not easily understandable from its parts, thus making it challenging for residents and visitors to navigate and comprehend the city's spatial organization.

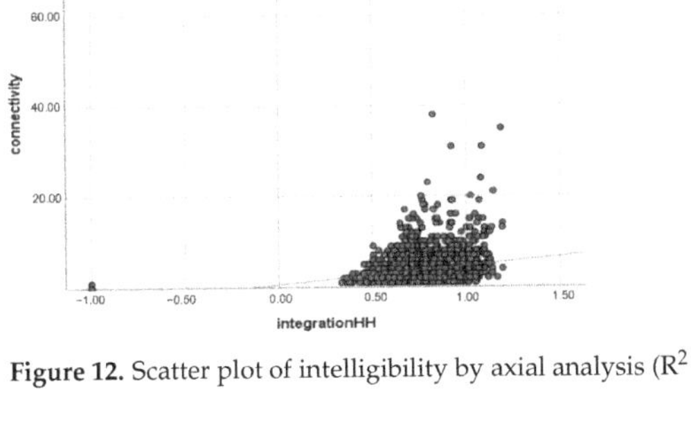

Figure 12. Scatter plot of intelligibility by axial analysis ($R^2 = 0.048$).

The scatter plot indicates a moderate alignment between the city's global and local dynamics (Figure 13). The correlation between local integration (radius 3) and global integration (radius n) assesses the alignment of the dynamics. This level of synergy suggests that some crucial axes and spaces are functioning effectively as convergence points for both local and global movements.

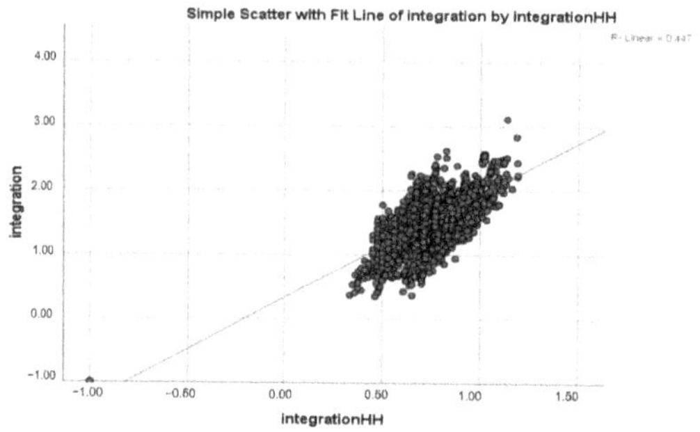

Figure 13. Scatter plot of synergy by axial analysis ($R^2 = 0.447$).

The moderate synergy suggests that individual neighborhoods still disrupt the overall flow of movement throughout the city, though the urban system works well together as a dynamic whole. The findings confirm that Bechar's mixed urban structure experiences urban segregation and socioeconomic differentiation, impacting sustainability, socioeconomic structures, and urban interaction.

The study identified areas of both segregation and integration, assessed connectivity and navigational challenges, and evaluated the urban layout's support for social and commercial activities using space syntax measures. The study also highlighted the spatial disparities between well-connected central areas and marginalized peripheries. Synergy assessed connectivity and navigational challenges and evaluated the urban layout's support for social and commercial activities. It also highlighted spatial disparities between well-connected central areas and marginalized peripheries.

4. Discussion

Like many Algerian cities, the case of Bechar is enlightening in explaining extensive urbanization and challenges in reclaiming the urban form. The axial and syntactic analysis of Bechar reveals its fragmented urban structure, shaped by colonial and post-independence developments. The city extends linearly over 17 km and contrasts starkly with its historical center, the ksar, a disconnected fragment nowadays. The master plan's proposal for north–south expansion potentially leads to further linear development. New peripheral extensions also show disconnection from the central core, contrasting with the National Agency for Territorial Development's (ANAT) aim to densify the urban fabric and control urban growth. The well-equipped central district, the hub of administrative, financial, and commercial activities, contrasts other neighborhoods with limited services, leading to urban segregation and unequal access to resources and opportunities [93,94]. The peripheries, characterized by a mix of legal and illegal housing and poorly accessible public facilities, have led to urban sprawl, resulting in a complex and fragmented city. Spatial analysis shows that Bechar's structure is illegible, with poor orientation and accessibility. Low synergy values indicate a lack of cohesive dynamics between global and local levels, leading to isolated neighborhood spaces, contrasting with a crucial regulatory goal of the master plan: rationalizing urban spaces and reducing network management costs. While the central areas have well-integrated road systems, peripheral areas remain segregated.

The primary integrated roads align with regulatory plans and include major roads connecting southern, northern, and western neighborhoods, structuring and connecting the city center with surrounding areas. Increased fragmentation correlates with decreased intelligibility, especially in the city center. This fragmentation and the irregular connections between the center and periphery contribute to Bechar's low intelligibility. Controllable spaces, characterized by inaccessibility due to fewer connections, isolate residents and detach these areas from the rest of the city, aligning with segregated zones where broader city interactions are restricted [57,70].

Urban planning regulations, including the master and land use plans, have allowed us to understand and identify the limiting factors of their implementation. The master plan sets fundamental orientations to organize and control land use for sustainable and balanced development. It includes plans defining short-, medium-, and long-term orientations, dividing the territory into specific zones with regulated uses. Key aspects include construction standards, environmental preservation, infrastructure improvement, and urban rehabilitation. Public participation is encouraged to ensure that citizens' needs are considered. Lastly, the regulations support economic and social development by creating economic activity zones and promoting access to affordable housing and social services. Its logical conclusion is found in the adoption of a land use plan. It is difficult to conceive a land use plan that does not rely on a clear statement from the master plan. Similarly, a land use plan is effective when it allows for implementing the master plan's options.

This research aimed to clarify assumptions about the impacts of spatial fragmentation on social and economic structures and space syntax measures in Bechar. By addressing related questions, this paper contributes to urban sustainability and space syntax by providing further insights into the actual distribution of urban space. It offers actionable insights for enhancing accessibility, promoting equitable resource distribution, reducing urban sprawl, and fostering social interactions. Additionally, the study reveals that the area's level of spatial intelligibility influences residential activities at urban scale more than spatial structure (grid-like or organic).

The results of the space syntax analysis in this research highlight the strengths and challenges of the urban space in Bechar. They provide a framework for evaluating design proposals to achieve specific design goals, such as increasing or decreasing certain land uses in an area. In this study, the axial lines method describes the spatial configurations of the case studies. This approach examines the typological distances between lines representing street networks [68]. It is crucial for planning authorities and urban planners to carefully consider the planning and design of road networks, as these are inherently static and hard to alter. This consideration is especially pertinent to new urban developments and is consistent with other research in the field [89,95,96]. Practically, the findings of this study offer an analytical framework for evaluating master plan proposals for new city areas, allowing the exploration and modification of urban space based on the street network.

The findings in this study provide valuable insights to predict the optimal locations for amenities and facilities in new urban developments. By implementing strategic urban design, planning authorities can effectively reduce the demand for public lighting, propose well-structured streets to mitigate the heat island effect, and decrease fuel consumption, thus enhancing urban sustainability. Key design elements, such as accessibility to facilities, sunlight exposure, life cycle analyses, and the impact of communication systems on urban design, are crucial in these strategies [95,97].

In Bechar, it is recommended that planning authorities develop guidelines that advocate for highly intelligible urban designs. These guidelines can serve as indicators to provide feedback on policy implementation regarding urban design and land-use distribution. While the study offers a broad understanding of urban space in Bechar, it also highlights the spatial distribution's role in promoting or hindering human behaviors and economic activities. The quality of urban space is influenced by various physical and economic factors, such as the integration and accessibility of an area, pedestrian movement, and the availability of public transportation. Achieving sustainable mobility necessitates a flexible paradigm and policy measures to enhance urban sustainability, including the high-quality implementation of innovative schemes and gaining public acceptability through active involvement [96,98].

This research focuses on the urban spatial configurations of Bechar and their impact on the city's functionality and sustainability. The study, however, has some limitations. First, it relies on the availability of satellite imagery and field surveys. These limit the ability to capture all dimensions of the urban environment, potentially affecting the accuracy of the analysis. Regardless of how powerful the spatial syntax tool is, it may not fully account for all socioeconomic and cultural factors influencing urban development. Hence why, the additional dimensions are used to interpret the results. The findings are specific to Bechar and may not directly apply to other cities with different geographical, cultural, and socioeconomic contexts. Therefore, comparative studies with other regions are needed to generalize the results. The research provides a snapshot of Bechar's current urban form, not accounting for historical transformations or future changes beyond the data collected. This problem highlights the need for longitudinal studies to understand urban growth dynamics over time. The use of DepthmapX and QGIS 3.30.3 software for spatial analysis has inherent limitations related to software capabilities and user expertise, with potential variations in application leading to differences in analytical outcomes. While examining specific spatial attributes, the study does not extensively explore the role of the Network Accessibility and Connectivity Hierarchy (NACH) or other influencing factors like land prices, population density, and public transport in land-use distribution. Future research should address these factors to build on this study and enhance our understanding of sustainable urban planning in arid regions.

5. Conclusions

The application of spatial syntax to Bechar's urban fabric reveals significant spatial disparities and connectivity issues linked to the history of colonization and post-independence expansion. It highlights a central urban dichotomy: highly integrated central areas versus segregated peripheral neighborhoods, reflecting spatial and socioeconomic disparities. The fragmented urban structure becomes less intelligible from the center to the peripheries, with centrally organized regular grids sharply contrasting with the irregular and disconnected layouts of peripheral areas. These peripheries, characterized by unplanned sprawl and a mix of legal and illegal housing, lack coherent planning for accessibility and integration. As a result, Bechar is a complex and compartmentalized city where navigation is challenging, particularly for newcomers. These findings support the idea that a city's urban form influences urban sustainability.

By using spatial syntax attributes, the study quantifies the spatial structure. It evaluates the impact of spatial fragmentation on social and economic structures and processes in the urban space, enabling proactive measures for urban sustainability. Hillier and Hanson [70] noted that the spatial configuration of cities directly impacts social structure and community interactions and functioning.

In principle, Bechar's urban planning rules aim to coherently structure land use, promote infrastructure, and encourage balanced growth. These guidelines integrate master plans with local specifics. Despite their potential, the regulations have failed to control urbanization, leading to chaos due to uncoordinated expansions. These regulations, dating from the 1990s, are considered outdated and ineffective, lacking an integrated approach and future vision, resulting in traffic congestion and insufficient implementation.

Overall, Bechar's urban dynamics are marked by spatial fragmentation, environmental deterioration, and infrastructure vulnerability. Urban policy failures are due to technical issues, lack of skills, and the complexity of social and political relations. Effective management requires greater participation, flexible legal tools, and a democratic institutional framework. Sustainable urban development is impossible without inclusive and participatory planning processes [26]. Land-use plans that divide cities into sectors show that limited visions are insufficient for future planning. Poor city organization, unprepared land, adequate facilities, and incompatible activities hinder sustainable socioeconomic and urban development, threatening current and future generations. Urban planning actions lack coherent codes and adequate field monitoring, preventing the achievement of socioeconomic goals.

In response to these findings, there is an urgent need for strategic urban planning to address Bechar's spatial and social ruptures. Effective solutions include improving connectivity between central and peripheral areas, fostering neighborhood integration, and enhancing the intelligibility of the city's layout. These measures are crucial for the city's physical, social, and economic fabric and to improve residents' quality of life. As Lefebvre [99] emphasized, the right to the city is not merely the right to access what already exists but the right to modify it according to our desires. Investigating urban form and segregation through space syntax spatial attributes can guide energy-efficient urban design strategies. Key strategies include urban canopy application, street orientation, and optimized building ratios. These strategies significantly influence energy use and urban sustainability. Integrating renewable energy potential (considering microclimatic effects on cooling demands) and mitigating air pollution is crucial. Holistic urban planning should focus on the collective impact of buildings, emphasizing courtyards, renewable energy integration, and energy-efficient urban blocks [100,101]. These strategies can help reduce energy requirements for cooling and manage commercial land use impacts, offering a pathway toward achieving urban sustainability goals through thoughtful design and planning. By combining these approaches, Bechar can address its current challenges and move towards a more sustainable and integrated urban future.

Critics argue that the Space Syntax approach (SSA) has several limitations. Ratti [102] from MIT points out that SSA lacks precision, is overly subjective, and fails to represent space accurately. According to Ratti, the approach does not capture cultural norms affecting building usage and oversimplifies spatial analysis by limiting it to two dimensions. Similarly, Osman and Suliman [103] argue that SSA does not adequately consider cultural differences in space utilization. These critiques highlight the need for comprehensive methodologies that incorporate cultural nuances and three-dimensional aspects of urban spaces.

To conclude, developing sustainable cities in the Saharan-arid regions requires a study of sustainability indicators that balance socioeconomic, environmental, and socio-cultural parameters. Following a strategic framework adapted to community needs, it is possible to achieve sustainable development. New urban planning measures and renewed debate are necessary. These require awareness and mobilization from researchers, public authorities, local communities, and all stakeholders to improve future generations' living conditions. The future of urban sustainability depends on creating inclusive, resilient, and adaptive urban environments [104].

Author Contributions: Methodology, M.E.F.M. and W.Z.; Software, M.E.F.M.; Validation, X.W.; Formal analysis, M.E.F.M. and W.Z.; Investigation, M.E.F.M.; Resources, M.E.F.M. and R.W.B.; Writing—original draft, M.E.F.M., X.W. and R.W.B.; Writing—review & editing, M.E.F.M. and R.W.B.; Visualization, M.E.F.M.; Supervision, X.W. All authors have read and agreed to the published version of the manuscript.

Funding: This research received no external funding.

Data Availability Statement: Data is contained within the article.

Conflicts of Interest: The authors declare no conflict of interest.

References

1. Fuentes, L.; Ramirez, M.I.; Rodriguez, S. Urban Sustainability and Perceived Satisfaction in Neoliberal Cities. *Cities* **2022**, *126*, 103647. [CrossRef]
2. Maclaren, V.W. Urban Sustainability Reporting. *J. Am. Plan. Assoc.* **2007**, *4363*, 184–202. [CrossRef]
3. Hamlili, F.Z.; Dakhia, A.; Biara, R.W. Unravelling the User Perception: Evaluating Kenadsa's Heritage Office Building Performance Through Perceptual Dimensions and Behaviors. *Indones. J. Soc. Sci. Res.* **2024**, *5*, 84–95.
4. Messafri, N.; Biara, R.W.; Farhi, A. Residential Subdivisions, What Potential for Quality Production?: Case of the City of Bechar, Algeria. *A | Z ITU J. Fac. Archit.* **2022**, *19*, 557–568. [CrossRef]

5. Benslimane, N.; Biara, R.W. Le mode de production spatial au niveau de la ville de bechar. *Rev. d'Archit.* **2022**, *2*, 60–80.
6. Dechaicha, A.; Daikh, A.; Alkama, D. Monitoring and Landscape Quantification of Uncontrolled Urbanisation in Oasis Regions: The Case of Adrar City in Algeria. *J. Contemp. Urban Aff.* **2021**, *5*, 209–219. [CrossRef]
7. Dizdaroglu, D. The Role of Indicator-Based Sustainability Assessment in Policy and the Decision-Making Process: A Review and Outlook. *Sustainability* **2017**, *9*, 1018. [CrossRef]
8. Kadri, S. Espaces Publics Urbains: Les Limites Réelles Des Rues. Le Cas De La Ville De Bechar 2020. Doctoral Thesis, USTO University, Oran, Algeria, 2020.
9. Zatir, S. Les Plans de l'Urbanisme et Leurs Applications Sur Le Terrain Cas Du PDAU. Doctoral Thesis, USTO University, Oran, Algeria, 2018.
10. Kaddouri, M. L'adaptation Des Formes Urbaines Interscalaires Des Villes Sahariennes Face Aux Inondations. Cas de La Ville de Béchar. Master's Thesis, Mohamed Khider University of Biskra, Biskra, Algérie, 2019.
11. Zatir, S.; Laabaci, H.; Biara, R.W.; Bouannini, M. Planning Sustainable Urban Development in Bechar City. *IJSET* **2017**, *5*, 46–70.
12. Hamidi, A. L'Évolution d'Un Quartier Périphérique En Centre d'Animation: Debdaba (Bechar). *Insaniyat Rev. Algérienne Anthropol. Sci. Soc.* **2011**, *51–52*, 185–196.
13. Biara, R.W.; Hamouine, A.; Nabou, M. The Future of the Ksar in Sustainable Development. *Energy Procedia* **2012**, *18*, 35–42. [CrossRef]
14. Benslimane, N.; Biara, R.W. The Urban Sustainable Structure of the Vernacular City and Its Modern Transformation: A Case Study of the Popular Architecture in the Saharian Region. *Energy Procedia* **2019**, *157*, 1241–1252. [CrossRef]
15. Dechaicha, A. La Soutenabilité de La Ville Au Sahara Entre Compacité et Tendances d'étalement Urbain. Cas d'étude: Les Grandes Villes Du Bas-Sahara Algérien. Doctoral Thesis, Mohamed Khider University of Biskra, Biskra, Algérie, 2020.
16. Chandio, I.A.; Aftab, M.I.R.; Talpur, H.; Shaikh, F.A. Urban Land Use Planning Trend and Sustainable Challenges in socio-economic developmentn. *Mehran Univ. Res. J. Eng. Technol.* **2018**, *37*, 397–404. [CrossRef]
17. Baig, F.; Rana, I.A. Determining Factors Influencing Residents' Satisfaction Regarding Urban Livability in Pakistan. *Int. J. Community Well-Being* **2019**, *2*, 91–110. [CrossRef]
18. Napiah, M.; Teknologi, U. Transportation Planning Survey Methodologies for the Proposed Study of Physical and Socio- Economic Development Study of Physical and Socio-Economic Development of Deprived Rural Regions: A Review. *Mod. Appl. Sci.* **2014**, *6*, 1. [CrossRef]
19. Khahro, S.H.; Memon, Z.A.; Gungat, L.; Razuhanafi, M.; Yazid, M.; Rahim, A.; Mubaraki, M.; Yusoff, N.I. Low-Cost Pavement Management System for Developing Countries. *Sustainability* **2021**, *13*, 5941. [CrossRef]
20. Khahro, S.H.; Aftab, M.; Talpur, H.; Bhellar, M.G.; Das, G.; Shaikh, H. GIS-Based Sustainable Accessibility Mapping of Urban Parks: Evidence from the Second Largest Settlement of Sindh, Pakistan. *Sustainability* **2023**, *15*, 6228. [CrossRef]
21. Aftab, M.; Talpur, H.; Chandio, I.; Das, G.; Shah, M.H. Accessibility and Usability of Sustainable Green Buildings: A Case Study On Accessibility and Usability of Sustainable Green Buildings: A Case Study On Hyderabad. *Eng. Sci. Technol. Int. Res. J.* **2021**, *5*, 96–101.
22. Hyderabad, D. Energy Crisis and Household's Perception about Solar Energy Acceptance: District Hyderabad, Pakistan. *SINDH Univ. Res. J.-Sci. Ser.* **2017**, *49*, 601. [CrossRef]
23. Khahro, S.H.; Subramaniam, U.; Ghosh, A. Efficiency for Energy, Economy and Environment (3E) Benefit. *Energies* **2021**, *14*, 4487. [CrossRef]
24. Jawed, A.; Talpur, M.A.H.; Chandio, I.A.; Mahesar, P.N. Impacts of In-Accessible and Poor Public Transportation System on Urban Environment: Evidence from Hyderabad, Pakistan. *Eng. Technol. Appl. Sci. Res.* **2019**, *9*, 3896–3899. [CrossRef]
25. *United Nations World Urbanization Prospects*; United Nations: New York, NY, USA, 2018; Volume 12, ISBN 9789211483192.
26. UN-Habitat World Cities Report 2016: Urbanization and Development–Emerging Futures. *Publ. UN-Habitat* **2016**. [CrossRef]
27. El-Aby, M.F. Towards sustainable urban development in arid regions: Masdar city as a case study. *J. Al-Azhar Univ. Eng. Sect.* **2017**, *12*, 199–211. [CrossRef]
28. Ryser, S. The Anti-Politics Machine of Green Energy Development: The Moroccan Solar Project in Ouarzazate and Its Impact on Gendered Local Communities. *Land* **2019**, *8*, 100. [CrossRef]
29. Kopenhagen, K. The Bicycle Account 2018: Copenhagen City of Cyclists. *Cph. Tech. Environ. Adm. Cph.* **2019**, 32.
30. NYC Economic Development Corporation High Line Economic Impact Study. Available online: https://www.nycedc.com/project/high-line (accessed on 10 March 2024).
31. Cortes, V.; Urazan, C.F.; Escobar, D.A. Accesibilidad Territorial Urbana Ofrecida Por La Red de TransMilenio Según Condición Socioeconómica. Caso Bogotá (Colombia). *Rev. Espac. ISSN* **2020**, *798*, 1015.
32. Tan, B.A.; Gaw, L.Y.-F.; Masoudi, M.; Richards, D.R. Nature-Based Solutions for Urban Sustainability: An Ecosystem Services Assessment of Plans for Singapore's First "Forest Town". *Front. Environ. Sci.* **2021**, *9*, 610155. [CrossRef]
33. Golden, J.S.; Kaloush, K.E. Mesoscale and Microscale Evaluation of Surface Pavement Impacts on the Urban Heat Island Effects. *Int. J. Pavement Eng.* **2006**, *7*, 37–52. [CrossRef]
34. Cartwright, A.; Marrengane, N. Urban Governance and Turning African Cities around: City of Johannesburg. *Partnersh. Afr. Soc. Gov. Res. Work. Pap.* **2016**, *17*, 756–771.
35. Gao, K.; Haddad, S.; Paolini, R.; Feng, J.; Altheeb, M.; Mogirah, A.A.; Moammar, A.B.; Santamouris, M. The Use of Green Infrastructure and Irrigation in the Mitigation of Urban Heat in a Desert City. *Build. Simul.* **2024**, *17*, 679–694. [CrossRef]

36. Mondal, T. Do We Understand Sustainability? An Architectural Insight into the Origin of the Concept. *Int. J. Arch. Plan.* **2021**, *2*, 9–14. [CrossRef]
37. Mazzoncini, R.; Somaschini, C.; Longo, M. *The Infrastructure for Sustainable Mobility*; Springer Nature: Cham, Switzerland, 2020; ISBN 9783030410711.
38. Lambin, J.-J. Capitalism and Sustainable Development. *Symphonya. Emerg. Issues Manag.* **2009**, *2*, 3–9. [CrossRef]
39. José, G. *Vargas-Hernández, E.P.O.-Q. Sustainable Resilience in Urban Land Use*; IGI Global: Hershey, PA, USA, 2022; Volume 24. [CrossRef]
40. Jabareen, Y. *The Risk City*; Lect. Notes Energy; Springer: Dordrecht, The Netherlands, 2015. Available online: https://link.springer.com/ (accessed on 23 December 2023).
41. Oksman, V.; Väätänen, A.; Ylikauppila, M. Co-Creation of Sustainable Smart Cities: Users, Participation and Service Design. In Proceedings of the 8th International Conference on Mobile Ubiquitous Computing, Systems, Services and Technologies, Ubicomm 2014, Rome, Italy, 24–28 August 2014; International Academy, Research, and Industry Association IARIA; pp. 189–195.
42. Zhang, X.; Li, H. Urban Resilience and Urban Sustainability: What We Know and What Do Not Know? *Cities* **2018**, *72*, 141–148. [CrossRef]
43. Shafiei-Dastjerdi, M.; Ghaffari, A.; Lak, A. Unveiling the Concept of Spatial Resilience in Urban Design: A Systematic Qualitative Review. *Mon. Sci. J. Bagh-E Nazar* **2022**, *19*, 69–80.
44. Ugalde, C.; Rigatti, D.; Zampieri, F.; Braga, A. Intelligibility in Large Spatial Systems. In Proceedings of the 7th International Space Syntax Symposium, Stockholm, Sweden, 8–11 June 2009; pp. 111–119.
45. Bibri, S.E.; Krogstie, J. Smart Sustainable Cities of the Future: An Extensive Interdisciplinary Literature Review. *Sustain. Cities Soc.* **2017**, *31*, 183–212. [CrossRef]
46. Beatley, T.; Collins, R. Smart Growth and beyond: Transitioning to a Sustainable Society. *Va. Environ. Law J.* **2000**, *19*, 287–322.
47. Sev, A. How Can the Construction Industry Contribute to Sustainable Development ? *Sustain. Dev.* **2009**, *173*, 161–173. [CrossRef]
48. Hillier, B.; Iida, S. Network and Psychological Effects in Urban Movement. In *Proceedings of the International Conference on Spatial Information Theory, COSIT 2005, 14–18 September 2005, Ellicottsville, NY, USA*; Springer: Berlin, Germany, 2005; pp. 475–490.
49. Karimi, K. A Configurational Approach to Analytical Urban Design: 'Space Syntax' Methodology. *Urban Des. Int.* **2012**, *17*, 297–318. [CrossRef]
50. Baran, P.K.; Rodríguez, D.A.; Khattak, A.J. Space Syntax and Walking in a New Urbanist and Suburban Neighbourhoods. *J. Urban Des.* **2008**, *13*, 5–28. [CrossRef]
51. Chiaradia, A.; Hillier, B.; Schwander, C.; Wedderburn, M. Compositional and Urban Form Effects on Centres in Greater London. *Proc. Inst. Civ. Eng. Des. Plan.* **2012**, *165*, 21–42. [CrossRef]
52. Hillier, B. Spatial Sustainability in Cities Organic Patterns and Sustainable Forms Note 1. *R. Inst. Technol.* **2009**, 1–20.
53. Ma, Y.; Zhang, P.; Zhao, K.; Zhou, Y.; Zhao, S. A Dynamic Performance and Differentiation Management Policy for Urban Construction Land Use Change in Gansu, China. *Land* **2022**, *11*, 942. [CrossRef]
54. Sminkey, P.V.; Le Doux, J. Case Management Ethics: High Professional Standards for Health Care's Interconnected Worlds. *Prof. Case Manag.* **2016**, *21*, 193–198. [CrossRef]
55. Mebirouk, H. La Fragmentation Socio-Spatiale à Annaba, Un Paradigme de Crise Urbaine. In *International Conference, The Cities Challenged by Sustainable Development*; University of Sfax: Sfax, Tunisia, 2022.
56. Santos, C.M. Metrópole Corporativa Fragmentada: O Caso de São Paulo de Milton Santos. São Paulo Nobel Secr. Estado da Cult. 1990. Available online: https://www.google.com.hk/url?sa=t&source=web&rct=j&opi=89978449&url=https://revistas.unifacs.br/index.php/rde/article/download/1896/1486&ved=2ahUKEwjmj7ycv4yHAxUWg44IHXSUDFsQFnoECBEQAQ&usg=AOvVaw0vHu8CR1iHSeUL3KBEtWZ1 (accessed on 15 March 2021).
57. Navez-Bouchanine, F. *La Fragmentation en Question: Des Villes Entre Fragmentation Spatiale et Fragmentation Sociale?* L'Harmattan: Paris, France, 2002; ISBN 2747521702.
58. Bénit-Gbaffou, C. Unbundled Security Services and Urban Fragmentation in Post-Apartheid Johannesburg. *Geoforum* **2008**, *39*, 1933–1950. [CrossRef]
59. Jaglin, S. Villes Disloquées? Ségrégations et Fragmentation Urbaine En Afrique Australe/Broken-up Cities: Segregation and Urban Fragmentation in Southern Africa. In Proceedings of the Annales de géographie; JSTOR: New York, NY, USA, 2001; pp. 243–265.
60. Séguin, A.M. Les Quartiers: Des Lieux de Fragmentation. *Cah. Geogr. Que.* **2011**, *55*, 69–73.
61. Vidal, L. La Notion de Fragmentation Dans Les Recherches Sur La Ville. Le Cas de La Communauté Scientifique Brésilienne. *Rapp. Mission. Rendu Au Pir-Villes (Cnrs) Numéro D'ordre* **1993**, *9*, 99.
62. Paquot, T. *Ville Fragmentée Ou Urbain Éparpillé? Navez-Bouchanine, Françoise. La Fragm. en Quest. des Villes entre Fragentation Spat. Fragm. Soc.*; L'Harmattan: Paris, France, 2002; pp. 114–118.
63. Chouadra, S. De La Fragmentation à La Recomposition: Cas de La Ville Algérienne. In Proceedings of the Penser la Ville—Approches Comparatives, Khenchela, Algeria, 25–26 October 2008; p. 14.
64. Mansouri, Z. Cohabitation Entre l'architecture Traditionnelle et Moderne Pour Un Modèle d'habitat Adapté à l'aspect Climatique et Social Des Villes Sahariennes:Cas d'étude La Ville de Bechar. Ph.D. Thesis, Université Mohamed Khider–Biskra, Biskra, Algeria, 2018.

65. Khodakarami, L.; Pourmanafi, S.; Mokhtari, Z.; Soffianian, A.R.; Lotfi, A. Urban Sustainability Assessment at the Neighborhood Scale: Integrating Spatial Modellings and Multi-Criteria Decision Making Approaches. *Sustain. Cities Soc.* **2023**, *97*, 104725. [CrossRef]
66. Yamu, C.; Nes, A. Van Bill Hillier's Legacy: Space Syntax—A Synopsis of Basic Concepts, Measures, and Empirical Application. *Sustainability* **2021**, *13*, 3394. [CrossRef]
67. Alalouch, C.; Al-Hajri, S.; Naser, A.; Al Hinai, A. The Impact of Space Syntax Spatial Attributes on Urban Land Use in Muscat: Implications for Urban Sustainability. *Sustain. Cities Soc.* **2019**, *46*, 101417. [CrossRef]
68. Lamíquiz, P.J.; López-Domínguez, J. Effects of Built Environment on Walking at the Neighbourhood Scale. A New Role for Street Networks by Modelling Their Configurational Accessibility? *Transp. Res. Part A Policy Pract.* **2015**, *74*, 148–163. [CrossRef]
69. De Koning, R.E.; Van Nes, A.; Ye, Y.; Roald, H.J. Strategies for Integrated Densification with Urban Qualities: Combining Space Syntax with Building Density, Land Usage, Public Transport and Property Rights in Bergen City. In Proceedings of the XI SSS: 11th International Space Syntax Symposium 2017, Lisbon, Portugal, 3–7 July 2017; Instituto Superior Técnico: Lisbon, Portugal, 2017; pp. 51–56.
70. Hillier, B.; Hanson, J. *The Social Logic of Space*; Cambridge University Press: Cambridge, UK, 1984; ISBN 9780521233651.
71. Jiang, B.; Claramunt, C. Integration of Space Syntax into GIS: New Perspectives for Urban Morphology. *Trans. GIS* **2002**, *6*, 295–309. [CrossRef]
72. Hillier, B. La Morphologie de l'espace Urbain: L'évolution de l'approche Syntaxique. *Archit. Comport. Behav.* **1987**, *3*, 205–216.
73. Hillier, B. Between Social Physics and Phenomenology. In Proceedings of the Fifth Space Syntax Symposium, Delft, The Netherlands, 13–17 June 2005.
74. Hillier, B. *Space Is the Machine: A Configurational Theory of Architecture*; CreateSpace Independent Publishing Platform: Scotts Valley, CA, USA, 2007.
75. Hillier, B. *Space Is the Machine UK Cambridge*; Cambridge University Press: Ulverston, LA, USA, 1996.
76. Liu, X.; Jiang, B. Defining and Generating Axial Lines from Street Center Lines for Better Understanding of Urban Morphologies. *Int. J. Geogr. Inf. Sci.* **2012**, *26*, 1521–1532. [CrossRef]
77. Mazouz, S. Fabrique de La Ville En Algérie et Pérennisation d'un Modèle: Le Cas de La Nouvelle Ville Ali Mendjeli à Constantine. Courrier du Savoir-N°15 23–30. 2013. Available online: https://www.academia.edu/106332400/Fabrique_De_La_Ville_en_Algerie_et_Perennisation_D_Un_Modele_Le_Cas_De_La_Nouvelle_Ville_Ali_Mendjeli_a_Constantine (accessed on 4 May 2023).
78. Femmam, N.; Mazouz, S. Analysis of Legibility in Urban Public Spaces. Case of El-Alia North-East Neighborhood in Biskra, Algeria. *J. Appl. Eng. Sci. Technol.* **2018**, *4*, 203–211.
79. Al-Sayed, K.; Turner, A.; Hillier, B.; Iida, S.; Penn, A. *Space Syntax Methodology*; Bartlett School of Architecture, UCL: London, UK, 2014.
80. Dalton, N.S.C. Synergy, Inteligibility and Revelation in Neighbourhood Places. 2011. Available online: https://discovery.ucl.ac.uk/id/eprint/1334117/ (accessed on 4 May 2023).
81. Van Nes, A. Space Syntax in Theory and Practice. *Geod. Integr. Des. Geospat. Sci.* **2014**, 237–257. [CrossRef]
82. Penn, A. Space Syntax and Spatial Cognition: Or Why the Axial Line? *Environ. Behav.* **2003**, *35*, 30–65. [CrossRef]
83. Penn, A.; Hillier, B.; Banister, D.; Xu, J. Configurational Modelling of Urban Movement Networks. *Environ. Plan. B Plan. Des.* **1998**, *25*, 59–84. [CrossRef]
84. Koohsari, M.J.; Sugiyama, T.; Mavoa, S.; Villanueva, K.; Badland, H.; Giles-Corti, B.; Owen, N. Street Network Measures and Adults' Walking for Transport: Application of Space Syntax. *Health Place* **2016**, *38*, 89–95. [CrossRef]
85. Musterd, S. Segregation and Integration: A Contested Relationship. *J. Ethn. Migr. Stud.* **2003**, *29*, 623–641. [CrossRef]
86. Zlatkovic, M.; Zlatkovic, S.; Sullivan, T.; Bjornstad, J.; Shahandashti, S.K.F. Assessment of Effects of Street Connectivity on Traffic Performance and Sustainability within Communities and Neighborhoods through Traffic Simulation. *Sustain. Cities Soc.* **2019**, *46*, 101409. [CrossRef]
87. Hiller, B. The Common Language of Space: A Way of Looking at the Social, Economic and Environmental Functioning of Cities on a Common Basis. *Sp. Syntax Lab.* **2005**, *11*.
88. Hillier, B. Cities as Movement Economies. *Urban Des. Int.* **1996**, *1*, 41–60. [CrossRef]
89. Laouar, D.; Mazouz, S.; Teller, J. Spatial Accessibility and Urban Fragmentation in Colonial Cities. The Case of Annaba City. *Cybergeo Eur. J. Geogr.* **2019**, 111–123. [CrossRef]
90. Lynch, K. *The Image of the City*; MIT Press: Cambridge, MA, USA, 1960; Volume 208.
91. Dalton, N. Configuration and Neighbourhood: Is Place Measurable? *Sp. Syntax Spat. Cogn. Work.* **2006**, *6*, 53–64.
92. Kim, Y.O. The Role of Spatial Configuration in Spatial Cognition. In Proceedings of the Third International Space Syntax Symposium, Atlanta, GA, USA, 7–11 May 2001; University of Michigan: Ann Arbor, MI, USA, 2001; pp. 41–49.
93. Laouar, D.; Mazouz, S. La Carte Axiale, Un Outil d' Analyse de l' Accessibilité Spat Iale: Cas de La Ville d'Annaba The Axial Map as an Analysis' Tool for the Spatial Accessibility: Case of Annaba City. *Rev. Sci. Technol. Synthèse* **2017**, *35*, 111–123.
94. Préteceille, E. La ségrégation sociale a-t-elle augmenté ?La métropole parisienne entre polarisation et mixité. *Sociétés Contemp.* **2006**, *62*, 69–93. [CrossRef]
95. Jenks, M. *Future Forms and Design for Sustainable Cities*; Routledge: London, UK, 2005; ISBN 075066309X.
96. Newman, P.; Kenworthy, J. *Sustainability and Cities: Overcoming Automobile Dependence*; Island Press: Washington, DC, USA, 1999; ISBN 1559636602.

97. Beatley, T. *Handbook of Biophilic City Planning and Design*; Island Press: Washington, DC, USA, 2016.
98. Banister, D. The Sustainable Mobility Paradigm. *Transp. Policy* **2008**, *15*, 73–80. [CrossRef]
99. Lefebvre, H. From the Production of Space. In *Theatre and Performance Design*; Routledge: London, UK, 2012; pp. 81–84.
100. Futcher, J.; Mills, G.; Emmanuel, R.; Korolija, I. Creating Sustainable Cities One Building at a Time: Towards an Integrated Urban Design Framework. *Cities* **2017**, *66*, 63–71. [CrossRef]
101. Sarralde, J.J.; Quinn, D.J.; Wiesmann, D.; Steemers, K. Solar Energy and Urban Morphology: Scenarios for Increasing the Renewable Energy Potential of Neighbourhoods in London. *Renew. Energy* **2015**, *73*, 10–17. [CrossRef]
102. Ratti, C. Space Syntax: Some Inconsistencies. *Environ. Plan. B Plan. Des.* **2004**, *31*, 487–499. [CrossRef]
103. Osman, K.M.; Suliman, M. The Space Syntax Methodology: Fits and Misfits. *Archit. Comport. Behav.* **1994**, *10*, 189–204.
104. De Moerloose, S. *World Bank Environmental and Social Conditionality as a Vector of Sustainable Development*; Schulthess Juristische Medien: Zürich, Switzerland, 2020; ISBN 3725587833.

MDPI AG
Grosspeteranlage 5
4052 Basel
Switzerland
Tel.: +41 61 683 77 34

Buildings Editorial Office
E-mail: buildings@mdpi.com
www.mdpi.com/journal/buildings

www.ingramcontent.com/pod-product-compliance
Lightning Source LLC
Chambersburg PA
CBHW061811210926
51850CB00006B/672
9783725855711